Operator Theory: Advances and Applications
Vol. 137

Editor:
I. Gohberg

Introduction to the Theory of Toeplitz Operators with Infinite Index

Translated from the Russian by Andrei Iacob

Vladimir Dybin
Sergei M. Grudsky

Springer Basel AG

Authors:

Vladimir Dybin
Mechanics and Mathematics Department
Rostov State University
Zorge Str. 5
344090 Rostov-on-Don
Russia
e-mail: dybin@math.rsu.ru
grudsky@aaanet.ru

Sergei M. Grudsky
Mechanics and Mathematics Department
Rostov State University
Zorge Str. 5
344090 Rostov-on-Don
Russia
e-mail: grudsky@aaanet.ru

and

Departamento de Matemáticos
CINVESTAV del I.P.N.
Apartado Postal 14-470
07000 Mexico, D.F.
Mexico
e-mail: grudsky@math.cinvestav.mx

2000 Mathematics Subject Classification 30D50, 45E10, 47B35

A CIP catalogue record for this book is available from the
Library of Congress, Washington D.C., USA

Bibliographic information published by Die Deutsche Bibliothek
Die Deutsche Bibliothek lists this publication in the Deutsche Nationalbibliografie; detailed bibliographic data is available in the Internet at <http://dnb.ddb.de>.

ISBN 978-3-0348-9476-0 ISBN 978-3-0348-8213-2 (eBook)
DOI 10.1007/978-3-0348-8213-2

Originally published by Birkhäuser Verlag Basel in 2002
Softcover reprint of the hardcover 1st edition 2002

Printed on acid-free paper produced from chlorine-free pulp. TCF ∞
Cover design: Heinz Hiltbrunner, Basel

ISBN 978-3-0348-9476-0

9 8 7 6 5 4 3 2 1 www.birkhauser-science.com

To our families selflessly
and worthily enduring
the difficulties of modern
Russian mathematical life

Contents

Preface

> We offer the reader of this book some specimens of "infinity" that we seized from the "mathematical jungle" and trapped within the solid cage of analysis

The creation of the theory of singular integral equations in the mid 20th century is associated with the names of N.I. Muskhelishvili, F.D. Gakhov, N.P. Vekua and their numerous students and followers and is marked by the fact that it relied principally on methods of complex analysis.

In the early 1960s, the development of this theory received a powerful impulse from the ideas and methods of functional analysis that were then brought into the picture. Its modern architecture is due to a constellation of brilliant mathematicians and the scientific collectives that they produced (S.G. Mikhlin, M.G. Kreĭn, B.V. Khvedelidze, I. Gohberg, I.B. Simonenko, A. Devinatz, H. Widom, R.G. Douglas, D. Sarason, A.P. Calderón, S. Prössdorf, B. Silbermann, and others). In the ensuing period, the Fredholm theory of singular integral operators with a finite index was completed in its main aspects in wide classes of Banach and Fréchet spaces.

At the same time, as early as the beginning of the 1960s, N.V. Govorov constructed examples of Riemann boundary value problems with infinite index. As the study of singular integral operators with discontinuous coefficients progressed, operators that are not Φ-operators, but rather $\Phi_{\pm}$-operators with infinite index – in the sense that their kernels and cokernels are infinite-dimensional – have been encountered more and more frequently. In connection with this, the beginning of the 1980s saw the onset of an intensive development of the constructive theory of singular integral operators with infinite index, including the theory of one-sided invertibility and the description of classes of solutions and of solvability conditions of the corresponding equations.

The main goal of this book is to harvest the fruits of this development, primarily as it relates to the results and interests of the authors. Attempting to give the book an introductory format, we have chosen as a demonstration model the Toeplitz operator, which lies at the intersection of the one-dimensional theories of singular integral equations and convolution-type equations, because in the space L_2 it is isometrically similar to the well-known Wiener-Hopf operator. With great satisfaction we acknowledge that as we worked on this the book our leading light was I. Gohberg and N.Ya. Krupnik's monograph *Introduction to the theory of singular integral operators*, a book that has already guided three generations of "singularchiks" (the Russian nickname for those who work with singular imtegral operators).

We are sincerely grateful to I. Gohberg, whose remarks and advice have profoundly influenced the form of our book, and to I.B. Simonenko, who gave us access to unpublished materials on the Φ-theory of one-dimensional singular integral equations. We are also greatly indebted to A. Böttcher for his many extremely useful comments on several parts of this book. On final stage of the work the second author was supported by CONACYT (cátedras patrimoniales project N 5758/010286).

Last but not least, we wish to express our deep gratitude to O.N. Grudskaya for her outstanding typesetting work.

July 2002, Rostov-on-Don, Mexico City

Introduction

The *Toeplitz operator* $T(a)$ on the real line $\mathbb{R}$ induced by a function a given on $\mathbb{R}$ is defined by

$$T(a)f = P^+(af), \tag{1}$$

where P^+ is the analytic projector of the space $L_p(\mathbb{R})$, $1 < p < \infty$, onto the Hardy subspace $H^p(\Pi_+)$ for the upper complex half-plane Π_+. The function a is called the *symbol* of the operator $T(a)$. To study such operators in the Hardy space $H^p(\Pi_+)$ we will assume from now on that $a \in L_\infty(\mathbb{R})$. In the simplest case the symbol a admits a representation

$$a(x) = a^+(x)\left(\frac{x-i}{x+i}\right)^\kappa a^-(x), \tag{2}$$

where

$$(a^+)^{\pm 1} \in H^\infty(\Pi_+), \quad (a^-)^{\pm 1} \in H^\infty(\Pi_-), \quad \kappa \in \mathbb{Z}, \tag{3}$$

which is known as a *canonical factorization.* The rational function

$$r(x) = \left(\frac{x-i}{x+i}\right)^\kappa \tag{4}$$

plays a determining role in the description of the main qualitative characteristics of the operator $T(a)$. Specifically, for $\kappa > 0$ the operator $T(a)$ is left-invertible and $\dim \operatorname{coker} T(a) = \kappa$, for $\kappa < 0$ it is right-invertible and $\dim \ker T(a) = -\kappa$, and for $\kappa = 0$ it is invertible. As a consequence, the index of the operator $T(a)$ is equal to $-\kappa$ and is also equal to minus the topological index of the function r, which is defined by

$$\operatorname{ind}_{\mathbb{R}} r = \frac{1}{2\pi} \arg r(x)\big|_{-\infty}^{\infty}.$$

The theory of singular integral equations with infinite index is a natural development of the theory of singular integral equations with finite index. At its foundations lie the deep investigations by N. V. Govorov [1]–[9] on the theory of the Riemann boundary value problem with infinite index and the works of I. Gohberg and I. A. Fel′dman [1]–[3] on integro-difference equations. In 1968–69, these authors discovered that the apparatus of factorization, which at that time allowed one to construct the Φ-theory of singular integral and convolution-type equations in broad classes of Banach spaces, is also applicable to the more general case in which the symbol of the Wiener-Hopf operator under study has infinite index. In the aforementioned works of I. Gohberg and I. A. Fel′dman the role of the function r in representation (2), (4) is played by the meromorphic function

$$h(x) = \exp(i\sigma x)r(x), \qquad \sigma \in \mathbb{R} \setminus \{0\},$$

and it is readily verified that its topological index is equal to $+\infty$ or $-\infty$, depending on the sign of σ.

The development of a constructive theory of singular integral equations with infinite index in $L_p(\Gamma)$ was inspired, on one hand, by a number of qualitative investigations ($\Phi_\pm$-theory) of various classes of convolution-type equations with infinite index and, on the other hand, by applications, first and foremost to problems in the mechanics of continua and diffraction theory.

As it has been always the case in the history of the theory of singular integral equations, adopting a new direction required the involvement of new results of complex analysis.

First, it turned out that the technique of inner functions and, in particular, of Blaschke products, is well adapted for modelling rich classes of functional discontinuities of the second kind of "infinite index" type ("whirl points" in the terminology of N. V. Govorov).

And second, the model subspaces in the functional models of B. Szökefalvi-Nagy and C. Foias for the nonclassical spectral theory together with the interpolation theory in these subspaces are perfectly suited to give effective descriptions of the kernel and the image of singular integral operators with infinite index.

Let us outline the monograph chapter by chapter.

In Chapter 1, after an exposition of the requisite auxiliary material (Sections 1.1–1.4), we demonstrate on the simplest examples how a Toeplitz operator may acquire an infinite index and how the problems arising in this way can be solved.

In the first example (Section 1.5) we trace the effects of passing from a finite to an infinite Blaschke product.

In the second and third examples (Sections 1.6 and 1.7) we take as the symbol of the operator $T(a)$ the function

$$a(x) = \exp(i\sigma x) - c, \tag{5}$$

where $\sigma \in \mathbb{R} \setminus \{0\}$ and $c \in \mathbb{C}$.

For $|c| > 1$ the operator $T(a)$ is invertible in the space $H^2(\Pi_+)$. For $|c| < 1$, it is right-invertible and $\dim\ker T(a) = \infty$ if $\sigma < 0$, and left-invertible and $\dim\operatorname{coker} T(a) = \infty$ if $\sigma > 0$. Moreover, the subspaces $\ker T(a)$ for $\sigma < 0$ and $\ker T^*(a)$ for $\sigma > 0$ depend on the Wiener-Paley class $\mathcal{L}_2^{|\sigma|}$, which is isomorphic to the Hilbert space $L_2(-|\sigma|, |\sigma|)$. This allows us to describe the subspaces $\ker T(a)$ and $\operatorname{im} T(a)$ in terms of classical orthonormal bases consisting of entire functions of exponential type.

If now $|c| = 1$, then a function $a(x)$ of the form (5) has a periodic degeneration on the real line $\mathbb{R}$, and consequently the operator $T(a)$ is not normally solvable in the space $H^2(\Pi_+)$. However, it is an unboundedly invertible operator in $H^2(\Pi_+)$. Accordingly, we describe the linear manifold $\operatorname{im} T(a)$ and give a construction of an unbounded inverse operator for $T(a)$, thus anticipating ideas that will be pursued in more depth in Chapter 6.

In Chapter 2 we elaborate an apparatus for studying the one-sided invertibility of Toeplitz operators with infinite index in the space $L_p^+(\Gamma, \varrho)$, $1 < p < \infty$. Our technique for the investigation of an operator $T(a)$ whose symbol has whirl points breaks the procedure into the following steps: study first the case of a symbol on the real line $\mathbb{R}$ with a single discontinuity at infinity, then use appropriate isometric mappings to go over to the case of an arbitrary circle in the complex plane, and finally transfer the discontinuity to a smooth closed curve. By virtue of this approach, in writing this book a problem we had to deal with from the very beginning was how to choose the underlying family of weights. Regardless of the fact that a large part of the results presented in this chapter are valid in the general case of the Hunt-Muckenhoupt-Wheeden A_p-weights, we were forced, in order to avoid complicating the technique, to confine ourselves to the scale of power weights, which is invariant under the steps described above.

The investigation of Toeplitz operators with infinite index is based on the generalization of the concept of the (p, ϱ)-factorization of a bounded function. In Sections 2.1–2.6 we present known results on (p, ϱ)-factorization, the Φ-theory of Toeplitz operators, and inner-outer factorization. In particular, in Section 2.4 we provide a description of the class $\operatorname{fact}(p, \varrho)$.

In Section 2.7 we introduce the class $\operatorname{fact}(\infty, p, \varrho)$, $1 < p < \infty$. A function $a \in L_\infty(\Gamma)$ belongs to this class if it is representable in the form

$$a = bh \quad \text{or} \quad a = bh^{-1}, \tag{6}$$

where

(1) $b \in \operatorname{fact}(p, \varrho)$, $1 < p < \infty$;

(2) $h \in L_\infty^+(\Gamma)$, $h^{-1} \in L_\infty(\Gamma)$.

The following theorem describes the main result of Chapter 2 under simplified assumptions about the function b.

Theorem 0.1 *Let $a \in \operatorname{fact}(\infty, p, \varrho)$ and let the function b in* (6) *admit a canonical factorization* (*see* (2)) *with index zero.*

Then, if $a = bh^{-1}$, the operator $T(a)$ is right-invertible in the space $L_p^+(\Gamma, \varrho)$, $\dim\ker T(a) = \infty$, *and the operator*

$$[T(a)]^{-1} = \frac{h}{b^+} P^+ \frac{1}{b^-} I,$$

where $b = b^+b^-$ is the canonical factorization of the function b, is a right inverse of $T(a)$. For a function φ to belong to $\ker T(a)$ it is necessary and sufficient that

$$\varphi = \frac{h}{b^+}P^-\frac{b^+}{h}\psi, \quad \textit{where} \quad \psi \in \ker T(h^{-1}).$$

If $a = bh$, then the operator $T(a)$ is left-invertible in $L_p^+(\Gamma, \varrho)$,

$$\dim \operatorname{coker} T(a) = \infty,$$

and the operator

$$[T(a)]^{-1} = P^+\frac{1}{b^+h}P^+\frac{1}{b^-}I$$

is a left inverse for $T(a)$. For a function f to belong to $\operatorname{im} T(a)$, it is necessary and sufficient that

$$\int_\Gamma \psi_j(t)f(t)dt = 0, \quad j = 1, 2, \ldots,$$

where

$$\psi_j = \frac{1}{b^-}P^-\frac{(t-z_0)^{-j}}{hb^+} \in L_q^-(\Gamma, \varrho^{1-q}), \quad q = p/(p-1),$$

and z_0 is a fixed point in the domain $\mathcal{D}_\Gamma^-$.

In Sections 2.8–2.10 we give examples of whirl points (almost periodic, power and logarithmic whirls) and also obtain formulas for the argument of a Blaschke product and that of an arbitrary inner function. In what follows, the first of these formulas (Theorem 2.8) plays a very important role in the modelling of a broad class of discontinuities of "infinite-index" type, while the second of these formulas allows us to show that the unboundedness of the function $\arg a(t)$ is not always a sufficient reason for the operator $T(a)$ to fail being a Φ-operator (Propositions 2.10 and 2.11).

Chapter 3 is devoted to the study of the subspace $\ker T(h^{-1})$, which is used in Theorem 0.1 to describe the kernel of the operators $T(a)$ with symbols in the class $\operatorname{fact}(\infty, p, \varrho)$.

A function h that satisfies the conditions

$$h \in L_\infty^+(\Gamma), \quad h^{-1} \in L_\infty(\Gamma),$$

is called a *model function* on the curve Γ, and the operator $T(h^{-1})$, acting in $L_p^+(\Gamma, \varrho)$, and the subspace $\mathcal{M} = \ker T(h^{-1})$ are called the *model operator* and the *model subspace* (in the space $L_p^+(\Gamma, \varrho)$) generated by h.

There are two methods of describing model subspaces: in terms of the boundary behavior of their elements, which are analytic or meromorphic functions in the domains $\mathcal{D}_\Gamma^\pm$, and in terms of expansions of their elements in series with respect to special bases.

The first method (which may be termed *analytic*) is a generalization and further development of the method used in complex analysis to describe various classes of entire functions by means of estimates of their behavior at infinity. It enables us to study the functional nature of the elements of model subspaces.

The second method (which may be referred to as *constructive*) was elaborated in numerous investigations on complex interpolation by means of entire and meromorphic functions. This method provides us with tools to solve a number of constructive problems for Toeplitz operators (construction of inverse operators, among them Moore-Penrose inverses, the formulation of problems for Toeplitz equations that guarantee existence and uniqueness of the solution as well as continuous dependence on data, and so on).

In Section 3.1 we obtain some integral representations for the elements of a model subspace $\mathcal{M}(h)$, which allow us to study their analytic properties. Depending on the analytic properties of the model function $h(z)$, the elements of the model subspace may admit analytic continuation to a domain larger than their original domain of definition. As a consequence, the subspace $\mathcal{M}(h)$ turns out to be stable under certain deformations of the curve Γ near whirl points. Such deformations and the resulting curves will be termed *admissible*.

In Section 3.2 we give a description of the admissible deformations. In what follows, as a rule, we use circles or lines as admissible curves. In connection with this, we study model subspaces on the real line $\mathbb{R}$ in more detail in Section 3.3. In Section 3.4 we obtain a description of the subspace $\mathcal{M}(h)$ in terms of the boundary behavior of its elements for smooth curves.

Sections 3.5–3.10 deal with various bases in model subspaces of the space $L_p(\mathbb{R}, \varrho)$, $1 < p < \infty$, generated by the inner functions $B(x)$ and $\exp(ix\sigma)$. Therein we solve an interpolation problem which guarantees that any element of the model subspace has an expansion in a Lagrange interpolation series. In recent investigations this problem turned out to play a very important role in the nonelliptic theory of Toeplitz operators.

Chapter 4 is devoted to Toeplitz operators whose symbols have the best known form of discontinuities of oscillating type: almost periodic, semi-almost periodic, and whirl points of power order.

Symbols with *almost periodic discontinuities* (a.p. discontinuities, for short) are considered in Sections 4.1–4.5. Understanding of Toeplitz operators with such symbols is, in a sense, the key to the realm of Toeplitz operators with infinite index. We will use just these symbols to demonstrate the technique described above and to uncover the essence of our approach.

In Section 4.1 we study Toeplitz operators on $\mathbb{R}$ with uniformly almost periodic symbols. Section 4.2 gives a description of the model subspaces on a bounded smooth contour Γ that are generated by model functions of the form

$$h(t) = \exp(a_0(t - t_0)^{-1}), \quad a_0 \in \mathbb{C}, \quad t_0 \in \Gamma.$$

In Section 4.3 we introduce the class of symbols with *standard almost periodic discontinuities*. A function $a \in GL_\infty(\Gamma)$ is said to have a standard almost periodic discontinuity in a point $t_0 \in \Gamma$ if it is continuous on $\Gamma \setminus \{t_0\}$ and admits a representation

$$a(t) = \exp(a_0(t - t_0)^{-1})b(t)$$

with $a_0 \in \mathbb{C}$ and $b \in \operatorname{fact}(p, \varrho)$. For Toeplitz operators whose symbols have a finite number of standard almost periodic discontinuities, we fully implement the program of study outlined in chapters 2 and 3. In addition, in Section 4.4 we give versions of well-posed problems for the corresponding Toeplitz equations, in which the existence and uniqueness of the solution and its continuous dependence on data are guaranteed. The final Section 4.5 of this chapter treats Toeplitz operators whose symbols have general a.p. discontinuities.

In Sections 4.6–4.9 we study Toeplitz operators on $\mathbb{R}$ whose symbols exhibit different almost periodic behaviors when $x \to -\infty$ and $x \to \infty$. Such discontinuities are called *semi-almost periodic.* A model function for this class of discontinuities is connected with the integral sine function $\operatorname{Si}(z)$, well known in the theory of special functions. In Theorems 4.23–4.26 we construct a generalized factorization of the symbol and establish a theory of normal solvability of the operator $T(a)$, providing also a description of the inverse operators and of the subspaces $\ker T(a)$ and $\operatorname{im} T(a)$.

In Sections 4.10–4.11 we deal with smooth closed contours. We there consider points of one- and two-sided whirling of power character, which means that

$$\arg a(t) = O(|t - t_0|^{\lambda_\pm}), \quad 0 < \lambda_\pm < 1,$$

when $t \to t_0 \pm 0$, $t_0 \in \Gamma$.

Chapter 5, which occupies an important place in the book, is designed to prove how broad the classes of Toeplitz operators with oscillating symbols are that can be studied with the help of (∞, p, ϱ)-factorization. The approach proposed there is based on the factorization of so-called u-periodic functions.

Let $u(t)$ be an inner function on the unit circle Γ_0 and $a(t) \in L_\infty(\Gamma_0)$. Then the composition $a(u(t))$ is called an *u-periodic function.* The justification for this terminology is that, for the simplest inner function on the real line, $u(\theta) = \exp(i\theta)$, $\theta \in \mathbb{R}$, the composition $a(\exp(i\theta))$ is an ordinary 2π-periodic function. To make clear the idea underlying the factorization of u-periodic functions, let us consider the analogue of representation (2) for the circle Γ_0:

$$a(t) = a^+(t)t^\kappa a^-(t), \tag{7}$$

where

$$(a^+)^{\pm 1} \in H^\infty(D_+), \quad (a^-)^{\pm 1} \in H^\infty(D_-), \quad \kappa \in \mathbb{Z},$$

and D_+ and D_- denote the interior and the exterior of Γ_0, respectively. Making the substitution $t = u(\tau)$ in (7), we obtain the equality

$$a(u(\tau)) = a^+(u(\tau))u^\kappa(\tau)a^-(u(\tau)),$$

which is an (∞, p, ϱ)-factorization of the function $a(u(\tau))$, because

$$u \in H^{\infty}(D_+), \quad (a^+)^{\pm 1} \in H^{\infty}(D_+), \quad (a^-)^{\pm 1} \in H^{\infty}(D_-).$$

We wish to point out that the theory of u-factorization is readily extended to the case of matrix functions. In connection with this, Sections 5.1 and 5.2 are devoted to constructing a theory of normal solvability of matrix Toeplitz operators with infinite index. The foundation of this theory is constituted by the notion of the generalized factorization of matrix functions, which is the analogue of the (∞, p, ϱ)-factorization of functions considered in Chapter 2.

In Section 5.3 we present a theory of Toeplitz operators with u-periodic matrix symbols. This makes extensive use of the properties of the classical composition operator

$$(\gamma_u a)(t) = a(u(t)).$$

As a first example of application of the technique of u-factorization, in Section 5.4 we consider the case when the argument of the symbol of the Toeplitz operator tends to infinity in a logarithmic fashion near a discontinuity point.

Sections 5.6–5.8 occupy the central place in Chapter 5. Here we obtain sufficient conditions on the function $\arg a(t)$ under which $a(t)$ admits an (∞, p, ϱ)-factorization of the form

$$a(t) = B(t)d(t), \tag{8}$$

where $B(t)$ is a Blaschke product whose argument is close to the argument of $a(t)$ and $d \in \text{fact}(\infty, p, \varrho)$. These conditions are so flexible that they cover the cases when the function $\arg a(t)$ is of power, power-logarithmic, exponential, super-exponential, and many other types.

In Section 5.8 we single out situations in which representation (8) can be replaced by

$$a(t) = U(t)c(t), \tag{9}$$

where $U(t)$ is some inner function connected with the Blaschke product $B(t)$ and the function $c(t)$ is continuous. In the last two sections of Chapter 5, 5.9 and 5.10, a representation of the form (9) is used to study scalar and matrix Toeplitz operators with generalized almost periodic symbols.

Chapter 6 is devoted to Toeplitz operators whose symbols have zeros and the degeneracy of the symbol has polynomial or locally-polynomial character.

In Section 6.2 we discuss one variant of normalization of linear operators in a reflexive Banach space. We assume that the linear operator A can be written as

$$A = BC,$$

where B is unboundedly invertible and C is a normally solvable operator. In this case, by introducing the Banach space $DX = \operatorname{im} B$ equipped with a new norm, one achieves that the operator $A : X \to DX$ is normally solvable. In the process we also describe a general class of linear operators that admit such a representation.

In Section 6.3 we carry out the normalization and construct a theory of one-sided invertibility of the operator $T(a)$ in the space $L_p(\Gamma, \varrho)$ under the assumption that its symbol can be written as

$$a = wb,$$

where w is a polynomial with complex coefficients with all roots on Γ, and $b \in \mathrm{fact}(\infty, p, \varrho)$. At the same time we show how generalized solutions arise in the normalization process of a Toeplitz operator with vanishing symbol.

In Section 6.4 we consider Toeplitz operators in the space $L_p^+(\mathbb{R})$, $1 < p < \infty$, under the assumption that their symbols have degeneracies of the form $x - x_k$ on an arbitrary separated infinite subset of the real line.

We next introduce in an axiomatic manner the class of *generators of zeros*, i.e., inner functions $w \in L_\infty(\mathbb{R})$ whose sets of zeros lie on $\mathbb{R}$, with the property that the operator $T(w)$ is unboundedly invertible in $L_p^+(\mathbb{R})$, $1 < p < \infty$. Armed with this concept, we carry out the normalization of operators $T(a)$ with symbols a of the form $a = wb$, where $b \in \mathrm{fact}(\infty, p, \varrho)$, and we construct a theory of the one-sided invertibility of such operators and provide a constructive description of the space $DL_p^+(\mathbb{R})$ (Theorem 6.12).

Finally, in Section 6.5 we examine the most important examples of generators of zeros that are encountered in the literature. These are special entire functions of exponential type, namely, sine-type functions (as introduced by B. Ya. Levin) with real zeros, and special meromorphic functions, connected with Blaschke products, whose sets of zeros lie on lines parallel to the real line.

Chapter 1

Examples of Toeplitz Operators with Infinite Index

In this chapter, after introducing the necessary definitions and studying the simplest properties of Toeplitz operators, we will show on a number of examples how a Toeplitz operator may turn out to have an infinite index.

Auxiliary material

1.1 The space $L_p(\mathbb{R}, \varrho)$ and the operator $S_{\mathbb{R}}$

Let $\varrho(x)$ be a weight function on the real line $\mathbb{R}$ of the form

$$\varrho(x) = |x+i|^{\alpha_0} \sum_{k=1}^{m} |x - x_k|^{\alpha_k}, \quad x_k \in \mathbb{R}, \tag{1.1}$$

$$-1 < \alpha_k, \alpha < p-1, \quad \alpha = \sum_{k=0}^{m} \alpha_k.$$

A measurable function $f(x)$ on $\mathbb{R}$ is said to belong to the space $L_p(\mathbb{R}, \varrho)$, $1 \le p < \infty$, if

$$\|f\|_{p,\varrho} = \left(\int_{\mathbb{R}} |f(x)|^p \varrho(x) dx \right)^{1/p} < \infty. \tag{1.2}$$

The space $L_p(\mathbb{R}, \varrho)$, $1 \le p < \infty$, equipped with the norm (1.2), is a Banach space. As usual, we denote $L_p(\mathbb{R}) = L_p(\mathbb{R}, 1)$. The dual of the space $L_p(\mathbb{R}, \varrho)$, $1 < p < \infty$, is $L_q(\mathbb{R}, \varrho^{1-q})$, where $q = p/(p-1)$.

We will denote by $S_{\mathbb{R}}$ the Cauchy-Lebesgue singular integral operator, defined by the formula

$$(S_{\mathbb{R}}f)(x) = \frac{1}{\pi i}\int_{\mathbb{R}} \frac{f(\tau)d\tau}{\tau - x}, \quad x \in \mathbb{R},$$

where the integration is understood in the sense of the principal value. The operator $S_{\mathbb{R}}$ coincides with the well-known Hilbert transform and is bounded in the space $L_p(\mathbb{R}, \varrho)$, $1 < p < \infty$, provided ϱ has the form (1.1).

1.2 The classes $L_p^{\pm}(\mathbb{R}, \varrho)$

Let us introduce the analytic projectors

$$P^{\pm} = P_{\mathbb{R}}^{\pm} = \frac{1}{2}(I \pm S_{\mathbb{R}}). \tag{1.3}$$

The projectors $P^{\pm}$ are orthogonal in the sense that $P^{+}P^{-} = P^{-}P^{+} = 0$. Now let us define the spaces $L_p^{\pm}(\mathbb{R}, \varrho)$ by

$$L_p^{\pm}(\mathbb{R}, \varrho) = P_{\mathbb{R}}^{\pm}(L_p(\mathbb{R}, \varrho)).$$

From the Sokhotskiĭ-Plemelj formula, which connects the boundary values of the Cauchy-type integral

$$I(x+iy) = \frac{1}{2\pi i}\int_{\mathbb{R}} \frac{f(\tau)d\tau}{\tau - x - iy}, \quad y \neq 0, \quad f \in L_p(\mathbb{R}, \varrho),$$

with the singular integral operator $S_{\mathbb{R}}$, namely,

$$\lim_{y \to 0 \pm 0} I(x+iy) = \pm\frac{1}{2}f(x) + \frac{1}{2}(S_{\mathbb{R}}f)(x), \quad x \in \mathbb{R}, \tag{1.4}$$

it follows that the space $L_p^{\pm}(\mathbb{R}, \varrho)$ consists of functions that admit analytic continuation to the half-plane $\Pi_{+} = \{x+iy : y > 0\}$ and $\Pi_{-} = \{x+iy : y < 0\}$, respectively. Equality (1.4) holds almost everywhere on $\mathbb{R}$ when the point $z = x+iy$ tends to x along any path that is not tangent to $\mathbb{R}$ at x, with the path remaining either in Π_{+} or in Π_{-}. From definition (1.3) and the boundedness of the projectors $P^{\pm}$ it follows that the space $L_p(\mathbb{R}, \varrho)$ admits the direct sum decomposition

$$L_p(\mathbb{R}, \varrho) = L_p^{+}(\mathbb{R}, \varrho) \oplus L_p^{-}(\mathbb{R}, \varrho). \tag{1.5}$$

Let us denote by $R^{\pm}(\mathbb{R})$ the algebras of rational functions vanishing at infinity and having their poles in the respective half-plane $\Pi_{\mp}$. Then for any p, $1 < p < \infty$[1],

$$R^{\pm}(\mathbb{R}) \subset L_p^{\pm}(\mathbb{R}, \varrho) \quad \text{and} \quad \operatorname{clos} R^{\pm}(\mathbb{R}) = L_p^{\pm}(\mathbb{R}, \varrho).$$

Here are a few criteria for a function to belong to the spaces $L_p^{\pm}(\mathbb{R}, \varrho)$.

[1] Here and in what follows **clos** stand for closure.

Proposition 1.1 *Let $f \in L_p(\mathbb{R},\varrho)$, $1 < p < \infty$. The following assertions are equivalent:*

(1) $f \in L_p^{\pm}(\mathbb{R},\varrho)$;

(2) $\displaystyle \frac{1}{2\pi i}\int\limits_{\mathbb{R}} \frac{f(\tau)d\tau}{\tau - z} = \pm f(z)$ *for all* $z \in \Pi_{\pm}$;

(3) $\displaystyle \int\limits_{\mathbb{R}} \frac{f(\tau)d\tau}{\tau - z} = 0$ *for all* $z \in \Pi_{\mp}$;

(4) $\displaystyle \int\limits_{\mathbb{R}} \frac{f(\tau)d\tau}{(\tau - z_0)^n} = 0$ *for all* $n \in \mathbb{N}$, $z_0 \in \Pi_{\mp}$.

Assume that the weight $\varrho(x)$ has the form (1.1). Let us introduce the functions

$$\varrho_p^{\pm}(z) = [(z \pm i)^{\alpha_0/p}]_{\pm} \prod_{k=1}^{m} [(z - x_k)^{\alpha_k/p}]_{\pm}, \tag{1.6}$$

where the branches of the multi-valued functions $(z \pm i)^{\alpha_0/p}$, $(z - x_k)^{\alpha_k/p}$ are chosen so that the functions $\varrho_p^{\pm}(z)$ are analytic in the respective half-planes $\Pi_{\pm}$.

The following result establishes a connection between the spaces $L_p^{\pm}(\mathbb{R},\varrho)$ and $L_p^{\pm}(\mathbb{R})$.

Proposition 1.2 *For f to belong to $L_p^{\pm}(\mathbb{R},\varrho)$ it is necessary and sufficient that $\varrho_p^{\pm} f \in L_p^{\pm}(\mathbb{R})$, $1 < p < \infty$.*

A function $f(z)$, $z = x + iy$, which is analytic in $\Pi_{\pm}$ is said to belong to the *Hardy space* $H^{p,\varrho}(\Pi_{\pm})$, $1 \le p < \infty$, if

$$\|f\|_{p,\varrho} = \sup_{y>0} \left(\int_{\mathbb{R}} |\varrho_p^{\pm}(x \pm iy) f(x \pm iy)|^p dx \right)^{1/p} < \infty. \tag{1.7}$$

As usual, one denotes $H^{p,1}(\Pi_{\pm}) = H^p(\Pi_{\pm})$, $1 \le p < \infty$. We let $H^{\infty}(\Pi_{\pm})$ denote the spaces of all bounded analytic functions in $\Pi_{\pm}$, equipped with the norms

$$\|f\|_{\infty} = \sup_{y>0} \sup_{x\in\mathbb{R}} |f(x \pm iy)|.$$

When $1 < p < \infty$, the spaces $L_p^{\pm}(\mathbb{R},\varrho)$ coincide with the Hardy spaces $H^{p,\varrho}(\Pi_{\pm})$, in the sense that if $f(x) \in L_p^{\pm}(\mathbb{R},\varrho)$, then its analytic continuation $f(z)$ to the domain $\Pi_{\pm}$ is a function of class $H^{p,\varrho}(\Pi_{\pm})$. Conversely, if $f(z) \in H^{p,\varrho}(\Pi_{\pm})$, $z = x + iy$, then for $z \to x$, $z \in \Pi_{\pm}$, the nontangential limit values $f(x)$ exist for almost all points $x \in \mathbb{R}$ and form a function of the class $L_p^{\pm}(\mathbb{R},\varrho)$. The indicated correspondence is an isometry of Banach spaces. In view of this property, from now on, whenever needed and without specially mentioning it, we will identify a function $f(x)(\in L_p^\pm(\mathbb{R},\varrho))$ with its analytic continuation $f(z)(\in H^{p,\varrho}(\Pi_\pm))$. Moreover, for uniformity purposes we will use the notations

$$L_p^\pm(\mathbb{R},\varrho) = H^{p,\varrho}(\Pi_\pm), \quad 1 \le p < \infty,$$

$$L_\infty^\pm(\mathbb{R}) = H^\infty(\Pi_\pm).$$

Note that $L_p^+(\mathbb{R},\varrho) \cap L_p^-(\mathbb{R},\varrho) = \{0\}$, $1 \le p < \infty$.

Let us denote by $L_\infty(\mathbb{R})$ the algebra of measurable essentially bounded functions $f(x)$ on $\mathbb{R}$, equipped with the norm

$$\|f\|_\infty = \operatorname*{ess\,sup}_{x\in\mathbb{R}} |f(x)|.$$

The algebra $L_\infty(\mathbb{R})$ does not decompose into the direct sum of its subalgebras $L_\infty^\pm(\mathbb{R})$; however,

$$\operatorname{clos}(L_\infty^+(\mathbb{R}) + L_\infty^-(\mathbb{R})) = L_\infty(\mathbb{R}).$$

We conclude this section by describing the dual spaces of the spaces $L_p^\pm(\mathbb{R},\varrho)$.

Proposition 1.3 *For $1 < p < \infty$ the space $(L_p^\pm(\mathbb{R},\varrho))^*$ is isometrically isomorphic to the quotient space $L_q(\mathbb{R},\varrho^{1-q})/L_q^\mp(\mathbb{R},\varrho^{1-q})$, $q = p/(p-1)$. Analogously, $(L_1^\pm(\mathbb{R}))^* \cong L_\infty(\mathbb{R})/L_\infty^\mp(\mathbb{R})$.*

A functional $G \in (L_q(\mathbb{R},\varrho^{1-q})/L_q^\mp(\mathbb{R},\varrho^{1-q}))$ is understood in the following sense: there exists a unique function $g \in L_q^\pm(\mathbb{R},\varrho^{1-q})$ such that for any function $f \in L_p^\pm(\mathbb{R},\varrho)$

$$G(f) = \int_{\mathbb{R}} f(x)\overline{g(x)}dx.$$

Then

$$\|G\| = \operatorname{dist}(g, L_q^\mp(\mathbb{R},\varrho^{1-q})).$$

If now $G \in (L_\infty(\mathbb{R})/L_\infty^\mp(\mathbb{R}))$, then there exists a unique function $g \in L_\infty^\pm(\mathbb{R})$ such that for any function $f \in L_1^\pm(\mathbb{R})$

$$G(f) = \int_{\mathbb{R}} f(x)\overline{g(x)}dx,$$

and then $\|G\| = \operatorname{dist}(g, L_\infty^\mp(\mathbb{R}))$.

1.3 Normally solvable operators

Let X be a Banach space and $\mathcal{L}(X)$ be the algebra of bounded linear operators in X. An operator $A \in \mathcal{L}(X)$ is said to be *normally solvable* if its image (range) is closed in X

$$\operatorname{im} A = \operatorname{clos} \operatorname{im} A.$$

We put $\ker A = \{x \in X : Ax = 0\}$ and $\operatorname{coker} A = X/\operatorname{clos}\operatorname{im} A$. A normally solvable operator A is called a *Φ_+-operator* if $\alpha = \dim\ker A < \infty$, a *$\Phi_-$-operator* if $\beta = \dim\operatorname{coker} A < \infty$, and a *$\Phi$-operator* if both $\ker A$ and $\operatorname{coker} A$ are finite-dimensional. The number

$$\alpha - \beta = \dim\ker A - \dim\operatorname{coker} A$$

is called the *index* of the operator A and is denoted by $\operatorname{ind} A$.

Note that $\dim\operatorname{coker} A = \dim\ker A^*$. Φ-operators (resp., $\Phi_\pm$-operators) are also called *Fredholm* (resp., *semi-Fredholm*) operators.

An operator $A \in \mathcal{L}(X)$ is said to be *left* (resp., *right*) *invertible* if there exists an operator $B \in \mathcal{L}(X)$ such $BA = I$ (resp., $AB = I$). The operator B is called a *left* (resp., *right*) *inverse* of A and will be denoted by A^{-1}.

An operator $A \in \mathcal{L}(X)$ is left invertible if and only if its kernel is trivial, $\ker A = \{0\}$, and its image $\operatorname{im} A$ is closed and has a direct complement in X. If these conditions are satisfied and $A_0^{-1} \in \mathcal{L}(X)$ is one of the left inverses of A, then the general form of the left inverses of A is $A_0^{-1}P$, where P is an arbitrary bounded projector in X with the property $\operatorname{im} P = \operatorname{im} A$.

An operator $A \in \mathcal{L}(X)$ is right invertible if and only if $\operatorname{im} A = X$ and $\ker A$ has a direct complement in X. If these conditions are satisfied and $A_0^{-1} \in \mathcal{L}(X)$ is one of the right inverses of A, then the general form of the right inverses of A is PA_0^{-1}, where P is an arbitrary bounded projector in X with the property that $\ker P = \ker A$.

An operator A is invertible if and only if it is simultaneously left and right invertible.

Any left invertible operator is a Φ_+-operator ($\alpha = 0$) and any right invertible operator A is a Φ_--operator ($\beta = 0$).

We will also need the notion of generalized invertibility. A normally solvable operator A is said to be *generalized invertible* (or to *have a generalized inverse*) if there exists an operator $B \in \mathcal{L}(X)$ such that $ABA = A$. In this case the operator B is called a *generalized inverse* of A.

An operator $A \in \mathcal{L}(X)$ is generalized invertible if and only if its image $\operatorname{im} A$ is closed and the subspaces $\ker A$ and $\operatorname{im} A$ have direct complements in X. If these conditions are satisfied and $A_0^{-1} \in \mathcal{L}(X)$ is one of the generalized inverses of the operator A, then the general form of the generalized inverses of A is $B = QA_0^{-1}P$, where P and Q are arbitrary bounded projectors in X with the properties $\operatorname{im} P = \operatorname{im} A$, $\ker Q = \ker A$. Note that the two subspaces $\ker A$ and $X/\operatorname{im} A$ may be infinite-dimensional in this case. It is also clear that one-sided invertible operators are generalized invertible.

The definitions given above extend in a natural way to bounded linear operators acting from a space X into a space Y when X and Y have a structure that is more general than that of a Banach space, for instance, when they are Fréchet spaces, i.e., complete locally convex metrizable linear spaces.

1.4 Toeplitz operators

A Toeplitz operator in the space $L_p^+(\mathbb{R}, \varrho)$, $1 < p < \infty$, is defined by the rule

$$T(a) = P_{\mathbb{R}}^+ aI. \tag{1.8}$$

The function $a \in L_\infty(\mathbb{R})$ is called the *symbol* of the operator $T(a)$. As usual, aI denotes the operator of multiplication by the function $a(t)$. When $p = 2$ and $\varrho \equiv 1$,

$$\|T(a)\|_{L_2(\mathbb{R})} = \|a\|_{L_\infty(\mathbb{R})}, \tag{1.9}$$

while in the general case

$$\|T(a)\|_{L_p(\mathbb{R},\varrho)} \le c_p \|a\|_{L_\infty(\mathbb{R})}, \quad \text{where} \quad c_p = \|P_{\mathbb{R}}^+\|_{L_p(\mathbb{R},\varrho)}. \tag{1.10}$$

The adjoint operator $T^*(a)$, which acts in the space $L_q^+(\mathbb{R}, \varrho^{1-q})$, $q = p/(p-1)$, is given by

$$T^*(a) = T(\overline{a}).$$

Let us discuss a number of basic properties of Toeplitz operators.

Proposition 1.4 *Let $a \in L_\infty(\mathbb{R})$ and $a \not\equiv 0$. Then at least one of the numbers* $\dim\ker T(a)$ *and* $\dim\operatorname{coker} T(a)$ *is equal to zero.*

Proof. Let $f \in \ker T(a)$ in $L_p^+(\mathbb{R}, \varrho)$ and $g \in \ker T(\overline{a})$ in $L_q^+(\mathbb{R}, \varrho^{1-q})$, $q = p/(p-1)$. Then

$$\begin{aligned} P^+ af &= 0 \iff af = f_1, \text{ where } f_1 \in L_p^-(\mathbb{R}, \varrho), \\ P^+ \overline{a} g &= 0 \iff \overline{a} g = g_1, \text{ where } g_1 \in L_q^-(\mathbb{R}, \varrho^{1-q}). \end{aligned}$$

Consequently, $f_1\overline{g} = f\overline{g}_1$. Since $\overline{g}_1 \in L_q^+(\mathbb{R}, \varrho^{1-q})$, it follows that $f\overline{g}_1 \in L_1^+(\mathbb{R})$.
Indeed, using the Hölder inequality we have

$$\begin{aligned} &\int_{\mathbb{R}} |f(x+iy)\overline{g_1(x+iy)}|dx \\ &= \int_{\mathbb{R}} |f(x+iy)\varrho_p^+(x+iy)||\overline{g_1(x+iy)}[\varrho_p^+(x+iy)]^{-1}|dx \\ &\le \left(\int_{\mathbb{R}} |f(x+iy)\varrho_p^+(x+iy)|^p dx\right)^{1/p} \\ &\times \left(\int_{\mathbb{R}} |\overline{g_1(x+iy)}[\varrho_p^+(x+iy)]^{-1}|^q dx\right)^{1/q} \\ &\le \|f\|_{p,\varrho}\|g_1\|_{q,\varrho^{1-q}} < \infty, \end{aligned}$$

for all for $y > 0$, where $\varrho_p^+(x+iy)$ has the form (1.6). But then

$$\sup_{y>0} \int_{\mathbb{R}} |f(x+iy)\overline{g_1(x+iy)}|dx < \infty,$$

i.e., $f\overline{g}_1 \in L_1^+(\mathbb{R})$, as claimed. In a similar way one verifies that $f_1\overline{g} \in L_1^-(\mathbb{R})$. Since $L_1^+(\mathbb{R}) \cap L_1^-(\mathbb{R}) = \{0\}$, we conclude that $f_1\overline{g} = f\overline{g}_1 = 0$ almost everywhere. Now suppose that $f \neq 0$ almost everywhere, and hence $g_1 = 0$ almost everywhere on $\mathbb{R}$. By assumption, there exists a set $M \subset \mathbb{R}$, $\operatorname{meas} M > 0$, on which $\overline{a} \neq 0$. But then $g = 0$ on M, that is (see, for example, Privalov [1], p. 292), $g = 0$ almost everywhere on $\mathbb{R}$. In much the same manner the condition $g \neq 0$ implies that $f = 0$ almost everywhere on $\mathbb{R}$. □

A consequence of this result is that a normally solvable Toeplitz operator in $L_p^+(\mathbb{R}, \varrho)$, $1 < p < \infty$, is necessarily a Φ_+-, Φ_-- or Φ-operator; moreover, in the case of the space $H^2(\Pi_+)$ such an operator is at least one-sided invertible, because in a Hilbert space any subspace has a direct complement.

The next assertion is established in an analogous manner.

Proposition 1.5 *Let $a \in L_\infty(\mathbb{R})$, $a(t) \not\equiv 0$, and assume that there exists a set $M \subset \mathbb{R}$, $\operatorname{meas} M > 0$, on which $a = 0$ almost everywhere. Then* $\dim\ker T(a) = \dim\operatorname{coker} T(a) = 0$.

The following result describes a necessary condition for the normal solvability of Toeplitz operators.

Proposition 1.6 *Let $a \in L_\infty(\mathbb{R})$ and $a \not\equiv 0$. In order for a Toeplitz operator $T(a)$ of the form* (1.8) *to be normally solvable in the space $L_p^+(\mathbb{R}, \varrho)$, $1 < p < \infty$, it is necessary that*

$$\operatorname*{ess\,inf}_{x\in\mathbb{R}} |a(x)| > 0. \tag{1.11}$$

Proof. We proceed by reductio ad absurdum. Suppose that (1.11) does not hold, but the operator $T(a)$ is normally solvable. Then there exist a set $M \subset \mathbb{R}$, $\operatorname{meas} M > 0$, and a function $a_\varepsilon \in L_\infty(\mathbb{R})$, such that $a_\varepsilon = 0$ almost everywhere on M and $\|a - a_\varepsilon\|_{L_\infty(\mathbb{R})} < \varepsilon$, where ε is an arbitrarily small fixed positive number. By Proposition 1.4, $T(a)$ is a Φ_+- or Φ_--operator. Since the sets of all Φ_+-operators and all Φ_--operators, acting in a given Banach space, are stable under small perturbations in the operator norm (see Gohberg and Krupnik [2], Theorem 15.4), inequality (1.10) implies that $T(a_\varepsilon)$ is also a Φ_+- or Φ_--operator. By Proposition 1.5, $T(a_\varepsilon)$ is invertible, and consequently the equation $T(a_\varepsilon)f = g$ is solvable for every $g \in L_p^+(\mathbb{R}, \varrho)$.

Take $g(x) = 1/(x+i)$, and let $f \in L_p^+(\mathbb{R}, \varrho)$ be the solution of the equation $(T(a_\varepsilon)f)(x) = 1/(x+i)$. Then

$$a_\varepsilon(x)f(x) = (P^- a_\varepsilon f)(x) + \frac{1}{x+i}$$

for almost all $x \in \mathbb{R}$. The function

$$h(z) = (P^- a_\varepsilon f)(z) + \frac{1}{z+i}$$

is meromorphic in Π_- and has zero nontangential limit values almost everywhere on M. By a theorem of N. N. Luzin and I. I. Privalov (Privalov [1], p. 292), $h(x) = 0$ almost everywhere on $\mathbb{R}$, i.e.,

$$(P^- a_\varepsilon f)(x) = -\frac{1}{x+i}$$

almost everywhere on $\mathbb{R}$, which is impossible. □

In general, the composition of two Toeplitz operators is not a Toeplitz operator. For this reason the next result plays an important role in the algebraic theory of Toeplitz operators.

Proposition 1.7 *Let $a^\pm \in L^\pm_\infty(\mathbb{R})$ and $a \in L_\infty(\mathbb{R})$. Then*

$$T(a)T(a^+) = T(aa^+), \quad T(a^-)T(a) = T(a^-a).$$

Proof. The proof of this assertion is elementary and uses only the definitions of the Toeplitz operator and the projectors $P^\pm$. □

A central place in the theory of Toeplitz operator is occupied by the concept of factorization of symbols.

At first let us consider the most simple variant of it.

We say that a function $a \in L_\infty(\mathbb{R})$ admits a *standard factorization* if it can be represented in the form

$$a(x) = a^-(x)\left(\frac{x-i}{x+i}\right)^\kappa a^+(x), \quad x \in \mathbb{R}, \tag{1.12}$$

where $\kappa \in \mathbb{Z}$ and

$$(a^+)^{\pm 1} \in L^+_\infty(\mathbb{R}), \quad (a^-)^{\pm 1} \in L^-_\infty(\mathbb{R}). \tag{1.13}$$

The number κ is called the *index* of the factorization.

Theorem 1.1 *Let the function $a \in L_\infty(\mathbb{R})$ admit a standard factorization of the form (1.12), (1.13). If $\kappa = 0$, then the operator $T(a)$ given by (1.8) is invertible in $L^+_p(\mathbb{R}, \varrho)$, and its inverse is given by*

$$[T(a)]^{-1} = \frac{1}{a^+}P^+\frac{1}{a^-}I.$$

If $\kappa < 0$, then $T(a)$ is right invertible in $L^+_p(\mathbb{R}, \varrho)$, $\dim\ker T(a) = -\kappa$, *and a right inverse is given by*

$$[T(a)]^{-1} = \left(\frac{x+i}{x-i}\right)^\kappa \frac{1}{a^+}P^+\frac{1}{a^-}I.$$

Furthermore,

$$\ker T(a) = \operatorname{span}\{[a^+(x)(x+i)^j]^{-1},\ j = 1, 2, \ldots, |\kappa|\}.$$

If $\kappa > 0$, then $T(a)$ is left invertible in $L_p^+(\mathbb{R}, \varrho)$, $\dim \operatorname{coker} T(a) = \kappa$, *and a left inverse is given by*

$$[T(a)]^{-1} = P^+ \left(\frac{x+i}{x-i}\right)^\kappa \frac{1}{a^+} P^+ \frac{1}{a^-} I.$$

Also, for f to belong to $\operatorname{im} T(a)$ *it is necessary and sufficient that*

$$\int_{\mathbb{R}} \frac{f(x)}{a^-(x)(x-i)^k} dx = 0, \quad k = 1, 2, \ldots, \kappa.$$

The proof of a more general version of this theorem will be given later in Chapter 2 (see Section 2.1).

All the definitions and properties of Toeplitz operators introduced above on the real line $\mathbb{R}$ carry over, with the requisite modifications, to the case of those Jordan curves for which the Cauchy-Lebesgue singular integral operator generates a bounded operator.

Examples of operators with infinite index

1.5 Blaschke products

Let us consider a finite Blaschke product on the real line:

$$B_n^-(x) = \prod_{k=1}^n \nu_k \frac{x - z_k}{x - \overline{z}_k}, \quad \nu_k = \frac{1 + z_k^2}{|1 + z_k^2|}, \tag{1.14}$$

constructed from distinct zeros $z_k \in \Pi_-$. Clearly

$$\arg B_n^-(x)\big|_{-\infty}^{\infty} = -2\pi n$$

and so

$$\operatorname{ind}_{\mathbb{R}} B_n^- := \frac{1}{2\pi} \arg B_n^-(x)\Big|_{-\infty}^{\infty} = -n. \tag{1.15}$$

Let us study the operator $T(B_n^-)$ in $L_p^+(\mathbb{R}, \varrho)$ without resorting to Theorem 1.1. Since $B_n^- \in H^\infty(\Pi_-)$, $\overline{B_n^-} = (B_n^-)^{-1} \in H^\infty(\Pi_+)$, the operator $T(B_n^-)$ is right invertible, and $T(\overline{B_n^-})$ is a right inverse for it.

We claim that

$$\ker T(B_n^-) = \operatorname{span}\{(x - z_k)^{-1},\ k = 1, \ldots, n\}. \tag{1.16}$$

Indeed,

$$\frac{B_n^-(x)}{x - z_k} \in L_p^-(\mathbb{R}, \varrho) \Longrightarrow (x - z_k)^{-1} \in \ker T(B_n^-).$$

Conversely, if $f \in \ker T(B_n^-)$, then one can choose numbers $c_k \in \mathbb{C}$ such that

$$g(x) = B_n^-(x)f(x) - \sum_{k=1}^{n} \frac{c_k}{x - \overline{z}_k} \in L_p^+(\mathbb{R}, \varrho).$$

But $\overline{z}_k \in \Pi_+$, and so $(x - \overline{z}_k)^{-1} \in L_p^-(\mathbb{R}, \varrho)$, and since $P^+B_n^- f = 0$, we see that $P^+g = 0$. This means that $g \in L_p^-(\mathbb{R}, \varrho)$, and hence that $g \equiv 0$. We conclude that

$$f(x) = \sum_{k=1}^{n} \frac{c_k}{(x - \overline{z}_k)B_n^-(x)} = \sum_{k=1}^{n} \frac{\widetilde{c}_k}{x - z_k}, \quad \widetilde{c}_k \in \mathbb{C},$$

and so $f \in \operatorname{span}\{(x - z_k)^{-1}, k = 1, \ldots, n\}$, as claimed.

If $B_n^+(x)$ has the form (1.14), where $z_k \in \Pi_+$, i.e., it is a finite Blaschke product for the upper half-plane, then

$$\operatorname{ind}_{\mathbb{R}} B_n^+ = n. \tag{1.17}$$

Since $B_n^+ \in H^\infty(\Pi_+)$, $\overline{B_n^+} = (B_n^+)^{-1} \in H^\infty(\Pi_-)$, it follows that the operator $T(B_n^+)$ is left invertible and that the operator $T(\overline{B_n^+})$ is a left inverse for it. It remains to verify that

$$\operatorname{im} T(B_n^+) = [\operatorname{span}\{(x - \overline{z}_k)^{-1}, k = 1, \ldots, n\}]^\perp.$$

Indeed,

$$\operatorname{im} T(B_n^+) = [\ker T^*(B_n^+)]^\perp.$$

But $T^*(B_n^+) = T(\overline{B_n^+})$, and the function $\overline{B_n^+}(x)$ is a finite Blaschke product with the zeros $\overline{z}_k$, $k = 1, \ldots, n$, which lie in Π_-. Therefore,

$$\ker T^*(B_n^+) = \operatorname{span}\{(x - \overline{z}_k)^{-1}, k = 1, \ldots, n\}.$$

From this it follows that for f to belong to $\operatorname{im} T(B_n^+)$ it is necessary and sufficient that the following orthogonality conditions be satisfied:

$$\int_{\mathbb{R}} \frac{f(x)dx}{x - z_k} = 0, \quad k = 1, \ldots, n. \tag{1.18}$$

Now let $n \to \infty$ in (1.14), requiring that the sequences $\{\overline{z}_k\}$, if $z_k \in \Pi_-$, and $\{z_k\}$, if $z_k \in \Pi_+$, $z_k \neq z_j$ for $k \neq j$, satisfy the Blaschke condition

$$\sum_{k=1}^{\infty} \frac{\operatorname{Im} z_k}{1 + |z_k|^2} < \infty. \tag{1.19}$$

Then the infinite products

$$B^\pm(x) = \prod_{k=1}^{\infty} \nu_k \frac{x - z_k}{x - \overline{z}_k}, \quad z_k \in \Pi_\pm, \tag{1.20}$$

converge uniformly on any compact subset of $\Pi_\pm$ to functions $B^\pm$ that belong to the classes $H^\infty(\Pi_\pm)$ (see Garnett [1], Chap. II, §2). The functions $B^\pm(x)$ are called *Blaschke products* for the respective half-plane $\Pi_\pm$, and (1.15) and (1.17) justify to define

$$\operatorname{ind}_{\mathbb{R}} B^\pm = \frac{1}{2\pi} \arg B^\pm(x)\big|_{-\infty}^{\infty} = \pm\infty.$$

Thus, infinite Blaschke products provide the first examples (and very important ones for the entire theory) of symbols with infinite index. As a result of the above passage to the limit $n \to \infty$, the operators $T(B^-)$ and $T(B^+)$ acquire an infinite-dimensional kernel and cokernel, respectively. The reasoning in the finite-dimensional case is readily extended to the limit case, and equality (1.16) and conditions (1.18) admit straightforward generalizations.

Theorem 1.2 *Let the Blaschke products $B^\pm$ have the form* (1.20). *Then the operators $T(B^+)$ and $T(B^-)$ are left and respectively right invertible in the space $L_p^+(\mathbb{R}, \varrho)$, $1 < p < \infty$, and the operators $T(\overline{B^+})$ and $T(\overline{B^-})$ are the corresponding one-sided inverses. Moreover,*

$$\dim\ker T(B^-) = \dim\operatorname{coker} T(B^+) = \infty,$$

and

$$\ker T(B^-) = \operatorname{clos\,span}\{(x - z_k)^{-1}, k \in \mathbb{N}\}, \quad z_k \in \Pi_-. \tag{1.21}$$

Also, in order that $f \in \operatorname{im} T(B^+)$ it is necessary and sufficient that the following orthogonality condition be satisfied:

$$\int_{\mathbb{R}} \frac{f(x)dx}{x - z_k} = 0, \quad k \in \mathbb{N}, \quad z_k \in \Pi_+. \tag{1.22}$$

Proof. Since $\overline{B^\pm} = (B^\pm)^{-1} \in H^\infty(\Pi_\mp)$, using Proposition 1.7 one readily obtains the assertions on the invertibility of the operators $T(B^\pm)$. Now let us prove equality (1.21).

When $z_k \in \Pi_-$, $B^-(x)(x - z_k)^{-1} \in L_p^-(\mathbb{R}, \varrho)$, and consequently $(x - z_k)^{-1} \in \ker T(B^-)$. Therefore,

$$\mathcal{L} = \operatorname{clos\,span}\{(x - z_k)^{-1}, k \in \mathbb{N}\} \subset \ker T(B^-).$$

Conversely, if $f \in \ker T(B^-)$, then $P^+B^-f = 0$, i.e., $B^-f \in L_p^-(\mathbb{R}, \varrho)$. Assuming that $f \notin \mathcal{L}$ and applying the Hahn-Banach theorem one finds a function $g \in L_q^+(\mathbb{R}, \varrho^{1-q})$, $q = p/(p-1)$, such that

$$\int_{\mathbb{R}} f(x)\overline{g(x)}dx = 1, \quad g(\overline{z}_k) = \frac{1}{2\pi i}\int_{\mathbb{R}} \frac{g(x)}{x - \overline{z}_k}dx = 0, \quad k \in \mathbb{N}.$$

Constructing the Blaschke product for Π_+ with the zeros $\overline{z}_k$, $k \in \mathbb{N}$, we observe that it coincides with $\overline{B^-}$.

Next, using F. Riesz's theorem (see Garnett [1], Chap. II, §2) and Proposition 1.2, we obtain

$$g = \overline{B^-} g_1, \quad \text{where} \quad g_1 \in L_q^+(\mathbb{R}, \varrho^{1-q}).$$

Therefore,

$$h = f\overline{g} = B^- f \overline{g_1} \in H^1(\Pi_-),$$

because Hölder's inequality yields the estimate

$$\begin{aligned}
&\sup_{y<0} \int_{\mathbb{R}} |h(x+iy)| dx \\
&= \sup_{y<0} \int_{\mathbb{R}} |B^-(x+iy) f(x+iy) \varrho_p^-(x+iy)| \cdot |\overline{g}_1(x+iy)(\varrho_p^-(x+iy))^{-1}| dx \\
&\leq \|f\|_{L_p(\mathbb{R},\varrho)} \|g_1\|_{L_q(\mathbb{R},\varrho^{1-q})},
\end{aligned}$$

where the weight ϱ_p^- is given by (1.6).

But then

$$\int_{\mathbb{R}} f(x)\overline{g(x)} dx = \int_{\mathbb{R}} h(x) dx = 0 \Longrightarrow f \in \mathcal{L} \Longrightarrow \ker T(B^-) = \mathcal{L}.$$

To establish (1.22), we observe that the linear manifold $\operatorname{im} T(B^+)$ is closed in $L_p^+(\mathbb{R}, \varrho)$. Hence, by Banach's theorem, $f \in \operatorname{im} T(B^+)$ if and only if $\langle f, g \rangle = 0$ for all $g \in \ker T^*(B^+)$. Since $T^*(B^+) = T(\overline{B^+})$, then, as shown above, $\ker T^*(B^+) = \operatorname{clos span}\{(x - \overline{z}_k)^{-1},\ k \in \mathbb{N}\}$, and therefore condition (1.22) is necessary and sufficient for f to lie in $\operatorname{im} T(B^+)$. □

Let us remark that under some supplementary conditions the system of rational fractions $\{(x - z_k)^{-1}\}_{k=1}^{\infty}$ is a basis in the space $\ker T(B^-)$. This question will be discussed in more detail in the last section of this chapter.

Theorem 1.2 already allows us to consider a class of symbols that play an important role in the ensuing constructions.

Let the function $a \in GL_\infty(\mathbb{R})$ admit one of the two representations

$$a(x) = a^-(x) B^{\pm}(x) a^+(x), \quad x \in \mathbb{R}, \tag{1.23}$$

where $B^{\pm}(x)$ are Blaschke products of the form (1.20), and put

$$a_0(x) = \overline{B^{\pm}(x)} a(x) = a^-(x) a^+(x) \quad , x \in \mathbb{R},$$

which is a standard factorization of the function $a_0(x)$.

Theorem 1.3 *Under assumptions* (1.23), *the operators*

$$T(a^- B^+ a^+) \qquad \text{and} \qquad T(a^- B^- a^+)$$

are left and respectively right invertible in the space $L_p^+(\mathbb{R}, \varrho)$ *and the operators*

$$[T(a)]^{-1} = P^+ \frac{\overline{B^+}}{a^+} P^+ \frac{1}{a^-} I \qquad \text{and} \qquad [T(a)]^{-1} = \frac{\overline{B^-}}{a^+} P^+ \frac{1}{a^-} I$$

are the corresponding one-sided inverses. Moreover,

$$\dim\ker T(a^- B^- a^+) = \dim\operatorname{coker} T(a^- B^+ a^+) = \infty,$$

and

$$\ker T(a^- B^- a^+) = \operatorname{clos\,span}\{[a^+(x)(x - z_k)]^{-1},\ k \in \mathbb{N}\}, \quad z_k \in \Pi_-. \tag{1.24}$$

Also, for f to belong to $\operatorname{im} T(a^- B^+ a^+)$ it is necessary and sufficient that the following orthogonality condition be satisfied:

$$\int_{\mathbb{R}} \frac{f(x)dx}{a^-(x)(x - z_k)} = 0, \quad k \in \mathbb{N}, \quad z_k \in \Pi_+. \tag{1.25}$$

Proof. Using Proposition 1.7, one verifies that, in the case of upper signs,

$$\begin{aligned} [T(a)]^{-1} T(a) &= P^+ \frac{\overline{B^+}}{a^+} P^+ \frac{1}{a^-} P^+ a^- B^+ a^+ I \\ &= P^+ \frac{\overline{B^+}}{a^+} P^+ \frac{1}{a^-} a^- B^+ a^+ I = P^+ \frac{\overline{B^+}}{a^+} B^+ a^+ I = P^+, \end{aligned}$$

and similarly $T(a)[T(a)]^{-1} = P^+$ in the case of lower signs.

Further, by Proposition 1.7 and Theorem 1.1,

$$T(a) = T(a^- B^- a^+) = T(B^-) T(a^- a^+),$$

where the operator $T(a^- a^+)$ is invertible, with $[T(a^- a^+)]^{-1} = \frac{1}{a^+} P^+ \frac{1}{a^-} I$. Consequently,

$$\begin{aligned} \ker T(a) &= [T(a^- a^+)]^{-1} (\ker T(B^-)) \\ &= \frac{1}{a^+} P^+ \frac{1}{a^-} I(\operatorname{clos\,span}\{(x - z_k)^{-1}, k \in \mathbb{N}\}). \end{aligned}$$

But when $z_k \in \Pi_-$

$$\begin{aligned} & P^+ \left[\frac{1}{a^-(x)} \cdot \frac{1}{x - z_k} \right] \\ &= P^+ \left[\left(\frac{1}{a^-(x)} - \frac{1}{a^-(z_k)} \right) \frac{1}{x - z_k} + \frac{1}{a^-(z_k)} \cdot \frac{1}{x - z_k} \right] \\ &= \frac{1}{a^-(z_k)} \cdot \frac{1}{x - z_k}, \end{aligned}$$

and so

$$\ker T(a^- B^- a^+) = \operatorname{clos\,span}\{[a^+(x)(x - z_k)]^{-1}, k \in \mathbb{N}\}.$$

Similarly,

$$T(a^- B^+ a^+) = T(a^- a^+) T(B^+),$$

and consequently $f \in \operatorname{im} T(a^- B^+ a^+)$ if and only if

$$\int_{\mathbb{R}} \frac{1}{x - z_k} \cdot \frac{1}{a^+(x)} \left(P^+ \frac{f}{a^-} \right)(x) dx = \int_{\mathbb{R}} \frac{f(x)}{a^-(x)} \left[P^- \frac{1}{a^+(t)(t - z_k)} \right](x) dx = 0,$$

$k \in \mathbb{N}$. But for $z_k \in \Pi_+$

$$\begin{aligned} P^- \left[\frac{1}{a^+(t)} \cdot \frac{1}{t - z_k} \right] &= P^- \left[\left(\frac{1}{a^+(t)} - \frac{1}{a^+(z_k)} \right) \frac{1}{t - z_k} + \frac{1}{a^+(z_k)} \cdot \frac{1}{t - z_k} \right] \\ &= \frac{1}{a^+(z_k)} \cdot \frac{1}{t - z_k}. \end{aligned}$$

We see that the preceding orthogonality conditions are equivalent to conditions (1.25). □

1.6 An elementary singular function

As a second example let us consider the Toeplitz operator with the symbol

$$a(x) = \exp(i\sigma x) - c, \quad \sigma \in \mathbb{R} \setminus \{0\}, \quad c \in \mathbb{C}. \tag{1.26}$$

Notice that for $\sigma > 0$, $a \in H^\infty(\Pi_+)$, while for $\sigma < 0$, $a \in H^\infty(\Pi_-)$. The function $a(x)$, being continuous at each finite point of the real line $\mathbb{R}$, has a discontinuity at infinity. Let us analyze the behavior of the function $\arg a(x)$ for different values of the parameters σ and c. Clearly, the numerical range of the function $a(x)$ when x runs through $\mathbb{R}$ is the circle of radius 1 centered at the point $z_0 = c$. Leaving aside for the moment the case $|c| = 1$, when the function $a(x)$ degenerates on $\mathbb{R}$, let us assume at first that $|c| > 1$ (Figure 1.1).

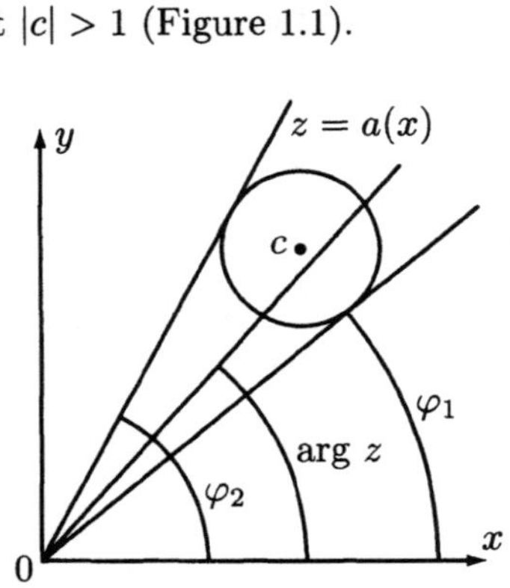

Figure 1.1

In this case any continuous branch of the function $\arg a(x)$ is a $(2\pi/|\sigma|)$-periodic function which changes from φ_1 to φ_2, where $0 < |\varphi_1 - \varphi_2| < \pi$. Despite the fact that the topological index of $a(x)$, defined by the formula

$$\operatorname{ind}_{\mathbb{R}} a(x) = \lim_{M_1, M_2 \to \infty} \left(\frac{1}{2\pi} \arg a(x) \Big|_{-M_1}^{M_2} \right),$$

fails to exist here, this case is intimately related to the index-zero case (see Theorem 1.5 below).

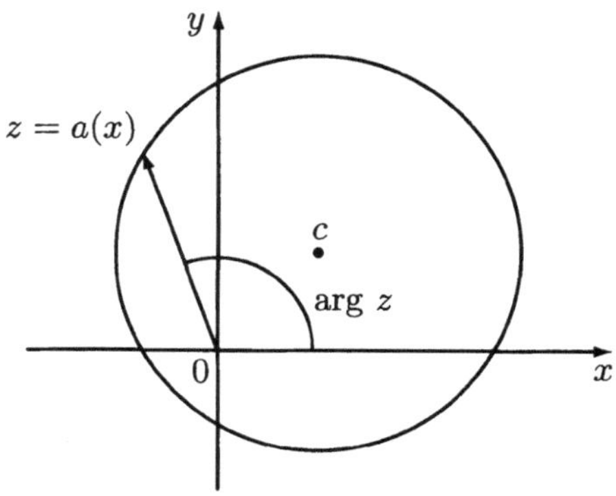

Figure 1.2

If $|c| < 1$ (Figure 1.2), then the continuous function $\arg a(x)$ is monotonic and its increment when x changes by $2\pi/\sigma$ is 2π. Hence, as $x \to +\infty$, $\arg a(x) \to +\infty$ if $\sigma > 0$ and $\arg a(x) \to -\infty$ if $\sigma < 0$. We put

$$\operatorname{ind}_{\mathbb{R}} a(x) = \begin{cases} +\infty, & \sigma > 0, \\ -\infty, & \sigma < 0. \end{cases}$$

Before dealing with the Toeplitz operator with the symbol (1.26), let us recall a few facts from the theory of the Fourier transformation.

The Fourier transformation $\mathcal{F}$ will be used in the form

$$f(x) = (\mathcal{F}g)(x) = \frac{1}{(2\pi)^{1/2}} \int_{\mathbb{R}} g(s) \exp(ixs) ds, \quad x \in \mathbb{R}.$$

An entire function $f(z)$ is called a *function of exponential type* if it obeys an estimate

$$|f(z)| \le A \exp(B|z|),$$

where the positive constants A and B do not depend on $z \in \mathbb{C}$. The infimum of all constants B for which this estimate holds is called the *type* of the function $f(z)$.

Let us denote by $\mathcal{L}_2^\sigma$, $\sigma > 0$, the set of functions of exponential type $\le \sigma$ that belong to the space $L_2(\mathbb{R})$. The class $\mathcal{L}_2^\sigma$ is a Hilbert space and coincides with the well-known Paley-Wiener space, which consists of the Fourier transforms of the square-integrable functions on $\mathbb{R}$ with support contained in $[-\sigma, \sigma]$. This assertion, known as the *Paley-Wiener theorem*, is part of the following theorem.

Theorem 1.4 *For the function $f(x)$ to belong to the space $H^2(\Pi_+)$, $H^2(\Pi_-)$, or $\mathcal{L}_2^\sigma$ it is necessary and sufficient that it admit the respective representation*

$$\begin{aligned} f(x) &= \frac{1}{(2\pi)^{1/2}} \int_0^\infty g(s) \exp(ixs) ds, \quad g \in L_2(0, \infty), \\ f(x) &= \frac{1}{(2\pi)^{1/2}} \int_{-\infty}^0 g(s) \exp(ixs) ds, \quad g \in L_2(-\infty, 0), \\ f(x) &= \frac{1}{(2\pi)^{1/2}} \int_{-\sigma}^\sigma g(s) \exp(ixs) ds, \quad g \in L_2(-\sigma, \sigma). \end{aligned}$$

Let us denote

$$\mathcal{L}_2^{\sigma\pm} = \mathcal{L}_2^\sigma \cap H^2(\Pi_\pm).$$

From Theorem 1.4 it follows that

$$\begin{aligned} f &\in \mathcal{L}_2^{\sigma+} \iff f(x) = \frac{1}{(2\pi)^{1/2}} \int_0^\sigma g(s) \exp(ixs) ds, \quad g \in L_2(0, \sigma), \\ f &\in \mathcal{L}_2^{\sigma-} \iff f(x) = \frac{1}{(2\pi)^{1/2}} \int_{-\sigma}^0 g(s) \exp(ixs) ds, \quad g \in L_2(-\sigma, 0). \end{aligned}$$

Now let us return to the Toeplitz operator with a symbol of the form (1.26).

Theorem 1.5 *If $|c| > 1$, then the Toeplitz operator $T(a)$ with the symbol $a(x)$ given by (1.26) is invertible in $H^2(\Pi_+)$.*

If $|c| < 1$, then $T(a)$ is one-sided invertible in $H^2(\Pi_+)$. More precisely, if $\sigma < 0$, then $T(a)$ is right invertible, $\dim \ker T(a) = \infty$, and $f \in \ker T(a)$ if and only if

$$f(x) = (1 - c\exp(-i\sigma x))^{-1} \varphi(x), \quad \text{where} \quad \varphi \in \mathcal{L}_2^{|\sigma|+}. \tag{1.27}$$

If $\sigma > 0$, then $T(a)$ is left invertible, $\dim \operatorname{coker} T(a) = \infty$, and $f \in \operatorname{im} T(a)$ if and only if

$$\int_{\mathbb{R}} f(x) \frac{\overline{\varphi(x)}}{1 - c\exp(-i\sigma x)} dx = 0, \quad \text{for all} \quad \varphi \in \mathcal{L}_2^{\sigma+}. \tag{1.28}$$

In all these cases, $T(a^{-1})$ is the corresponding one-sided or two-sided inverse to $T(a)$.

Proof. Let $|c| > 1$. If $\sigma > 0$, then $a \in H^\infty(\Pi_+)$ and $a^{-1} \in H^\infty(\Pi_+)$. If $\sigma < 0$, then $a, a^{-1} \in H^\infty(\Pi_-)$. Therefore, by Proposition 1.7, in both cases

$$T(a)T(a^{-1}) = T(a^{-1})T(a) = I.$$

If $|c| < 1$, then $T(a)$ admits the representation

$$T(a) = \begin{cases} T(\exp(i\sigma x))T(1 - c\exp(-i\sigma x)), & \text{if } \sigma < 0, \\ T(1 - c\exp(-i\sigma x))T(\exp(i\sigma x)), & \text{if } \sigma > 0, \end{cases}$$

where the operator $T(1 - c\exp(-i\sigma x))$ is invertible in $H^2(\Pi_+)$ and

$$T(1 - c\exp(-i\sigma x))]^{-1} = T((1 - c\exp(-i\sigma x))^{-1}).$$

At the same time, the operator $T(\exp(i\sigma x))$ is right (resp., left) invertible for $\sigma < 0$ (resp., $\sigma > 0$), because in this case $T(\exp(i\sigma x))T(\exp(-i\sigma x)) = I$ (resp., $T(\exp(-i\sigma x))T(\exp(ix\sigma)) = I$). This proves the validity of the second part of the theorem. Moreover, the corresponding inverse of the operator $T(a)$ is given by the formula

$$[T(a)]^{-1} = \begin{cases} T((1 - c\exp(-i\sigma x))^{-1})T(\exp(-i\sigma x)), & \text{if } \sigma < 0, \\ T(\exp(-i\sigma x))T((1 - c\exp(-i\sigma x))^{-1}), & \text{if } \sigma > 0, \end{cases}$$

i.e., $[T(a)]^{-1} = T(a^{-1})$.

Since for $\sigma < 0$ we have

$$\ker T(a) = (1 - c\exp(-i\sigma x))^{-1}(\ker T(\exp(i\sigma x))),$$

equality (1.27) follows from the fact that $\ker T(\exp(i\sigma x)) = \mathcal{L}_2^{|\sigma|+}$. To see this, let us use the representation of elements of $L_2(\mathbb{R})$ as Fourier integrals:

$$\varphi \in L_2(\mathbb{R}) \quad \Longleftrightarrow \quad \varphi(x) = \frac{1}{(2\pi)^{1/2}} \int_{\mathbb{R}} \psi(t)\exp(ixt)dt, \quad \text{where} \quad \psi \in L_2(\mathbb{R}).$$

Then the projectors $P^{\pm}$ act as follows:

$$(P^+\varphi)(x) = \frac{1}{(2\pi)^{1/2}} \int_0^{\infty} \psi(t)\exp(ixt)dt,$$

$$(P^-\varphi)(x) = \frac{1}{(2\pi)^{1/2}} \int_{-\infty}^{0} \psi(t)\exp(ixt)dt.$$

By the Paley-Wiener theorem (see Theorem 1.4)

$$\varphi \in \mathcal{L}_2^{|\sigma|+} \quad \Longleftrightarrow \quad \varphi(x) = \frac{1}{(2\pi)^{1/2}} \int_0^{|\sigma|} \psi(t)\exp(ixt)dt, \quad \text{where} \quad \psi \in L_2(0, |\sigma|).$$

Let $\varphi \in \mathcal{L}_2^{|\sigma|+}$. Then for $\sigma < 0$,

$$P^+(\exp(i\sigma x)\varphi) = \frac{1}{(2\pi)^{1/2}} P^+ \left(\int_0^{|\sigma|} \psi(t)\exp(ix(t+\sigma))dt \right)$$

$$= \frac{1}{(2\pi)^{1/2}} P^+ \left(\int_{\sigma}^{0} \psi(\tau - \sigma)\exp(ix\tau)d\tau \right) = 0,$$

and so $\varphi \in \ker T(\exp(i\sigma x))$. Thus, we have shown that $\mathcal{L}_2^{|\sigma|+} \subset \ker T(\exp(i\sigma x))$.

Let us verify that the opposite inclusion holds as well. Take

$$\varphi(x) = \frac{1}{(2\pi)^{1/2}} \int_0^\infty \psi(t) \exp(ixt)dt, \quad \text{where} \quad \psi \in L_2(0,\infty).$$

Then $\varphi \in \ker T(\exp(i\sigma x))$ implies $P^+(\exp(i\sigma x)\varphi) = 0$, or, equivalently,

$$P^+\left(\int_0^\infty \psi(t)\exp(ix(t+\sigma))dt\right) = P^+\left(\int_\sigma^\infty \psi(\tau-\sigma)\exp(ix\tau)d\tau\right)$$
$$= \int_0^\infty \psi(\tau-\sigma)\exp(ix\tau)d\tau = \exp(i\sigma x)\int_{|\sigma|}^\infty \psi(t)\exp(ixt)dt = 0.$$

Therefore,

$$\varphi(x) = \frac{1}{(2\pi)^{1/2}} \int_0^{|\sigma|} \psi(t)\exp(ixt)dt \in \mathcal{L}_2^{|\sigma|+},$$

and so $\ker T(\exp(i\sigma x)) \subset \mathcal{L}_2^{|\sigma|+}$.

Thus, equality (1.27) is established.

If now $\sigma > 0$, then

$$H^2(\Pi_+) = \operatorname{im} T(a) \oplus \ker T^*(a) \tag{1.29}$$

and since $T^*(a) = T(\overline{a})$, (1.27) yields

$$\ker T^*(a) = (1 - \overline{c}\exp(i\sigma x))^{-1} I(\mathcal{L}_2^{\sigma+}).$$

But then from (1.29) it follows that $f \in \operatorname{im} T(a)$ if and only if

$$\int_{\mathbb{R}} f(x)\overline{\left(\frac{\varphi(x)}{1-\overline{c}\exp(i\sigma x)}\right)}dx = \int_{\mathbb{R}} f(x)\frac{\overline{\varphi(x)}}{1-c\exp(-i\sigma x)}dx = 0$$

for all $\varphi \in \mathcal{L}_2^{\sigma+}$. □

Remark 1.1 Using Hilbert-space techniques, conditions (1.27) and (1.28) can be restated in a different form. To that end, let us consider the familiar orthonormal basis in $L_2(0,|\sigma|)$ given by

$$\varepsilon_k(t) = |\sigma|^{-1/2} i \exp\left(\frac{2\pi kit}{|\sigma|}\right), \quad k \in \mathbb{Z}.$$

Since the Fourier transformation $\mathcal{F}$ is an isomorphism of the Hilbert space $L_2(0,|\sigma|)$ onto the Hilbert space $\mathcal{L}_2^{|\sigma|+}$, the system of functions

$$e_k(x) = \frac{1}{(2\pi|\sigma|)^{1/2}} \cdot \frac{\exp(ix|\sigma|)-1}{x+2\pi k/|\sigma|} \tag{1.30}$$
$$= (\mathcal{F}\varepsilon_k)(x) = \frac{i}{(2\pi|\sigma|)^{1/2}} \int_0^{|\sigma|} \exp\left(\frac{2\pi kit}{|\sigma|} + ixt\right) dt$$

is an orthonormal basis in $\mathcal{L}_2^{|\sigma|+}$. Hence, $\varphi \in \mathcal{L}_2^{|\sigma|+}$ if and only if

$$\varphi = \sum_{k\in\mathbb{Z}} c_k e_k, \quad \text{where} \quad \{c_k\} \in l_2(\mathbb{Z}). \tag{1.31}$$

Therefore, equality (1.27) is equivalent to the equality

$$f(x) = (1 - c\exp(-i\sigma x))^{-1} \sum_{k\in\mathbb{Z}} c_k e_k(x), \quad \text{where} \quad \{c_k\} \subset l_2(\mathbb{Z}),$$

while condition (1.28) is equivalent to the orthogonality condition

$$\int_{\mathbb{R}} f(x) \frac{\overline{e_k(x)}}{1 - c\exp(-i\sigma x)} = 0, \quad k \in \mathbb{Z},$$

where e_k is given by (1.30).

Remark 1.2 The coefficients c_k of the series (1.31) can be computed explicitly. Namely, let $\Phi = \mathcal{F}^{-1}\varphi$. Then

$$\begin{aligned} c_k &= \langle \varphi, e_k \rangle = \langle \Phi, \varepsilon_k \rangle = \int_{\mathbb{R}} \Phi(t)\overline{\varepsilon_k(t)}dt \\ &= -i|\sigma|^{-1/2} \int_{\mathbb{R}} \Phi(t) \exp\left(-it2\pi k|\sigma|^{-1}\right) dt = -i\left(\frac{2\pi}{|\sigma|}\right)^{1/2} \varphi\left(-\frac{2\pi k}{|\sigma|}\right). \end{aligned}$$

From this it follows, in particular, that for any function φ in $\mathcal{L}_2^{|\sigma|+}$

$$\sum_{k\in\mathbb{Z}} \left|\varphi\left(-\frac{2\pi k}{|\sigma|}\right)\right|^2 < \infty,$$

and therefore the function φ has an expansion into a Lagrange interpolation series,

$$\varphi(x) = \frac{1}{i|\sigma|} \sum_{k\in\mathbb{Z}} \varphi(x_k) \frac{\exp(ix|\sigma|) - 1}{x - x_k}, \quad x_k = -\frac{2\pi k}{|\sigma|},$$

which converges in the square mean on $\mathbb{R}$.

1.7 Boundary degeneracy

Let us examine now the limiting case $|c| = 1$. Write $c = \exp(i\theta\sigma)$, where $\theta \in [0, 2\pi/\sigma)$. Then the function $a(x) = \exp(ix\sigma) - \exp(i\theta\sigma)$ vanishes at the points

$$x_k = \theta + 2\pi k/\sigma, \quad k \in \mathbb{Z}, \tag{1.32}$$

and so the normal solvability condition (1.11) for the operator $T(a)$ is violated. In this connection there arises the new problem of describing the linear manifold $\operatorname{im} T(a)$, the solution of which is given in Theorem 1.6 below.

First let us introduce several definitions.

Recall that by an element of the space $\operatorname{loc} L_1(\mathbb{R})$ one means a class of equivalent locally integrable functions, where two functions are said to be equivalent if they differ from one another on some set $M \subset \mathbb{R}$ of measure zero. Accordingly, the value of an integrable function at some given point t is not defined. Nevertheless, given an element f of $\operatorname{loc} L_1(\mathbb{R})$, one can associate a certain complex number $f(t_0)$ to f at almost every point t_0 of $\mathbb{R}$ (such points are referred to as the Lebesgue points of the function f). The number $f(t_0)$ is called the Lebesgue value of the function f at the point t_0. With Lebesgue values one can operate as with the values of ordinary, say continuous, functions. Let us give the precise definitions.

Let $f \in \operatorname{loc} L_1(\mathbb{R})$. A point $t_0 \in \mathbb{R}$ is called a *Lebesgue point* of the element f if there exists a complex number ζ_0 such that

$$\lim_{t \to t_0} \frac{1}{|t-t_0|} \int_{t_0}^{t} |f(\tau) - \zeta_0| d\tau = 0.$$

If t_0 is a Lebesgue point of $f \in \operatorname{loc} L_1(\mathbb{R})$, then

$$\zeta_0 = \lim_{t \to t_0} \frac{1}{t-t_0} \int_{t_0}^{t} f(\tau) d\tau,$$

because

$$\begin{aligned} \left| \frac{1}{t-t_0} \int_{t_0}^{t} f(\tau) d\tau - \zeta_0 \right| &= \frac{1}{|t-t_0|} \left| \int_{t_0}^{t} (f(\tau) - \zeta_0) d\tau \right| \\ &\leq \frac{1}{|t-t_0|} \int_{t_0}^{t} |f(\tau) - \zeta_0| d\tau. \end{aligned}$$

This implies that the number ζ_0 is unique whenever it exists. This number ζ_0 will be called the *Lebesgue value of the function f at the point t_0* and will be denoted by $f(t_0)$. It is known that, given a locally integrable function on $\mathbb{R}$, almost every point of $\mathbb{R}$ is a Lebesgue point of that function. Moreover, any point t_0 of continuity of an element f of $\operatorname{loc} L_1(\mathbb{R})$ is a Lebesgue point, and the Lebesgue value of f at t_0 coincides with the value at t_0 of the continuous function that represents the element f in a neighborhood of the point t_0.

Let us mention also that if $f(t) = (t-t_0)g(t)$, where $g \in \operatorname{loc} L_1(\mathbb{R})$, $t_0 \in \mathbb{R}$, then t_0 is a Lebesgue point of the function $f(t)$ and $f(t_0) = 0$. Indeed, in this case

$$\begin{aligned} \frac{1}{|t-t_0|} \int_{t_0}^{t} |f(\tau)| d\tau &= \frac{1}{|t-t_0|} \int_{t_0}^{t} |\tau - t_0||g(\tau)| d\tau \\ &\leq \int_{t_0}^{t} |g(\tau)| d\tau \to 0 \quad \text{as} \quad t \to t_0. \end{aligned}$$

From now on, each time we will speak about the value of an integrable function f at a point, it will be assumed that the point is a Lebesgue point of f, and the value of f at that point will be the corresponding Lebesgue value.

Let $f \in H^2(\Pi_+)$. Assume that the points x_k, $k \in \mathbb{Z}$, of the form (1.32) are Lebesgue points of the function $f(x)$ and that the values $\{f(x_k)\}$ form an element of the space $l_2(\mathbb{Z})$. Let us introduce the operator $\mathcal{Y}$ which takes a function f into a function $\mathcal{Y}f$ of class $\mathcal{L}_2^{|\sigma|+}$ according to the rule

$$(\mathcal{Y}f)(x) = \frac{1}{i|\sigma|} \sum_{k\in\mathbb{Z}} f(x_k) \frac{\exp(ix|\sigma|) - \exp(i\theta|\sigma|)}{x - x_k} \tag{1.33}$$

and the operator N which acts on the functions f of the above form by the formula

$$Nf = \frac{1}{a}(I - \exp(ix\sigma)\mathcal{Y})f. \tag{1.34}$$

Theorem 1.6 *Suppose that the function $a(x)$ has the form (1.26) with $c = \exp(i\theta\sigma)$ and $\theta \in [0, 2\pi/\sigma)$. Then the operator $T(a)$ has a nonclosed range in the space $H^2(\Pi_+)$ and $\dim\ker T(a) = \dim\operatorname{coker} T(a) = 0$ in $H^2(\Pi_+)$.*

If $\sigma < 0$, then for the function f to belong to $\operatorname{im} T(a)$ it is necessary and sufficient that the following conditions be satisfied:

(i) *the points $x_k = \theta + 2\pi k/\sigma$, $k \in \mathbb{Z}$, are Lebesgue points of $f(x)$;*

(ii) $\sum_{k\in\mathbb{Z}} |f(x_k)|^2 < \infty$;

(iii) $\dfrac{f(x) - \exp(ix\sigma)(\mathcal{Y}f)(x)}{a(x)} \in H^2(\Pi_+)$.

When these conditions are satisfied, the function $\varphi = Nf$, where the operator N is defined by (1.33), (1.34), is a solution of the equation $[T(a)]\varphi = f$.

If $\sigma > 0$, then $f \in \operatorname{im} T(a)$ if and only if $f/a \in H^2(\Pi_+)$. If this last condition is satisfied, then the function $\varphi = f/a$ is a solution of the equation $[T(a)]\varphi = f$.

Proof. Let $\sigma < 0$ and $\varphi \in \ker T(a)$, i.e., $P^+ a\varphi = 0$. Equivalently, $a\varphi = \varphi^-$, where $\varphi^- = P^- a\varphi \in H^2(\Pi_-)$. Explicitly, this reads

$$(\exp(ix\sigma) - \exp(i\theta\sigma))\varphi(x) = \varphi^-(x),$$

or

$$(1 - \exp(i(\theta - x)\sigma))\varphi(x) = \exp(-i\sigma x)\varphi^-(x),$$

where the left-hand side belongs to $H^2(\Pi_+)$. Applying the operator P^+ to both sides of the last equality we obtain

$$\begin{aligned}(1 - \exp(i(\theta - x)\sigma))\varphi(x) &= P^+\left(\exp(-ix\sigma)\int_{-\infty}^{0} \psi(t)\exp(ixt)dt\right) \\ &= P^+\left(\int_{-\infty}^{|\sigma|} \psi(t+\sigma)\exp(ixt)dt\right) = \int_0^{|\sigma|} \psi(t+\sigma)\exp(ixt)dt = \varphi_1(x) \in \mathcal{L}_2^{|\sigma|+},\end{aligned}$$

where $\psi = \mathcal{F}^{-1}\varphi^{-}$. Since $H^2(\Pi_+) \subset \operatorname{loc} L_1(\mathbb{R})$, the points x_k are Lebesgue points of the function $(1 - \exp(i(\theta - x)\sigma))\varphi(x)$, and the values of this function in the points x_k are equal to zero: $\varphi_1(\theta + 2\pi k/\sigma) = 0$ for all $k \in \mathbb{Z}$. Further, $\varphi_2(x) = \varphi_1(x - \theta) \in \mathcal{L}_2^{|\sigma|+}$ and $\varphi_2(2\pi k/\sigma) = 0$, $k \in \mathbb{Z}$. By Remark 1.2 and equality (1.31), $\varphi_2 \equiv 0$. Consequently, $\varphi_1 \equiv 0$, and so $\varphi = 0$ almost everywhere on $\mathbb{R}$. We conclude that $\ker T(a) = \{0\}$.

Now let $\sigma > 0$. Then $[T(a)]\varphi = 0$ is equivalent to $a(x)\varphi(x) = 0$ and implies that $\varphi = 0$ almost everywhere. Thus, $\ker T(a) = \{0\}$ in this case, too. Finally, the relation $T^*(a) = T(\overline{a})$ implies that $\operatorname{coker} T(a) = \{0\}$ in both cases $\sigma > 0$ and $\sigma < 0$.

Turning now to the description of the linear manifold $\operatorname{im} T(a)$ in the case $\sigma < 0$, we observe that if $f \in \operatorname{im} T(a)$, then

$$f(x) = P^+ \left[(\exp(ix\sigma) - \exp(i\theta\sigma))\varphi(x)\right]$$

for some $\varphi \in H^2(\Pi_+)$. From this it follows, by analogy with the case $f = 0$, that

$$\exp(-ix\sigma)f(x) = (1 - \exp(i(\theta - x)\sigma))\varphi(x) + \varphi_1(x),$$

where $\varphi_1 \in \mathcal{L}_2^{|\sigma|+}$. Clearly, the x_k's are Lebesgue points of the function $f(x)$ and $f(x_k) = \varphi_1(x_k)$. By Remark 1.2, $\{f(x_k)\} \in l_2(\mathbb{Z})$, and

$$\varphi_1(x) = \frac{1}{i|\sigma|} \sum_{k\in\mathbb{Z}} f(x_k) \frac{\exp(ix|\sigma|) - \exp(i\theta|\sigma|)}{x - x_k} = (\mathcal{Y}f)(x).$$

But then

$$(1 - \exp(i(\theta - x)\sigma))\varphi(x) = \exp(-ix\sigma)f(x) - (\mathcal{Y}f)(x),$$

and the requirement

$$\varphi(x) = \frac{\exp(-ix\sigma)f(x) - (\mathcal{Y}f)(x)}{1 - \exp(i(\theta - x)\sigma)} \in H^2(\Pi_+) \tag{1.35}$$

establishes the necessity of condition (iii).

The sufficiency of conditions (i), (ii), (iii) is verified by substituting a function φ of the form (1.35) in the equation $[T(a)]\varphi = f$:

$$\begin{aligned} P^+ a\varphi &= P^+ \left[(\exp(ix\sigma) - \exp(i\theta\sigma))\frac{\exp(-ix\sigma)f(x) - (\mathcal{Y}f)(x)}{1 - \exp(i(\theta - x)\sigma)}\right] \\ &= P^+ \left[f(x) - \exp(ix\sigma)(\mathcal{Y}f)(x)\right] = f, \end{aligned}$$

because $\exp(ix\sigma)\mathcal{Y}f \in H^2(\Pi_-)$.

In the case $\sigma > 0$ the requirement $f \in \operatorname{im} T(a)$ is equivalent to the existence of a function $\varphi \in H^2(\Pi_+)$ such that $a\varphi = f$, which is equivalent to the condition $f/a \in H^2(\Pi_+)$. □

Remark 1.3 Since for the operator $T(a)$ considered in Theorem 1.6 both

$$\dim \ker T(a) \qquad \text{and} \qquad \dim \operatorname{coker} T(a)$$

equal zero, it follows that $\operatorname{clos} \operatorname{im} T(a) = H^2(\Pi_+)$, and therefore $T(a)$ is unboundedly invertible in the space $H^2(\Pi_+)$. It is readily seen that in the case at hand the unbounded inverse operator $[T(a)]^{-1}$ coincides with the operator N given by (1.33), (1.34).

References and comments

1.1–1.3 This material is treated in a more complete fashion in, e.g., Gohberg and Krupnik [3], Chaps. I, II, IV.

1.4 There are several equivalent theories: that of Toeplitz and Wiener-Hopf operators, that of the Riemann boundary value problem, and that of singular integral equations. Among the extensive literature devoted to such problems, let us mention the monographs that conceptually are the closest to ours: Gohberg and Krupnik [3], Böttcher and Silbermann [2], and Litvinchuk and Spitkovsky [1].

Concerning the exact values of the norm of the singular integration operator in the space $L_p(\mathbb{R}, \varrho)$ (inequality (1.10)), see Krupnik [3].

Proposition 1.4 is due to Coburn [1]–[2].

Propositions 1.5 and 1.6 are due to Simonenko [2], [7]; see also Gohberg and Krupnik [3], Chap. VII, Theorem 4.1 and Lemma 4.1.

The standard factorization is historically the first factorization that allowed a constructive investigation of the operators mentioned above; see the works of Gahov (Gakhov) [2], Muskhelishvili [1], Mikhlin [1], Khvedelidze [1], M. G. Kreĭn [1].

In particular, Theorem 1.1 is a reformulation in operator-theoretic language of classical results of Gahov (see [2], §14) for the Riemann problem with Hölder continuous coefficients in a half-plane (see, e.g., Gohberg and Krupnik [3], Chap. III). Proposition 1.7, which is the basis of Theorem 1.1, is due to Hartman and Wintner [1].

1.5–1.7 Section 1.6 follows the work of Dybin [4], while Section 1.7 follows that of Gaponenko and Dybin [1]–[2]. A more general case of degeneration of the symbol on a set with almost-periodic accumulation points was considered by Dybin in [13], Chap. III, §4.

Chapter 2

Factorization and Invertibility

In this chapter we introduce the notion of the (p, ϱ)-factorization of a bounded measurable function and establish a connection between this notion and the Fredholmness of Toeplitz operators. After that we will propose a generalization of the theory of (p, ϱ)-factorization to the case of Toeplitz operators with infinite index.

(p, ϱ)-factorization and Φ-theory

2.1 The space $L_p(\Gamma, \varrho)$ and the operator S_Γ

Let Γ be a closed rectifiable Jordan curve in the complex plane $\mathbb{C}$. A function $\varrho : \Gamma \to [0, +\infty]$ is called a *weight* if $\varrho^{-1}(\{0, +\infty\})$ has (length) measure zero.

We will say that a measurable function $f(t)$ defined on Γ belongs to the space $L_p(\Gamma, \varrho)$, $1 \leq p < \infty$, if

$$\|f\|_{p,\varrho} = \left(\int_\Gamma |f(t)|^p \varrho(t) |dt| \right)^{\frac{1}{p}} < \infty.$$

The space $L_p(\Gamma, \varrho)$, $1 \leq p < \infty$, equipped with the above norm, is a Banach space. As usual, we write $L_p(\Gamma) = L_p(\Gamma, 1)$. The dual of the space $L_p(\Gamma, \varrho)$, $1 \leq p < \infty$, is $L_q(\Gamma, \varrho^{1-q})$, where $q = p/(p-1)$.

We let S_Γ denote the Cauchy-Lebesgue singular integral operator, defined by the formula

$$(S_\Gamma f)(t) = \frac{1}{\pi i} \int_\Gamma \frac{f(\tau) d\tau}{\tau - t}, \quad t \in \Gamma,$$

where the integral is understood in the sense of the principal value.

The curve Γ is called a *Carleson curve* (or an *Ahlfors-David curve*) if for every disk $c(z, r) := \{\zeta \in \mathbb{C} : |\zeta - z| < r\}$ the length of $\Gamma \cap c(z, r)$ is not larger than dr, where d is a positive constant that does not depend on z and r. The operator

S_Γ is bounded in $L_p(\Gamma)$, $1 < p < \infty$, if and only if Γ is a Carleson curve. On a Carleson curve the operator S_Γ is bounded in $L_p(\Gamma, \varrho)$, $1 < p < \infty$, if and only if the weight ϱ satisfies *condition* (A_p):

$$\sup_{z,r} \left(\frac{1}{r} \int_{\Gamma \cap c(z,r)} \varrho(t)|dt| \right) \left(\frac{1}{r} \int_{\Gamma \cap c(z,r)} (\varrho(t))^{-\frac{1}{p-1}} |dt| \right)^{p-1} = C_p < \infty. \qquad (A_p)$$

Ordinarily, we will assume that Γ is a smooth closed Jordan curve or the real line $\mathbb{R}$. In this case condition (A_p) reads

$$\sup_{|I|} \left(\frac{1}{|I|} \int_I \varrho(t)|dt| \right) \left(\frac{1}{|I|} \int_I (\varrho(t))^{-\frac{1}{p-1}} |dt| \right)^{p-1} = C_p < \infty,$$

where I is any arc on Γ and $|I|$ denotes the length of I.

As a rule, in what follows we will use one of B. V. Khvedelidze's *power weights*

$$\varrho(t) = \prod_{k=1}^{m} |t - t_k|^{\alpha_k}, \quad t_k \in \Gamma, \quad -1 < \alpha_k < p - 1. \qquad (2.1)$$

The constraints on the exponents α_k are necessary and sufficient for the operator S_Γ to be bounded in the space $L_p(\Gamma, \varrho)$, $1 < p < \infty$, with the weight (2.1).

Let us also give the formula for the operator S_Γ^* in the case of a smooth contour Γ. To this end let us denote by $\theta_\Gamma(t)$ the angle between the tangent to Γ at the point t and the positively oriented real axis. Then

$$S_\Gamma^* = -H_\Gamma S_\Gamma H_\Gamma,$$

where

$$(H_\Gamma f)(t) = \overline{\exp(i\theta_\Gamma(t)) f(t)}, \quad H_\Gamma^2 = I. \qquad (2.2)$$

2.2 Classes of bounded and continuous functions

We denote by $L_\infty(\Gamma)$ the algebra of measurable essentially bounded functions on Γ, equipped with the norm

$$\|f\|_\infty = \operatorname{ess\,sup}_{t \in \Gamma} |f(t)|.$$

Here are some important subalgebras of $L_\infty(\Gamma)$:

$PC(\Gamma)$ is the subalgebra of all functions that are continuous on Γ, except for a finite number of discontinuities of the first kind;

$C(\Gamma)$ is the subalgebra of all continuous functions on Γ;

$H_\mu(\Gamma)$, $0 < \mu \le 1$, is the subalgebra of all continuous functions on Γ that satisfy the Hölder condition

$$\sup_{t_1,t_2\in\Gamma} \frac{|f(t_1) - f(t_2)|}{|t_1 - t_2|^\mu} < \infty;$$

$R(\Gamma)$ is the subalgebra of all rational functions with no poles on Γ.

One has the chain of inclusions

$$R(\Gamma) \subset H_\mu(\Gamma) \subset C(\Gamma) \subset PC(\Gamma) \subset L_\infty(\Gamma) \subset L_p(\Gamma, \varrho) \subset L_1(\Gamma).$$

The subalgebras $C(\Gamma)$ and $R(\Gamma)$ are dense in $L_p(\Gamma, \varrho)$ and their closures in the norm of the space $L_p(\Gamma, \varrho)$ coincide with all of $L_p(\Gamma, \varrho)$:

$$\operatorname{clos} R(\Gamma) = \operatorname{clos} C(\Gamma) = L_p(\Gamma, \varrho).$$

Let Γ_0 be the unit circle (centered at the origin). We will denote by W the *Wiener algebra* of all continuous functions on Γ_0 that admit an absolutely convergent Fourier series expansion

$$f(t) = \sum_{k\in\mathbb{Z}} f_k t^k.$$

Endowed with the norm

$$\|f\|_W = \sum_{k\in\mathbb{Z}} |f_k|,$$

the algebra W becomes a Banach algebra.

2.3 The classes $L_p^{\pm}(\Gamma, \varrho)$

In what follows, unless otherwise stipulated, we will assume that the closed oriented Carleson curve Γ has no points of self-intersection and partitions the complex plane into two domains: a bounded one, $\mathcal{D}_\Gamma^+$, and an unbounded one, $\mathcal{D}_\Gamma^-$. The set of all curves with these properties will be denoted by $\mathcal{T}$. We also denote $\mathbb{D}^\pm = \mathcal{D}_{\Gamma_0}^\pm$.

The analytic projectors $P_\Gamma^\pm = \frac{1}{2}(I \pm S_\Gamma)$ and the spaces $L_p^\pm(\Gamma, \varrho)$ are introduced in the same manner as in the case of the line $\mathbb{R}$ (recall (1.3)). With these definitions, the decomposition of $L_p(\Gamma, \varrho)$ into the direct sum of the subspaces $L_p^+(\Gamma, \varrho)$ and $L_p^-(\Gamma, \varrho)$ is preserved for $1 < p < \infty$ and any weight $\varrho(t)$ satisfying condition (A_p).

When $1 < p < \infty$, the space $L_p^+(\Gamma)$ coincides with the *Smirnov space* $E_p(\mathcal{D}_\Gamma^+)$. Recall that a function $f(z)$, analytic in the domain $\mathcal{D}_\Gamma^+$, belongs to the space $E_p(\mathcal{D}_\Gamma^+)$, $1 \le p < \infty$, if there exists a nested sequence of smooth closed curves $\Gamma_n (\subset \mathcal{D}_\Gamma^+)$, $n = 1, 2, \ldots$, which converges to Γ as $n \to \infty$ in the sense that the domains $\mathcal{D}_n^+$ with boundaries Γ_n exhaust $\mathcal{D}_\Gamma^+$ (i.e., $\mathcal{D}_1^+ \subset \mathcal{D}_2^+ \subset \ldots \subset \mathcal{D}_\Gamma^+$ and $\bigcup_{n=1}^{\infty} \mathcal{D}_n^+ = \mathcal{D}_\Gamma^+$), and

$$\|f\|_{E_p} = \sup_n \left(\int_{\Gamma_n} |f(z)|^p |dz| \right)^{\frac{1}{p}} < \infty.$$

The statement that $L_p^+(\Gamma)$ and $E_p(\mathcal{D}_\Gamma^+)$ coincide means that if $f(t) \in L_p^+(\Gamma)$, then its analytic continuation $f(z)$ to the domain $\mathcal{D}_\Gamma^+$ is a function of class $E_p(\mathcal{D}_\Gamma^+)$ and that, conversely, if $f(z) \in E_p(\mathcal{D}_\Gamma^+)$, then for almost every $t \in \Gamma$ there exists the nontangential limit $f(t)$ as $z \to t$ $(z \in \mathcal{D}_\Gamma^+)$, and the resulting function belongs to $L_p^+(\Gamma)$.

The space $E_p(\mathcal{D}_\Gamma^-)$, $1 < p < \infty$, is introduced in a similar manner.

From the foregoing discussion it follows that for $1 < p < \infty$, $E_p(\mathcal{D}_\Gamma^-) = L_p^-(\Gamma) + \mathbb{C}$. From now on we will not use special notations for the Smirnov spaces, writing for the sake of a uniformity

$$L_p^+(\Gamma) = E_p(\mathcal{D}_\Gamma^+), \quad L_p^-(\Gamma) = E_p(\mathcal{D}_\Gamma^-) \ominus \mathbb{C}, \quad 1 \le p < \infty.$$

Notice that $L_p^+(\Gamma) \cap L_p^-(\Gamma) = \{0\}$, $1 \le p < \infty$.

Also, we will say that $f \in L_\infty^\pm(\Gamma)$ if the function $f(z)$ is analytic and bounded in the domain $\mathcal{D}_\Gamma^\pm$. Let us observe that, if the function $f(z)$ belongs to $L_\infty^\pm(\Gamma)$, then its nontangential limit $f(t)$ as $z \to t$ $(z \in \mathcal{D}_\Gamma^\pm)$ exists for almost all $t \in \Gamma$ and the resulting function is of class $L_\infty(\Gamma)$. However, the direct sum decomposition (1.5) is no longer valid, because the operator S_Γ is not bounded in $L_\infty(\Gamma)$.

In the particular case $\Gamma = \Gamma_0$, the space $E_p(\mathbb{D}^+)$ coincides with the Hardy space $H^p(\mathbb{D}^+)$.

By definition, a function $f(z)$ that is analytic in $\mathbb{D}^+$ belongs to the space $H^p(\mathbb{D}^+)$, $1 < p < \infty$, if

$$\|f\|_p = \sup_{0<r<1} \left\{ \frac{1}{2\pi} \int_0^{2\pi} |f(r\exp(i\theta))|^p d\theta \right\}^{\frac{1}{p}} < \infty,$$

and $f(z)$ belongs to $H^\infty(\mathbb{D}^+)$ if

$$\|f\|_\infty = \sup_{z\in\mathbb{D}^+} |f(z)| < \infty.$$

In the space $C(\Gamma)$ let us single out the subspaces $A^\pm(\Gamma)$ consisting of the functions that are analytic in $\mathcal{D}_\Gamma^\pm$ and continuous on $\overline{\mathcal{D}_\Gamma^\pm}$. Like $L_\infty(\Gamma)$, the space $C(\Gamma)$ is not invariant under the operator S_Γ. Therefore, it does not decompose into the direct sum of its subspaces $A^+(\Gamma)$ and $A^-(\Gamma) \ominus \mathbb{C}$. One only has the equality

$$C(\Gamma) = \operatorname{clos}(A^+(\Gamma) + A^-(\Gamma)).$$

In the spaces $H_\mu(\Gamma)$ and W the operator S_Γ *is* bounded, and consequently one has the decompositions

$$H_\mu(\Gamma) = H_\mu^+(\Gamma) \oplus H_\mu^-(\Gamma), \quad W = W^+ \oplus W^-,$$

where

$$\begin{aligned} H_\mu^{\pm}(\Gamma) &= P_\Gamma^{\pm}(H_\mu(\Gamma)) = H_\mu(\Gamma) \cap L_p^{\pm}(\Gamma, \varrho), \\ W^{\pm} &= P_{\Gamma_0}^{\pm}(W) = W \cap L_p^{\pm}(\Gamma_0, \varrho). \end{aligned}$$

Further, let us denote by $R^{\pm}(\Gamma)$ the algebras of rational functions whose poles lie in the respective domains $\mathcal{D}_\Gamma^{\mp}$ and which vanish at infinity in the case of the algebra $R^-(\Gamma)$. The following chains of inclusions hold:

$$R^+(\Gamma) \subset H_\mu^+(\Gamma) \subset A^+(\Gamma) \subset L_\infty^+(\Gamma) \subset L_p^+(\Gamma, \varrho) \subset L_1^+(\Gamma),$$

$$R^-(\Gamma) \subset H_\mu^-(\Gamma) \subset A^-(\Gamma) \ominus \mathbb{C} \subset L_\infty^-(\Gamma) \ominus \mathbb{C} \subset L_p^-(\Gamma, \varrho) \subset L_1^-(\Gamma).$$

Moreover, for $1 \le p < \infty$ we have $\operatorname{clos} R^{\pm}(\Gamma) = L_p^{\pm}(\Gamma, \varrho)$, and the classes $L_1^+(\Gamma)$ and $L_1^-(\Gamma)$ are maximally wide in the sense that they coincide with the set of all functions $f(z)$ that are analytic in $\mathcal{D}_\Gamma^+$ and $\mathcal{D}_\Gamma^-$ respectively (and satisfy $f(\infty) = 0$ in the second case) and possess a nontangential limit $f(t)$ for almost all points $t \in \Gamma$, such that the resulting function $f(t)$ is measurable and $f(z)$ is recovered from $f(t)$ by means of a Cauchy-type integral.

Propositions 1.1–1.3, appropriately modified, hold also for the space $L_p^{\pm}(\Gamma, \varrho)$. In particular, the function $\varrho_p^+(z)$ of Proposition 1.2 has now the form

$$\varrho_p^+(z) = \prod_{k=1}^{m} [(z - t_k)^{\frac{\alpha_k}{p}}]_+, \quad z \in \overline{\mathcal{D}_\Gamma^+}, \tag{2.3}$$

where the branches of the multi-valued functions $[(z - t_k)^{\frac{\alpha_k}{p}}]_+$ are chosen so that the function $\varrho_p^+(z)$ will be analytic in $\mathcal{D}_\Gamma^+$.

Let us also mention that most results of the theory of singular integral operators carry over from the case of the real line $\mathbb{R}$ to the case of the circle Γ_0 via fractional linear transformations that map $\mathbb{D}^{\pm}$ conformally onto the upper and lower complex half-planes, Π^+ and Π^-. Since the recipe for this transition will be needed in a somewhat more general situation, we will pause to discuss it in more detail.

Let $a_k \in \mathbb{C}$, $k = 1, 2, 3, 4$, and assume that $\Delta = a_1 a_4 - a_2 a_3 \neq 0$. Consider the following two fractional linear transformations, which are inverses to one another:

$$v(t) = \frac{a_1 t + a_2}{a_3 t + a_4}, \quad v^{-1}(x) = \frac{a_4 x - a_2}{a_1 - a_3 x}. \tag{2.4}$$

Let Γ_v denote the circle (or line) that the transformation v maps into $\mathbb{R}$, and consider the following pair of operators, which are inverses of one another:

$$(\mathcal{K} f)(t) = \frac{f(v(t))}{a_3 t + a_4}, \quad (\mathcal{K}^{-1} g)(x) = \frac{\Delta g(v^{-1}(x))}{a_1 - a_3 x}. \tag{2.5}$$

Proposition 2.1 *Let $1 < p < \infty$. Then we have the following assertions are true.*

(1) *If ϱ has the form* (1.1), *then*

$$\varrho_v(t) = |a_3 t + a_4|^{p-2} \varrho(v(t)) \in A_p(\Gamma_v).$$

(2) *The operator $\mathcal{K}$ effects a Banach space isomorphism $L_p(\mathbb{R}, \varrho) \cong L_p(\Gamma_v, \varrho_v)$.*

(3) $\mathcal{K}S = S_{\Gamma_v}\mathcal{K}$.

(4) *If $a_1, a_2, a_3, a_4 \in \mathbb{R}$, then $\Gamma_v = \mathbb{R}$ and $\mathcal{K}S = \operatorname{sign} \Delta S \mathcal{K}$.*

Proof. Assertion (2) results from of a change of variables in the Lebesgue integral. The operator identities can also be established by means of a change of variables for rational functions and then extended to $L_p(\mathbb{R}, \varrho)$ by continuity. □

From Proposition 2.1 it follows that

$$L_p^{\pm}(\mathbb{R}, \varrho) = P^{\pm}(L_p(\mathbb{R}, \varrho)) = \mathcal{K}^{-1}(L_p^{\pm}(\Gamma_v, \varrho_v)).$$

Let us also remark that

$$\mathcal{K}(L_2(\mathbb{R})) = L_2(\Gamma_v), \quad \mathcal{K}(L_p(\mathbb{R})) = L_p(\Gamma_v, \varrho_v),$$

$p \neq 2$, where $\varrho_v(t) = |a_3 t + a_4|^{p-2}$.

2.4 The class fact(p, ϱ)

A central role in the theory of singular integral operators is played by the notion of the (p, ϱ)-factorization, which generalizes the standard factorization and allows one to construct the Φ-theory of the Toeplitz operator

$$T(a) = P_\Gamma^+ aI \tag{2.6}$$

with a finite index in the space $L_p^+(\Gamma, \varrho)$.

Let $1 < p < \infty$. We say that the function $a \in L_\infty(\Gamma)$ admits a (p, ϱ)-*factorization with respect to the simple closed contour* Γ if it can be represented in the form

$$a(t) = a^-(t)(t - z_0)^\kappa a^+(t), \quad t \in \Gamma, \tag{2.7}$$

where κ is an integer, called the *index* of the factorization, $z_0 \in \mathcal{D}_\Gamma^+$, and the functions $a^\pm(t)$ satisfy the following conditions:

(1) $a^- \in L_p^-(\Gamma, \varrho) \oplus \mathbb{C}$, $1/a^- \in L_q^-(\Gamma, \varrho^{1-q}) \oplus \mathbb{C}$, $a^+ \in L_q^+(\Gamma, \varrho^{1-q})$, $1/a^+ \in L_p^+(\Gamma, \varrho)$, $q = p/(p-1)$;

(2) the operator $(1/a^+)S_\Gamma(1/a^-)I$ is bounded in $L_p(\Gamma, \varrho)$.

It is readily seen that a (p, ϱ)-factorization is unique up to a constant factor.

The set of all functions $a \in L_\infty(\Gamma)$ that admit a (p, ϱ)-factorization will be denoted by fact(p, ϱ). In the case of the space $L_p(\mathbb{R}, \varrho)$ the notion of the (p, ϱ)-factorization takes on the following form.

We say that the function $a \in L_\infty(\mathbb{R})$ admits a *(p, ϱ)-factorization with respect to the real line* $\mathbb{R}$ if it can be represented in the form

$$a(x) = a^-(x) \left(\frac{x-i}{x+i} \right)^\kappa a^+(x), \quad x \in \mathbb{R}, \tag{2.8}$$

where κ is an integer, called the *index* of the factorization, and the functions $a^\pm(x)$ are subject to the following conditions:

(1) $\dfrac{a^-(x)}{x-i} \in L_p^-(\mathbb{R}, \varrho), \quad \dfrac{1}{a^-(x)(x-i)} \in L_q^-(\mathbb{R}, \varrho^{1-q})$

$\dfrac{a^+(x)}{x+i} \in L_q^+(\mathbb{R}, \varrho^{1-q}), \quad \dfrac{1}{a^+(x)(x+i)} \in L_p^+(\mathbb{R}, \varrho), \quad q = p/(p-1);$

(2) the operator $(1/a^+)S(1/a^-)I$ is bounded in $L_p(\mathbb{R}, \varrho)$.

Before stating the main result of this section, let us point out the connection between the operator $T(a)$ in $L_p^+(\Gamma, \varrho)$ and the singular integral operator

$$A(a) = aP^+ + P^-, \tag{2.9}$$

acting in $L_p(\Gamma, \varrho)$ (or $L_p(\mathbb{R}, \varrho)$). The source of this connection is the operator identity

$$A(a) = T(a)P^+ + P^- + P^-aP^+, \tag{2.10}$$

from which it follows that

$$T(a) = P^+ A(a)|_{L_p^+(\Gamma, \varrho)}.$$

Moreover, the following proposition shows that the normal solvability theories for the operators $T(a)$ and $A(a)$ are equivalent.

Proposition 2.2 *Let $a \in L_\infty(\Gamma)$. Then for the operator $T(a)$ given by* (2.6) *to be invertible, right-invertible, left-invertible, semi-Fredholm, or normally solvable in the space $L_p^+(\Gamma, \varrho)$ it is necessary and sufficient that the operator $A(a)$ given by* (2.9) *have the same property in the space $L_p(\Gamma, \varrho)$. Furthermore, the corresponding one-sided or two-sided inverse operators $[T(a)]^{-1}$ and $[A(a)]^{-1}$ are connected through the relation*

$$[A(a)]^{-1} = [T(a)]^{-1}P^+ + P^- - P^-a[T(a)]^{-1}P^+. \tag{2.11}$$

Also,

$$\begin{aligned} \ker A(a) &= (P^+ - P^-aI)(\ker T(a)), \\ \operatorname{im} A(a) &= \operatorname{im} T(a) \oplus L_p^-(\Gamma, \varrho), \\ \dim \ker A(a) &= \dim \ker T(a), \\ \operatorname{codim} \operatorname{im} A(a) &= \operatorname{codim} \operatorname{im} T(a). \end{aligned}$$

Let us list a number of known facts concerning the operators $T^*(a)$ and $A^*(a)$. On a curve Γ, the operator $T^*(a)$ is given by the formula

$$T^*(a) = P_\Gamma^+ H_\Gamma (P_\Gamma^- a P_\Gamma^-) H_\Gamma, \tag{2.12}$$

while the operator $A^*(a)$ is given by the formula

$$A^*(a) = D_1 A(a^{-1}) D_2, \tag{2.13}$$

in which the invertible operators D_1 and D_2 have the form

$$D_1 = I + P^+ a P^-, \quad D_2 = (I - P^- a P^+)(a H_\Gamma), \tag{2.14}$$

and the operator H_Γ is given by (2.2). The adjoint operators $T^*(a)$ and $A^*(a)$ act in the spaces $L_q^+(\Gamma, \varrho^{1-q})$ and $L_q(\Gamma, \varrho^{1-q})$, respectively, where $q = p/(p-1)$.

Theorem 2.1 *Let $a \in L_\infty(\Gamma)$. Then operator $T(a)$ given by (2.6) is a Φ-operator in the space $L_p^+(\Gamma, \varrho)$, $1 < p < \infty$, if and only if $a \in \operatorname{fact}(p, \varrho)$. If representation (2.7) holds, then $\operatorname{ind} T(a) = -\kappa$, and for $\kappa = 0$ (resp., $\kappa > 0$, $\kappa < 0$) the operator $T(a)$ is invertible (resp., left-invertible, right-invertible); moreover, the formula*

$$[T(a)]^{-1} = P^+ \frac{1}{(t - z_0)^\kappa a^+} P^+ \frac{1}{a^-} I \tag{2.15}$$

gives the corresponding inverse operator. Further, for $\kappa < 0$, $\dim \ker T(a) = |\kappa|$, and

$$\ker T(a) = \operatorname{span}\left\{ \frac{t^{j-1}}{a^+},\ j = 1, 2, \ldots, |\kappa| \right\}, \tag{2.16}$$

while for $\kappa > 0$, $\dim \operatorname{coker} T(a) = \kappa$, and $f \in \operatorname{im} T(a)$ if and only if the following orthogonality conditions are satisfied:

$$\int_\Gamma f(\tau) \frac{1}{a^-(\tau)} \cdot \frac{1}{(\tau - z_0)^j}\, d\tau = 0, \quad j = 1, 2, \ldots, \kappa. \tag{2.17}$$

Proof. Suppose that $a \in \operatorname{fact}(p, \varrho)$. Then the operator $[T(a)]^{-1}$ given by (2.15) is bounded in $L_p^+(\Gamma, \varrho)$ thanks to condition (2) in the definition of the (p, ϱ)-factorization of the symbol a.

First let us assume that $\kappa \leq 0$, and hence $(t - z_0)^\kappa \in L_\infty^-(\Gamma)$. Let us show that the following equality holds in $L_p^+(\Gamma, \varrho)$:

$$T(a)[T(a)]^{-1} = I. \tag{2.18}$$

Let $f \in R^+(\Gamma)$. Then

$$\begin{aligned}
T(a)[T(a)]^{-1}(f) &= P^+ a^- (t - z_0)^\kappa a^+ P^+ \frac{1}{(t - z_0)^\kappa a^+} P^+ \frac{f}{a^-} \\
&= P^+ a^- P^+ \frac{f}{a^-} - P^+ a^- (t - z_0)^\kappa a^+ P^- \frac{1}{(t - z_0)^\kappa a^+} P^+ \frac{f}{a^-} \\
&= f - P^+ a^- P^- \frac{f}{a^-} - P^+ a^- (t - z_0)^\kappa a^+ P^- \frac{1}{(t - z_0)^\kappa a^+} P^+ \frac{f}{a^-}.
\end{aligned}$$

We claim that the second and third terms in the last expression are equal to zero.

Indeed, in view of condition (1) in the definition of the (p, ϱ)-factorization, $\frac{f}{a^-} \in L_q(\Gamma, \varrho^{1-q})$, that is, $P^- \frac{f}{a^-} \in L_q^-(\Gamma, \varrho^{1-q})$. But $a^- \in L_p^-(\Gamma, \varrho)$, and so

$$a^- P^- \frac{f}{a^-} \in L_1^-(\Gamma).$$

At the same time, by condition (2) in the definition of the (p, ϱ)-factorization,

$$a^- P^- \frac{f}{a^-} = \frac{a}{(t - z_0)^\kappa} \left(\frac{1}{a^+} P^- \frac{f}{a^-} \right) \in L_p(\Gamma, \varrho),$$

and so

$$a^- P^- \frac{f}{a^-} \in L_p^-(\Gamma, \varrho).$$

Therefore,

$$P^+ a^- P^- \frac{f}{a^-} = 0.$$

In a similar manner one can verify that

$$\frac{1}{(t - z_0)^\kappa a^+} P^+ \frac{f}{a^-} \in L_1^+(\Gamma) \cap L_p(\Gamma, \varrho),$$

and so

$$P^- \frac{1}{(t - z_0)^\kappa a^+} P^+ \frac{f}{a^-} = 0.$$

Thus, (2.18) holds in $R^+(\Gamma)$. Since $\operatorname{clos} R^+(\Gamma) = L_p^+(\Gamma, \varrho)$ and the operators $T(a)$ and $[T(a)]^{-1}$ are bounded in $L_p^+(\Gamma, \varrho)$, equality (2.18) extends by continuity to the whole space $L_p^+(\Gamma, \varrho)$.

In much the same way one proves that for $\kappa \geq 0$

$$[T(a)]^{-1} T(a) = I$$

in $L_p^+(\Gamma, \varrho)$.

Taking again $\kappa < 0$, we observe that, for $j = 1, 2, \ldots, |\kappa|$, the functions $t^{j-1}(a^+)^{-1}$ belong to $L_p^+(\Gamma, \varrho)$. At the same time,

$$[T(a)] \left(\frac{t^{j-1}}{a^+} \right) = P^+(a^-(t - z_0)^\kappa t^{j-1}) = 0, \quad j = 1, 2, \ldots, |\kappa|,$$

because $a^-(t - z_0)^\kappa t^{j-1} \in L_p^-(\Gamma, \varrho)$. It follows that

$$\operatorname{span} \left\{ \frac{t^{j-1}}{a^+}, \ j = 1, 2, \ldots, |\kappa| \right\} \subseteq \ker T(a).$$

To show that (2.16) holds it remains to verify that $\dim \ker T(a) = |\kappa|$. To that end, let us introduce the right-invertible Toeplitz operator $T_1 = T((t - z_0)^{-1})$

and show first that $\dim\ker T_1 = 1$. Indeed, let $f \in \ker T_1$. Then $P^+\frac{f(t)}{t-z_0} = 0$, or, equivalently,

$$P^+\frac{f(t)-f(z_0)}{t-z_0} = \frac{f(t)-f(z_0)}{t-z_0} = 0.$$

This means that $f(t) = c$, where $c \in \mathbb{C}$, i.e., $\dim\ker T_1 = 1$. Further, by Proposition 1.7,

$$T(a) = T((t-z_0)^\kappa)T(a^+a^-) = T_1^{|\kappa|}T(a^+a^-),$$

where the operator $T(a^+a^-)$ is invertible. Consequently,

$$\begin{aligned}\dim\ker T(a) &= \dim\ker T_1^{|\kappa|} = \operatorname{ind} T_1^{|\kappa|} \\ &= |\kappa| \operatorname{ind} T_1 = |\kappa| \dim\ker T_1 = |\kappa|.\end{aligned}$$

If now $\kappa > 0$ and $f \in \operatorname{im} T(a)$, then there exists a function $g \in L_p^+(\Gamma, \varrho)$ such that $[T(a)](g) = f$, or, equivalently,

$$P^+(a^-(t-z_0)^\kappa a^+ g) = f.$$

This can be rewritten as

$$a^-(t-z_0)^\kappa a^+ g = f + P^-(a^-(t-z_0)^\kappa a^+ g),$$

whence it follows that

$$g = \frac{1}{(t-z_0)^\kappa a^+} P^+ \frac{f}{a^-}.$$

Since $z_0 \in \mathcal{D}^+$ and the function

$$F(z) = \frac{1}{a^+(z)}\left(P^+\frac{f}{a^-}\right)(z) = \frac{1}{a^+(z)2\pi i}\int_\Gamma \frac{f(\tau)d\tau}{a^-(\tau)(\tau-z)}, \quad z \in \mathcal{D}^+,$$

is analytic in $\mathcal{D}_\Gamma^+$, the inclusion $g \in L_p^+(\Gamma, \varrho)$ is equivalent to the κ conditions

$$\left.\left(\int_\Gamma \frac{f(\tau)d\tau}{a^-(\tau)(\tau-z)}\right)^{(j)}\right|_{z=z_0} = j!\int \frac{f(\tau)d\tau}{a^-(\tau)(\tau-z_0)^{j+1}} = 0, \quad j = 0, 1, \ldots, \kappa-1,$$

which in turn are equivalent to conditions (2.17).

Now suppose that the operator $T(a)$ given by (2.6) is a Φ-operator of index $-\kappa$. Let us show that the symbol a belongs to $\operatorname{fact}(p, \varrho)$. To that end, we introduce the operator B by

$$B = \begin{cases} T((t-z_0)^{-\kappa})T(a), & \text{if } \kappa > 0, \\ T(a)T((t-z_0)^{-\kappa}), & \text{if } \kappa < 0, \end{cases}$$

where $z_0 \in \mathcal{D}^+$. By Proposition 1.7, B is a Toeplitz operator, $B = T(b)$, where

$$b = (t-z_0)^{-\kappa} a. \tag{2.19}$$

The operator $T((t-z_0)^{-\kappa})$ is a one-sided invertible Φ-operator and its index is equal to κ. Therefore, $T(b)$ is a Φ-operator and

$$\operatorname{ind} T(b) = \operatorname{ind} T((t-z_0)^{-\kappa}) + \operatorname{ind} T(a) = 0.$$

By Proposition 1.4, $T(b)$ is invertible, and by Proposition 1.6, $b \in GL_\infty(\Gamma)$.

Notice that the invertibility of the operator $T(b)$ in $L_p^+(\Gamma, \varrho)$ implies that of the operator $T(b^{-1})$ in $L_q^+(\Gamma, \varrho^{1-q})$. Indeed, by Proposition 2.2, the operator $A(b)$ is invertible in $L_p(\Gamma, \varrho)$. Therefore, $A^*(b)$ is invertible in $L_q(\Gamma, \varrho^{1-q})$. By formulas (2.13) and (2.14), the operator $A(b^{-1})$ is invertible in $L_q(\Gamma, \varrho^{1-q})$. Applying again Proposition 2.2, we conclude that the operator $T(b^{-1})$ is invertible in $L_q^+(\Gamma, \varrho^{1-q})$.

Next, consider the equation $[T(b)](f^+) = 1$ in $L_p^+(\Gamma, \varrho)$ and the equation $[T(b^{-1})](g^+) = 1$ in $L_q^+(\Gamma, \varrho^{1-q})$. They have unique solutions f^+ and g^+, and one can write the equalities

$$bf^+ = f^- + 1, \quad b^{-1}g^+ = g^- + 1,$$

where

$$\begin{aligned} &f^+ \in L_p^+(\Gamma, \varrho), \quad f^- = P^- b f^+ \in L_p^-(\Gamma, \varrho), \\ &g^+ \in L_q^+(\Gamma, \varrho^{1-q}), \quad g^- = P^- b^{-1} g^+ \in L_q^-(\Gamma, \varrho^{1-q}). \end{aligned}$$

Multiplying these equalities term by term, we deduce that

$$f^+g^+ = (f^- + 1)(g^- + 1)$$

almost everywhere on Γ. Since

$$f^+g^+ \in L_1^+(\Gamma), \quad (f^- + 1)(g^- + 1) \in L_1^-(\Gamma) \oplus \mathbb{C},$$

and

$$[(f^- + 1)(g^- + 1)](\infty) = 1,$$

we have

$$f^+g^+ = 1, \quad (f^- + 1)(g^- + 1) = 1$$

almost everywhere on Γ. It follows that

$$(f^+)^{-1} = g^+ \in L_q^+(\Gamma, \varrho^{1-q}), \quad (f^- + 1)^{-1} = g^- + 1 \in L_q^-(\Gamma, \varrho^{1-q}) \oplus \mathbb{C}.$$

Denoting

$$b^+ = (f^+)^{-1}, \quad b^- = f^- + 1,$$

we see that

$$b = b^+ b^-$$

almost everywhere on Γ, where the functions b^+ and b^- satisfy condition (1) of the definition of the (p, ϱ)-factorization.

Let us verify that condition (2) of the definition of the (p, ϱ)-factorization is also satisfied. Given any function $g \in L_p^+(\Gamma, \varrho)$, the equation $[T(b)](f^+) = g$ has a unique solution f^+ in $L_p^+(\Gamma, \varrho)$. Hence

$$b^+ b^- f^+ = f^- + g,$$

or, equivalently,

$$b^+ f^+ = \frac{f^-}{b^-} + \frac{g}{b^-}$$

almost everywhere on Γ, where $f^- = P^- b f^+ \in L_p^-(\Gamma, \varrho)$. But $b^+ f^+ \in L_1^+(\Gamma)$, $f^-/b^- \in L_1^-(\Gamma)$. Therefore, by a theorem of Privalov ([1], p. 190), the singular integrals $S(b^+ f^+)$ and $S(f^-/b^-)$ exist almost everywhere on Γ and $P^+(b^+ f^+) = b^+ f^+$, $P^+(f^-/b^-) = 0$. It follows that the singular integral $S(g/b^-)$ exists almost everywhere on Γ and that $f^+ = (1/b^+) P^+ (g/b^-)$. Since the operator $[T(b)]^{-1}$ is bounded in $L_p^+(\Gamma, \varrho)$ and $f^+ = [T(b)]^{-1}(g)$, the operator $(1/b^+)P^+(1/b^-)I$, and together with it the operator $(1/b^+)S(1/b^-)I = 2\,[(1/b^+)P^+(1/b^-)I - (1/b)I]$, is bounded in $L_p^+(\Gamma, \varrho)$. Now the (p, ϱ)-factorization of the function a follows from (2.19). □

2.5 A sufficient condition for (p, ϱ)-factorizability

In this section we give a sufficient condition for the membership of a function in the class $\mathrm{fact}(p, \varrho)$.

First of all, let us remark that any function that admits a standard factorization belongs to $\mathrm{fact}(p, \varrho)$.

A Banach algebra $\mathfrak{A} \subset C(\Gamma)$ is said to be *decomposable* if the operator S_Γ, acting in the Banach space $\mathfrak{A}$, is bounded. Examples of decomposable algebras are the algebra of Hölder continuous functions $H_\mu(\Gamma)$, $0 < \mu < 1$, and the Wiener algebra W. A Banach algebra $\mathfrak{A}$ is called an *R-algebra* if

$$\mathrm{clos}(R^+(\Gamma) + R^-(\Gamma)) = \mathfrak{A},$$

where the closure is taken in the norm of $\mathfrak{A}$.

Theorem 2.2 *Let $\mathfrak{A}$ be a decomposable R-algebra and let $a \in \mathfrak{A}$. Then for the operator $T(a)$ to be a Φ-operator in $L_p^+(\Gamma, \varrho)$ it is necessary and sufficient that*

$$a(t) \neq 0, \quad t \in \Gamma.$$

If the last condition is satisfied, then the function a admits a standard factorization with

$$(a^+)^{\pm 1} \in \mathfrak{A}^+, \quad (a^-)^{\pm 1} \in \mathfrak{A}^- \oplus \mathbb{C}, \quad \mathfrak{A}^\pm = P_\Gamma^\pm(\mathfrak{A}).$$

For the constructions given below we will need one more class of functions that admit a standard factorization.

Proposition 2.3 *Let the function $a \in GL_\infty(\Gamma)$ take positive values almost everywhere on Γ. Then a admits a standard factorization with index $\kappa = 0$. Moreover, the factors $a^\pm$ in this factorization are given by the formulas*

$$a^\pm = \exp\{P_\Gamma^\pm \log a\}. \tag{2.20}$$

This result allows us to reduce the question of the normal solvability of the operator $T(a)$ to the analogous question for the operator $T(a/|a|)$, whose symbol has modulus 1 almost everywhere on Γ (such functions are termed *unimodular*). Indeed, if the necessary normal solvability condition (1.11) is satisfied, then the function $|a|$ admits the standard factorization

$$|a| = d^- d^+,$$

where $(d^+)^{\pm 1} \in L_\infty^+(\Gamma)$, $(d^-)^{\pm 1} \in L_\infty^-(\Gamma)$. Hence, if we denote $a_0 = a/|a|$, then, by Proposition 1.7,

$$T(a) = T(d^-)T(a_0)T(d^+),$$

where the operators $T(d^\pm)$ are invertible and $[T(d^\pm)]^{-1} = T(1/d^\pm)$.

This argument clarifies the important role played by the study of Toeplitz operators with unimodular symbols of the form $a_0(t) = \exp(i\varphi(t))$, where φ is some real-valued function, which represents the argument of the function a.

Let us give some other sufficient conditions for membership in the class $\mathrm{fact}(p, \varrho)$.

Proposition 2.4 *Let $a = 1 + m \in L_\infty(\Gamma)$. If the number $\|m\|_\infty$ is sufficiently small, then the function a admits a (p, ϱ)-factorization with index $\kappa = 0$.*

In the case of the space $L_2(\Gamma_0)$ it suffices to require that $\|m\|_\infty < 1$.

Proposition 2.5 *Let $a \in \mathrm{fact}(p, \varrho)$. Then there exists an $\varepsilon > 0$ such that, for any function b satisfying $\|b\|_\infty < \varepsilon$, $a + b \in \mathrm{fact}(p, \varrho)$, and the factorization indices of the functions a and $a + b$ coincide.*

A rather broad subclass of the class $\mathrm{fact}(p, \varrho)$ is formed by the *sectorial* functions. Let us denote by $G_p(\Gamma, \varrho)$, $1 < p < \infty$, the set of all functions $a \in L_\infty(\Gamma)$ for which the following conditions are satisfied:

(1) $\operatorname*{ess\,inf}_{t \in \Gamma} |a(t)| > 0$,

(2) for each point $\tau \in \Gamma$ there exist an open arc $l(\tau) \subset \Gamma$ containing τ and functions $g_\tau^\pm(\tau) \in L_\infty^\pm(\Gamma)$ with the property that $1/g_\tau^\pm(t) \in L_\infty^\pm(\Gamma)$, such that values of the product $g_\tau^+(t)a(t)g_\tau^-(t)$ when t ranges over $l(\tau)$ lie in a sector with vertex at the origin and of angular opening at most $2\pi/\overline{p}$, where

$$\overline{p} = \max\left\{p, p(p-1)^{-1}, p(1+\alpha)^{-1}, p(p-(1+\alpha))^{-1}\right\},$$

$$\alpha = \begin{cases} \alpha_k, & \text{if} \quad t = t_k, \\ 0, & \text{if} \quad t \neq t_k, \quad k = 1, 2, \ldots, m, \end{cases}$$

and t_k are the nodes of the power weight (1.2) with exponents α_k. In the case $p = 2$ and $\varrho \equiv 1$, this subclass is maximally wide, because the angular opening of the sector figuring in its definition may be arbitrarily close to π. The functions in the class $G_p(\Gamma, \varrho)$ are called *locally sectorial*.

Theorem 2.3 (1) $G_2(\Gamma) = G_2(\Gamma, 1) = \text{fact}(2, 1)$.

(2) $G_p(\Gamma, \varrho) \subset \text{fact}(p, \varrho)$ *for all* $p \in (1, \infty)$ *and all* ϱ *of the form* (2.1).

(3) *For a continuous function* a *on* Γ *to be of class* $\text{fact}(p, \varrho)$ *for every admissible* p *and* ϱ *it is necessary and sufficient that* $a(t) \neq 0$ *for all* $t \in \Gamma$. *If this condition is satisfied, then the factorization index* κ *coincides with the topological index (winding number)* $\kappa(a)$ *of the function* a,

$$\kappa(a) = \frac{1}{2\pi} \arg a(t)\Big|_{\Gamma}. \tag{2.21}$$

Here $\arg a(t)|_\Gamma$ *denotes the increment of the function* $\arg a(t)$ *when the contour* Γ *is traversed in positive direction.*

Factorization and Toeplitz operators with infinite index

We remarked above that the existence of a (p, ϱ)-factorization for the symbol of a Toeplitz operator is equivalent to the operator being Fredholm (Theorem 2.1). Hence, it is only natural to attempt to generalize the concept of factorization in such a way that it will cover, to the extent to which this is possible, larger classes of symbols of normally solvable operators. Below we will introduce one such generalization.

But first let us recall a number of facts concerning the Riesz factorization that will be needed below.

2.6 Inner-outer factorization

A function $F(z)$ analytic in $\mathbb{D}^+$ is called an *outer function* if it can be represented in the form

$$F(z) = \lambda \exp\left(\frac{1}{2\pi} \int_{-\pi}^{\pi} \frac{\exp(i\theta) + z}{\exp(i\theta) - z} \log(k(\theta)) d\theta\right), \quad z \in \mathbb{D}^+,$$

where $k(\theta) \geq 0$, $\log(k(\theta)) \in L_1(-\pi, \pi)$, $|\lambda| = 1$.

Notice that in this case $|F(\exp(i\theta))| = k(\theta)$ almost everywhere on Γ_0, where $F(\exp(i\theta))$ are the nontangential limit values of the function $F(z)$ as $z \to \exp(i\theta)$.

A function $U(z)$ of class $H^\infty(\mathbb{D}^+)$ is called an *inner function* if $|U(t)| = 1$ almost everywhere on Γ_0.

A well-known example of inner function is the *Blaschke product*

$$B(z) = z^n \prod_{k=1}^{\infty} \frac{\overline{z}_k}{|z_k|} \cdot \frac{z_k - z}{1 - \overline{z}_k z}, \quad n \in \mathbb{N}, \quad z \in \mathbb{D}^+, \tag{2.22}$$

whose set of zeros $\{z_k\} \subset \mathbb{D}^+$ satisfies the *Blaschke condition*

$$\sum_{k=1}^{\infty} (1 - |z_k|) < \infty. \tag{2.23}$$

Condition (2.23) guarantees the uniform convergence of the infinite product (2.22) on all compact subsets of the disk $\mathbb{D}^+$. Let us note that the set of zeros $\{z_k\}$ of any nonzero function of class $H^p(\mathbb{D}^+)$, $p \geq 1$, satisfies condition the (2.23).

An inner function $E(z)$ is said to be *singular* if it can be written as

$$E(z) = \exp\left(- \int_{-\pi}^{\pi} \frac{\exp(i\theta) + z}{\exp(i\theta) - z} d\mu(\theta) \right), \quad z \in \mathbb{D}^+, \tag{2.24}$$

where μ is a nonnegative singular measure on Γ_0. In the case when the measure μ is concentrated at the point $t = 1$

$$E(z) = \exp\left(-\sigma \frac{1+z}{1-z} \right), \quad \text{where} \quad \sigma > 0.$$

Theorem 2.4 *Suppose that* $f \in H^1(\mathbb{D}^+)$ *and that* f *does not vanish identically. Then* f *admits the Riesz factorization*

$$f(z) = F(z)B(z)E(z), \quad z \in \mathbb{D}^+, \tag{2.25}$$

where $F(z)$ *is an outer function of class* $H^1(\mathbb{D}^+)$, $B(z)$ *is the Blaschke product constructed from the zeros of* $f(z)$, *and* $E(z)$ *is a singular inner function. The factorization* (2.25) *is unique. Moreover,*

(1) $f \in L_p^+(\Gamma_0, \varrho)$ *if and only if* $F \in L_p^+(\Gamma_0, \varrho)$, $1 < p < \infty$;

(2) $f \in H^\infty(\mathbb{D}^+)$, $f^{-1} \in L_\infty(\Gamma_0)$ *if and only if* $F^{\pm 1} \in H^\infty(\mathbb{D}^+)$.

Blaschke products and singular inner functions may be analytic in domains larger than $\mathbb{D}^+$.

Proposition 2.6 *A Blaschke product* $B(z)$ *of the form* (2.22) *with set of zeros* $\{z_k\}$ *converges uniformly on any compact subset of* $\mathbb{C} \setminus M$, *where* $M = \overline{\{\bar{z}_k^{-1}\}}$, *and hence has an analytic continuation to the domain* $\mathbb{C} \setminus M$. *Moreover,* $B^{-1}(z) \in H^\infty(\mathbb{D}^-)$.

Proposition 2.7 *A singular inner function* $E(z)$ *of the form* (2.24), *generated by a singular measure* μ, *admits an analytic continuation to the domain* $\mathbb{C} \setminus M$, *where* $M = \overline{\text{supp}(\mu)}$. *Moreover,* $E^{-1}(z) \in H^\infty(\mathbb{D}^-)$.

We will also need the following result.

Proposition 2.8 *Any inner function is a uniform limit in $\overline{\mathbb{D}^+}$ of Blaschke products.*

The results stated above can be reformulated for the case of the half-planes $\Pi_\pm$ by using fractional linear isomorphisms of the form (2.5). Here we will indicate only the form of the Blaschke product for Π_+ and the analogue of the Blaschke condition (2.23).

By definition, a Blaschke product for the upper half-plane is a function of the form

$$B^+(z) = \left(\frac{z-i}{z+i}\right)^n \prod_{k=1}^{\infty} \frac{|1+z_k^2|}{1+z_k^2} \cdot \frac{z-z_k}{z-\overline{z}_k}, \quad n \in \mathbb{N}, \quad z \in \Pi_+. \tag{2.26}$$

Here the set of zeros $\{z_k\}$ of $B^+(z)$ lies in Π_+ and the Blaschke condition is supposed to be satisfied:

$$\sum_{k=1}^{\infty} \frac{\operatorname{Im} z_k}{1+|z_k|^2} < \infty. \tag{2.27}$$

Condition (2.27) guarantees the uniform convergence of the infinite product (2.26) on every compact subset of Π_+. If the absolute magnitudes $|z_k|$ are bounded in common, then (2.27) takes on the form $\sum\limits_{k=1}^{\infty} \operatorname{Im} z_k < \infty$ and the convergence factors in the Blaschke product (2.26) are not needed, because the product

$$\prod_{k=1}^{\infty} \frac{z-z_k}{z-\overline{z}_k}$$

itself converges uniformly on compact subsets of Π_+.

In this monograph the reader will often encounter Blaschke products $B^+(z)$ of the form

$$B^+(z) = \prod_{k\in\mathbb{Z}} \frac{\overline{z}_k}{z_k} \cdot \frac{z-z_k}{z-\overline{z}_k}, \tag{2.28}$$

for which the set of zeros $\{z_k\}$ is contained in Π_+ and $z_k \to \infty$ as $|k| \to \infty$. The product (2.28) converges since $B^+(z) = \widetilde{B}^+(-\frac{1}{z})$, where

$$\widetilde{B}^+(z) = \prod_{k\in\mathbb{Z}} \frac{z-\zeta_k}{z-\overline{\zeta}_k}, \quad \zeta_k = -\frac{1}{z_k},$$

is a Blaschke product for Π_+ whose set of zeros $\{\zeta_k\}$ is bounded.

A Blaschke product $B^-(z)$ for the lower half-plane Π_- is defined in a similar manner. Notice that $1/B^+(z)$ is a Blaschke product for Π_-, while $1/B^-(z)$ is a Blaschke product for Π_+. Moreover, if $z \in \mathbb{R}$, then $1/B^\pm(z) = \overline{B^\pm(z)}$. Let us also remark that

$$B^+(z), 1/B^-(z) \in H^\infty(\Pi_+), \quad B^-(z), 1/B^+(z) \in H^\infty(\Pi_-).$$

The simplest singular inner functions for the half-planes $\Pi_\pm$ have the form $E^\pm(z) = \exp(\pm i\sigma z)$, $\sigma > 0$, and $E^\pm(z) \in H^\infty(\Pi_\pm)$, $\sigma > 0$.

2.7 The class fact(∞, p, ϱ) and one-sided invertibility

Let us remark at the outset that all the scalar operators studied below will be at least one-sided invertible. The following criterion for one-sided invertibility of operators $T(a)$ with unimodular symbol in the space $H^2(\mathbb{D}^+)$ leads in a natural manner to the infinite-index factorization introduced by us above.

Theorem 2.5 *Suppose that $|a(t)| = 1$ for almost every $t \in \Gamma$. Then the operator $T(a)$ is left (resp., right) invertible in $H^2(\mathbb{D}^+)$ if and only if*

$$\operatorname{dist}(a, H^\infty(\mathbb{D}^+)) < 1 \quad (\textit{resp.}, \ \operatorname{dist}(\overline{a}, H^\infty(\mathbb{D}^+)) < 1). \tag{2.29}$$

Proof. Let us consider the case of left invertibility. If (2.29) holds, then there exists a function $h \in H^\infty(\mathbb{D})^+$ such that $\|a-h\|_\infty < 1$. This implies that $\|1-\overline{a}h\|_\infty < 1$, whence $\|I - T(\overline{a}h)\|_2 < 1$. Therefore, the operator $T(\overline{a}h) = T(\overline{a})T(h)$ is invertible. It follows that the operator $T(\overline{a})$ is right-invertible, and therefore $T(a) = T^*(\overline{a})$ is left-invertible.

Conversely, suppose that $T(a)$ is left-invertible. Then there exists an $\varepsilon > 0$ such that

$$\|T(a)f\|_2 \geq \varepsilon \|f\|_2, \quad f \in H^2(\mathbb{D}^+).$$

Therefore,

$$\|f\|_2^2 = \|af\|_2^2 = \|P^+af\|_2^2 + \|P^-af\|_2^2 \geq \varepsilon^2\|f\|_2^2 + \|P^-af\|_2^2,$$

whence

$$\|P^-af\|_2^2 \leq (1-\varepsilon^2)\|f\|_2^2, \quad f \in H^2(\mathbb{D}^+).$$

This shows that for any $f \in H^2(\mathbb{D}^+)$ and $g \in H^2(\mathbb{D}^-)$

$$\left|\frac{1}{2\pi}\int_{\Gamma_0} a(t)f(t)\overline{g}(t)|dt|\right| = \left|\frac{1}{2\pi}\int_{\Gamma_0} (P^-af)(t)\overline{g}(t)|dt|\right|$$
$$\leq \|P^-af\|_2\|\overline{g}\|_2 \leq (1-\varepsilon^2)^{1/2}\|f\|_2\|\overline{g}\|_2.$$

Since any function $\psi \in H^1(\mathbb{D}^+)$ can be represented in the form $\psi = \psi_1\psi_2$, with $\psi_{1,2} \in H^2(\mathbb{D}^+)$ and $\|\psi\|_1 = \|\psi_1\|_2\|\psi_2\|_2$ (see, e.g., Garnett [1], p. 93), one has

$$\left|\frac{1}{2\pi}\int_{\Gamma_0} a(t)\psi(t)|dt|\right| = \left|\frac{1}{2\pi}\int_{\Gamma_0} a(t)\psi_1(t)\psi_2(t)|dt|\right|$$
$$\leq (1-\varepsilon^2)^{1/2}\|\psi_1\|_2\|\psi_2\|_2 = (1-\varepsilon^2)^{1/2}\|\psi\|_1.$$

Thus, the linear functional on $H^1(\mathbb{D}^+)$ defined by the function a has norm smaller than one. By Proposition 1.3, $\operatorname{dist}(a, H^\infty(\mathbb{D}^+)) < 1$.

It remains to observe that the right invertibility of the operator $T(a)$ is equivalent to the left invertibility of the operator $T^*(a) = T(\overline{a})$. □

As follows from the proof of Theorem 2.5, when the first condition (2.29) is satisfied, the operator $T(d)$, where $d = \bar{a}h$, is invertible in $H^2(\mathbb{D}^+)$. Hence, $T(d/|d|)$ is invertible (by Propositions 1.7 and 2.3). Therefore, $T^*(d/|d|) = T(\bar{d}/|d|) = T(|d|/d)$ is invertible, and thus so is $T(d^{-1}) = T(|d|^{-1}|d|/d)$. We conclude that $d^{-1} \in \text{fact}(2,1)$ (by Theorem 2.1). This shows that the left invertibility of the operator $T(a)$ implies that the following representation holds:

$$a = ch, \quad \text{where} \quad h \in H^\infty(\mathbb{D}^+) \cap GL_\infty(\Gamma_0), \quad c \in \text{fact}(2,1).$$

In much the same way, when the operator $T(a)$ is right-invertible in $H^2(\mathbb{D}^+)$ its symbol can be written in the form

$$a = ch^{-1}, \quad \text{where} \quad h \in H^\infty(\mathbb{D}^+) \cap GL_\infty(\Gamma_0), \quad c \in \text{fact}(2,1).$$

The foregoing constructions lead to the following definition. We say that a function $a \in L_\infty(\Gamma)$ *admits a generalized factorization with infinite index in the space* $L_p(\Gamma, \varrho)$ (or (p,ϱ)-factorization with infinite index) if it admits a representation

$$a = bh \quad \text{or} \quad a = bh^{-1}, \tag{2.30}$$

where

(1) $b \in \text{fact}(p, \varrho)$, $1 < p < \infty$;

(2) $h \in L_\infty^+(\Gamma) \cap GL_\infty(\Gamma)$.

The class of functions admitting a generalized factorization with infinite index in $L_p(\Gamma, \varrho)$ will be denoted by $\text{fact}_\Gamma(\infty, p, \varrho)$, or simply by $\text{fact}(\infty, p, \varrho)$.

Here are some properties of functions in the class $\text{fact}(\infty, p, \varrho)$.

1. $\text{fact}(p, \varrho) \subset \text{fact}(\infty, p, \varrho)$.

   It follows that the class of Toeplitz operators with symbols in $\text{fact}(\infty, p, \varrho)$ contains Φ-operators. The following condition excludes Φ-operators from this class: for any polynomial u with complex coefficients,

$$u/h \notin L_\infty^+(\Gamma). \tag{2.31}$$

   Indeed, suppose that condition (2.31) is not satisfied, i.e., there exists a polynomial u such that $v^+ = u/h \in L_\infty^+(\Gamma)$. With no loss of generality, we may assume that all the roots of u lie in the domain $\mathcal{D}^+(\Gamma)$ and are roots of the functions h. We may also without loss of generality assume that the orders of the zeros of u are greater than or equal to the corresponding orders of the zeros of h. Then $1/v^+ = h/u \in L_\infty^+(\Gamma)$. Hence, if $b \in \text{fact}(p, \varrho)$ then $bh, bh^{-1} \in \text{fact}(p, \varrho)$, and consequently $T(bh)$ and $T(bh^{-1})$ are Φ-operators in $L_p^+(\Gamma, \varrho)$.

   The fact that the condition (2.31) ensures that the subspaces $\ker T(bh^{-1})$ or $\operatorname{coker} T(bh)$ are infinite-dimensional will be shown below in Theorem 2.6.

2. *(p,ϱ)-factorization with infinite index does not enjoy the uniqueness property.*

   For example, if the first equality in (2.30) holds, then $a = \widetilde{b}\widetilde{h}$, where $\widetilde{h} = hd^+$, $\widetilde{b} = b(d^+)^{-1}$, $d^+ \in GL^+_\infty(\Gamma)$, is another factorization with infinite index in the space $L_p(\Gamma, \varrho)$. In this connection, choosing the function h in each concrete situation is an actual task, determined by the peculiarities of the function a under consideration.

3. *Let* $a \in \mathrm{fact}(\infty, p, \varrho)$ *and let condition* (2.31) *be satisfied. Then the function* h *in* (2.30) *can be chosen so that* $\mathrm{ind}\, b = 0$.

   Indeed, suppose, say, that $a = bh$, where $\mathrm{ind}\, b = \kappa \neq 0$, i.e., $b = b^+(t-z_0)^\kappa b^-$, $z_0 \in \mathcal{D}^+_\Gamma$. If $\kappa > 0$, then $a = \widetilde{b}\widetilde{h}$, where $\widetilde{b} = b^+b^- \in \mathrm{fact}(p,\varrho)$ and $\mathrm{ind}\,\widetilde{b} = 0$, $\widetilde{h} = (t-z_0)^\kappa h \in L^+_\infty(\Gamma)$. If $\kappa < 0$ and the function h has a polynomial divisor u of degree $|\kappa|$ in $L^+_\infty(\Gamma)$ such that the roots of u lie in $\mathcal{D}^+_\Gamma$, then $a = \widetilde{b}\widetilde{h}$, where $\widetilde{h} = hu^{-1} \in L^+_\infty(\Gamma)$, $\widetilde{b} = ub \in \mathrm{fact}(p,\varrho)$ and $\mathrm{ind}\,\widetilde{b} = 0$.

   Now suppose that $\kappa < 0$ and h has no polynomial divisor of the type indicated above in $L^+_\infty(\Gamma)$. Let us introduce a conformal mapping $z = \alpha(s)$ of the disk $\mathbb{D}^+$ onto the domain $\mathcal{D}^+_\Gamma$. The function $h(\alpha(s))$ lies in $H^\infty(\mathbb{D}^+)$ and, by Theorem 2.4, it admits a representation

   $$h(\alpha(s)) = F(s)B(s)E(s),$$

   where $F^{\pm 1} \in H^\infty(\mathbb{D}^+)$, B is a finite Blaschke product, and E is a singular inner function, which in view of condition (2.31) is not a constant. By Proposition 2.8, there exists a Blaschke product B_E such that $\|E - B_E\|_\infty < \varepsilon$ for sufficiently small ε. Since the function E has discontinuities of the second kind on Γ_0, it cannot be approximated arbitrarily close in the metric of $L_\infty(\Gamma_0)$ by continuous functions. Hence, by making ε smaller one can ensure that the function B_E has infinitely many zeros. We put

   $$\widetilde{h}(z) = F(\alpha^{-1}(z))B(\alpha^{-1}(z))B_E(\alpha^{-1}(z)), \quad b_1(t) = \frac{E(\alpha^{-1}(t))}{B_E(\alpha^{-1}(t))},$$

   and remark that for small ε the function bb_1 lies in $\mathrm{fact}(p,\varrho)$ (Proposition 2.5). Therefore, $a = \widetilde{b}\widetilde{h}$, where $\widetilde{h}$ has more than $|\kappa|$ zeros in $\mathcal{D}^+_\Gamma$, and $\widetilde{b} = bb_1$. The case $a = bh^{-1}$ is dealt with in a similar manner.

4. *Let* $a \in \mathrm{fact}_{\Gamma_0}(\infty, p, \varrho)$ (*or* $a \in \mathrm{fact}_{\mathbb{R}}(\infty, p, \varrho)$). *Then for the function* h *in* (2.30) *one can take some inner function* U.

   Indeed, let $a \in \mathrm{fact}(\infty, p, \varrho)$, $a = bh$ or $a = bh^{-1}$. By Theorem 2.4, $h = FU$, where F is outer, $F^{\pm 1} \in L^+_\infty$, and U is inner. It follows that $a = \widetilde{b}U$ or $a = \widetilde{b}\overline{U}$, where $\widetilde{b} = bF$ or $\widetilde{b} = bF^{-1}$, with $\widetilde{b} \in \mathrm{fact}(p,\varrho)$.

The following theorem contains the main results of this section.

Theorem 2.6 *Assume* $a \in \mathrm{fact}_\Gamma(\infty, p, \varrho)$, *condition* (2.31) *is satisfied, and* $\mathrm{ind}\, b = 0$. *If* $a = bh^{-1}$, *then the operator* $T(a)$ *is right-invertible in the space* $L_p^+(\Gamma, \varrho)$, $\dim \ker T(a) = \infty$, *and the operator*

$$[T(a)]^{-1} = \frac{h}{b^+} P^+ \frac{1}{b^-} I, \tag{2.32}$$

where $b = b^+ b^-$ *is a* (p, ϱ)*-factorization of the function* b, *is a right inverse of* $T(a)$. *For a function* φ *to belong to* $\ker T(a)$ *it is necessary and sufficient that*

$$\varphi = \frac{h}{b^+} P^- \frac{b^+}{h} \psi, \quad \text{where} \quad \psi \in \ker T(h^{-1}). \tag{2.33}$$

If $a = bh$, *then the operator* $T(a)$ *is left-invertible in the space* $L_p^+(\Gamma, \varrho)$, $\dim \operatorname{coker} T(a) = \infty$, *and the operator*

$$[T(a)]^{-1} = P^+ \frac{1}{b^+ h} P^+ \frac{1}{b^-} I \tag{2.34}$$

is a left inverse for $T(a)$. *For a function* f *to belong to* $\operatorname{im} T(a)$ *it is necessary and sufficient that*

$$\int_\Gamma \psi_j(t) f(t)\, dt = 0, \quad j = 1, 2, \ldots, \tag{2.35}$$

where

$$\psi_j = \frac{1}{b^-} P^- \frac{(t - z_0)^{-j}}{hb^+} \in L_q^-(\Gamma, \varrho^{1-q}), \quad q = p/(p-1),$$

and z_0 *is a fixed point in the domain* $\mathcal{D}_\Gamma^-$.

Proof. The operator equalities that follow are verified on the set $R^+(\Gamma)$, which is dense in $L_p^+(\Gamma, \varrho)$, and then extended by continuity to the closure. The verification process uses the identities

$$P^\pm \psi^\pm P^\pm f = \psi^\pm P^\pm f, \quad P^\pm \psi^\mp P^\pm f = P^\pm \psi^\mp f,$$

where $f \in L_p(\Gamma, \varrho)$, $\psi^+ \in L_q^+(\Gamma, \varrho^{1-q})$, $\psi^- \in L_q^-(\Gamma, \varrho^{1-q}) \oplus \mathbb{C}$, and $q = p/(p-1)$.

An operator $[T(a)]^{-1}$ of the form (2.32) is bounded on $L_p^+(\Gamma, \varrho)$ thanks to condition (2) of (2.7), and for the choice $a = bh^{-1}$ it is a right inverse for the operator $T(a)$, because in $L_p^+(\Gamma, \varrho)$ one has

$$T(a)[T(a)]^{-1} = P^+ \frac{b^- b^+}{h} P^+ \frac{h}{b^+} P^+ \frac{1}{b^-} I = P^+ b^- P^+ \frac{1}{b^-} I = P^+ = I.$$

Now let us show that a function φ of the form (2.33) lies in the subspace $\ker T(a)$. First,

$$P^- \varphi = P^- \frac{h}{b^+} P^- \frac{b^+}{h} \psi = P^- \psi = 0 \Longrightarrow \varphi \in L_p^+(\Gamma, \varrho).$$

Furthermore,

$$[T(a)]\varphi = P^+ \frac{ah}{b^+} P^- \frac{b^+}{h} \psi = P^+ b^- P^- \frac{b^+}{h} \psi = 0,$$

i.e., $\varphi \in \ker T(a)$.

Before verifying that every element of $\ker T(a)$ has the form (2.33), let us observe that if $\varphi \in \ker T(a)$, i.e., $P^+ a\varphi = 0$, then $a\varphi = \varphi^-$, where $\varphi^- = P^- a\varphi \in L_p^-(\Gamma, \varrho)$. Therefore, $\varphi = \varphi^-/a$, and so $P^-(\varphi^-/a) = 0$, or $\varphi^- \in \ker P^-(1/a)I$ in $L_p^-(\Gamma, \varrho)$. And conversely, if $\varphi^- \in \ker P^-(1/a)I$ in $L_p^-(\Gamma, \varrho)$, then $\varphi^- = a\varphi^+$, where $\varphi^+ \in \ker T(a)$.

Now let $\varphi \in \ker T(a)$. Then $\varphi = \varphi^-/a$, where

$$P^-(\varphi^-/a) = P^-(h\varphi^-/(b^+b^-)) = P^- h P^-(\varphi^-/(b^+b^-)) = 0,$$

and so $\psi^- = P^-(\varphi^-/(b^+b^-)) \in \ker P^- hI$ in $L_p^-(\Gamma, \varrho)$.

Applying the operator $b^- P^- b^+ I$ to both sides of the last equality, we conclude that $\varphi^- = b^- P^- b^+ \psi^-$. But $\psi = h\psi^- \in \ker T(h^{-1})$, and consequently $\varphi = (b^-/a)P^-(b^+\psi^-) = (h/b^+)P^-((b^+/h)\psi)$.

Let us show that $\dim \ker T(a) = \infty$. Assuming that this is not the case, i.e., $\dim \ker T(a) < \infty$, we infer that $T(a)$ is a Φ-operator, and hence $a \in \mathrm{fact}(p, \varrho)$, that is, $a = a^+(t - z_0)^{-\kappa} a^-$, where $z_0 \in \mathcal{D}_\Gamma^+$, $\kappa \geq 0$. But then

$$\psi^+(t)(t - z_0)^{-\kappa} = \psi^-(t)$$

almost everywhere on Γ, where

$$\psi^+ = ha^+(b^+)^{-1} \in L_1^+(\Gamma), \quad \psi^- = b^-(a^-)^{-1} \in L_1^-(\Gamma) \oplus \mathbb{C}.$$

Taking for $\kappa \geq 0$ a polynomial $u_{\kappa-1}$ of degree $\kappa - 1$ (we put $u_{-1}(t) \equiv 0$) with the property that $(t - z_0)^{-\kappa}(\psi^+(t) - u_{\kappa-1}(t)) \in L_1^+(\Gamma)$, we see that for almost every $t \in \Gamma$,

$$(t - z_0)^{-\kappa}(\psi^+(t) - u_{\kappa-1}(t)) = (t - z_0)^{-\kappa} u_{\kappa-1}(t) + \psi^-(t),$$

where the right-hand side lies in $L_1^-(\Gamma) \oplus \mathbb{C}$. But $L_1^+(\Gamma) \cap (L_1^-(\Gamma) \oplus \mathbb{C}) = \mathbb{C}$. Consequently, $\psi^+(t) = P_\kappa(t)$, where $P_\kappa(t)$ is some polynomial of degree at most κ. This shows that the fraction $P_\kappa(t)/h(t) = (b^+(t))^{-1} a^+(t)$ belongs to $L_1^+(\Gamma) \cap L_\infty(\Gamma)$. Therefore, $P_\kappa(t)/h(t) \in L_\infty^+(\Gamma)$, which contradicts (2.31).

Now let $a = bh$. By condition (2) of the definition of the factorization (2.7), the operator $[T(a)]^{-1}$ given by (2.34) is bounded in $L_p^+(\Gamma, \varrho)$. Moreover, it is a left inverse for the operator $T(a)$, because on $L_p^+(\Gamma, \varrho)$,

$$[T(a)]^{-1} T(a) = P^+ \frac{1}{hb^+} P^+ \frac{1}{b^-} P^+ b^- b^+ hI = P^+ \frac{1}{hb^+} P^+ b^+ hI = I.$$

The fact that the subspace $\operatorname{coker} T(a)$ is infinite-dimensional is established in the same way as for the subspace $\ker T(a)$ in the case where the operator $T(a)$ is right-invertible.

If $f \in \operatorname{im} T(a)$, then letting the operator $(1/b^+)P^+((1/b^-)I)$ act on both sides of the equation $[T(a)]\varphi = f$ we obtain

$$h\varphi = (1/b^+)P^+(f/b^-).$$

It follows that $f \in \operatorname{im} T(a)$ if and only if $(1/hb^+)P^+(f/b^-) \in L_p^+(\Gamma, \varrho)$. To show that f is in $\operatorname{im} T(a)$ if and only if (2.35) holds, it remains to apply the analogue of Proposition 1.1 for $L_p^+(\Gamma, \varrho)$: namely, for $z_0 \in \mathcal{D}_\Gamma^-$ and $j = 1, 2, \ldots$,

$$\int_\Gamma \frac{1}{(t - z_0)^j h(t) b^+(t)} \left(P^+ \frac{f}{b^-}\right)(t)dt = 0$$

$$\Longleftrightarrow \int_\Gamma \frac{1}{b^-(t)} \left(P^- \frac{1}{(t - z_0)^j hb^+}\right)(t)f(t)dt = 0.$$

Here we used the fact that in all cases the integrands are absolutely integrable on Γ and $\int\limits_\Gamma g(t)dt = 0$ whenever $g \in L_1^+(\Gamma)$ or $(tg) \in L_1^-(\Gamma)$.

Finally, let us show that $\psi_j = (1/b^-)P^- \frac{(t-z_0)^{-j}}{hb^+} \in L_q^-(\Gamma, \varrho^{1-q})$. From condition (1) of the definition of the factorization (2.7) it follows that $(1/b^+) \in L_p^+(\Gamma, \varrho)$, $(1/b^-) \in L_q^-(\Gamma, \varrho) \oplus \mathbb{C}$. Since $1/((t - z_0)^j h(t)) \in L_\infty(\Gamma)$, we have $\psi_j \in L_1^-(\Gamma)$. But the operator $(1/b^-)S(1/b^+)I$ is equal to $-H_\Gamma((1/b^+)S(1/b^-)I)^* H_\Gamma$, where the operator H_Γ is defined by equality (2.2), and hence it is bounded in $L_q(\Gamma, \varrho^{1-q}) = L_p^*(\Gamma, \varrho)$. Therefore, $\psi_j \in L_q(\Gamma, \varrho^{1-q})$. □

Remark 2.1 Let us return to the case $\Gamma = \Gamma_0$. Suppose that $a \in \operatorname{fact}(\infty, p, \varrho)$. By property 4 of the functions in $\operatorname{fact}(\infty, p, \varrho)$, $a = \widetilde{b}U$ or $a = \widetilde{b}\overline{U}$, where $\widetilde{b} \in \operatorname{fact}(p, \varrho)$ and U is an inner function. In this case, by Theorem 2.6, the description of the subspace $\ker T(a)$ reduces to that of a subspace of the form $\ker T(\overline{U})$. Subspaces of the form $\ker T(\overline{U})$ will be called *model subspaces*. Model subspaces are invariant under the operator $V^* = P^+\overline{t}I$, the dual of the shift operator $V = tI$ on the space $H^p(\mathbb{D}^+)$: if $f \in \ker T(\overline{U})$, which means that $P^+\overline{U}f = 0$, then $V^*f = P^+\overline{t}f \in \ker T(\overline{U})$. Indeed, by Proposition 1.7,

$$P^+\overline{U}P^+\overline{t}f = P^+\overline{U}\,\overline{t}f = P^+\overline{t}P^+\overline{U}f = 0.$$

Thus, Theorem 2.6 establishes a connection between the problem of describing the solutions of a Toeplitz equation with infinite index and the theory of model subspaces. The next chapter will be devoted to a detailed study of such subspaces.

Remark 2.2 The class of Toeplitz operators whose symbols admit a generalized factorization with infinite index in $L_p(\Gamma, \varrho)$ is rather broad. Indeed, it follows from Theorem 2.5 and Propositions 1.6 and 2.3 that in the case of the space $H^2(\mathbb{D}^+)$ this class coincides with the entire class of one-sided invertible and hence normally solvable Toeplitz operators (Proposition 1.4). Theorem 2.6 shows that representation (2.30) is sufficient for the one-sided invertibility of the operator $T(a)$ in $L_p^+(\Gamma, \varrho)$. However, at present the necessity of condition (2.30) in this case remains an open problem.

2.8 Examples of functions in fact(∞, p, ϱ)

Inner functions provide the simplest examples of symbols belonging to the class fact(∞, p, ϱ).

Example 2.1. In Section 1.6 we considered a symbol of the form

$$h(x) = \exp(i\sigma x), \quad \sigma \in \mathbb{R}. \tag{2.36}$$

This is the simplest singular inner function on $\mathbb{R}$ which belongs to the class $\text{fact}_{\mathbb{R}}(\infty, p, \varrho)$. The function $h(x)$ has a discontinuity at infinity. Its argument is

$$\arg h(x) = \sigma x, \quad x \in \mathbb{R}.$$

In what follows functions of the form (2.36) will play a crucial role in the construction of the theory of Toeplitz operators whose symbols have almost periodic discontinuities.

The next two examples are samples of functions h with a more complex behavior of their argument at infinity.

Example 2.2. Power whirls. Let

$$h(x) = \exp(ix^\lambda) + \exp\{ix^\lambda \exp(i\pi(1-\lambda))\}, \quad 0 < \lambda < 1, \tag{2.37}$$

where the branch of the multi-valued function $f(z) = z^\lambda$ is specified as follows: if $z = r\exp(i\varphi)$, then $f(z) = r^\lambda \exp(i\lambda\varphi)$, $-\pi/2 < \varphi \le 3\pi/2$. Since

$$\begin{aligned} h(z) &= \exp\{r^\lambda(-\sin\lambda\varphi + i\cos\lambda\varphi)\} \\ &\quad + \exp\{r^\lambda[-\sin(\lambda\varphi + \pi(1-\lambda)) + i\cos(\lambda\varphi + \pi(1-\lambda))]\}, \end{aligned}$$

it is readily seen that $h \in H^\infty(\Pi_+)$. Moreover, on $\mathbb{R}$ the function $h(x)$ is continuous and separated away from zero, and hence it satisfies condition (2) of the definition of the class $\text{fact}_{\mathbb{R}}(\infty, p, \varrho)$. Since the first term on the right-hand side of equality (2.37) tends to zero as $x \to -\infty$ while the second vanishes as $x \to +\infty$, we have

$$\arg h(x) = (\operatorname{sign} x)|x|^\lambda + o(x) \quad \text{as} \quad |x| \to \infty.$$

Example 2.3. Logarithmic whirls. Let

$$h(x) = h_+(x) + h_-(x) = \exp\{i\log^\nu x\} + \exp\{i\log^\nu(-x)\}, \quad \nu > 1. \tag{2.38}$$

To avoid discussing the behavior of the function h near the origin, we take for Γ the contour obtained from the real line $\mathbb{R}$ by means of a smooth deformation into Π_+ in a vicinity of the origin. This deformation will be described in detail later.

As a cut for the multi-valued functions in formula (2.38), we will take the negative imaginary semiaxis. For a power function, we pick the principal branch

($x^\nu > 0$ if $x > 0$), while the single-valued branches of the logarithmic function are specified by the conditions (in which $z = r\exp(i\varphi)$):

$$\log z = \log r + i\varphi, \quad \log(-z) = \log r + i(\varphi - \pi), \quad -\pi/2 < \varphi \le 3\pi/2.$$

Now let $z \in \mathcal{D}_\Gamma^+$. Then

$$\begin{aligned} |h_+(z)| &= |\exp\{i\log^\nu x\}| = \left|\exp\left\{i(\log^2 r + \varphi^2)^{\nu/2}\exp\left\{i\nu\arctan\frac{\varphi}{\log r}\right\}\right\}\right| \\ &= \exp\left\{-(\log^2 r + \varphi^2)^{\nu/2}\sin\left(\nu\arctan\frac{\varphi}{\log r}\right)\right\} \end{aligned}$$

and similarly,

$$\begin{aligned} |h_-(z)| &= |\exp\{-i\log^\nu(-z)\}| \\ &= \exp\left\{(\log^2 r + (\varphi-\pi)^2)^{\nu/2}\sin\left(\nu\arctan\frac{\varphi-\pi}{\log r}\right)\right\}. \end{aligned}$$

Since $\nu > 1$ and since as $r \to \infty$,

$$\sin\left(\nu\arctan\frac{\varphi}{\log r}\right) \sim \frac{\nu\varphi}{\log r}, \quad 0 < \varphi \le \pi,$$

the function $h_+(z)$ is bounded in $\overline{\Pi}_+$ outside some neighborhood of the origin. Moreover, it tends to zero along the rays $R_\varphi = \{z \in \mathbb{C} : \arg z = \varphi\}$, $0 < \varphi \le \pi$, and $|h_+(z)| = 1$ on the ray R_0. In much the same way, when $r \to \infty$,

$$\sin\left(\nu\arctan\frac{\varphi-\pi}{\log r}\right) \sim -\frac{\nu(\varphi-\pi)}{\log r}, \quad 0 \le \varphi < \pi,$$

and consequently the function $h_-(z)$ is bounded in $\overline{\Pi}_+$ outside some neighborhood of the origin. Moreover, it tends to zero along the rays R_φ, $0 \le \varphi < \pi$, and $|h_-(z)| = 1$ on the ray R_π.

Carrying out the aforementioned deformation of the line $\mathbb{R}$ in such a way that the resulting curve Γ does not pass through the zeros of the function $h(z) = h_+(z) + h_-(z)$, we can ensure that $h \in L_\infty^+(\Gamma)$. At the same time, $h^{-1} \in L_\infty(\Gamma)$, because $h(x) \ne 0$ for all nonzero $x \in \mathbb{R}$. Therefore, the function h satisfies condition (2) in the definition of the class $\mathrm{fact}_\Gamma(\infty, p, \varrho)$.

It is readily verified that

$$\arg h_+(x) = \log^\nu x, \quad x > 0, \quad \arg h_-(x) = -\log^\nu |x|, \quad x < 0,$$

and so

$$\arg h(x) = (\operatorname{sign} x)\log^\nu |x| + o(x) \quad \text{as} \quad |x| \to \infty, \quad \nu > 1.$$

Remark 2.3 If one subjects formula (2.38) to the complex shift $x \to x + i$, the resulting function $h(x + i)$ will be a model function on $\mathbb{R}$, and $\arg h(x + i)$ will preserve the asymptotics of $\arg h(x)$ when $|x| \to \infty$. The calculations needed to establish this fact are only slightly more difficult.

The transfer of discontinuous symbols from the real line to other curves presents a certain technical challenge. In the next two examples we outline one of the possible ways of achieving such a transfer, confining ourselves to symbols of the form (2.36).

Example 2.4. Let Γ_r be the circle of radius r centered at the point z_0 and let $t_0 = z_0 + r\exp(i\theta_0)$, $\theta_0 \in [0, 2\pi)$, be a point on Γ_r. The fractional linear transformation

$$\zeta = v(z) = t_0 - 2i(t_0 - z_0)(z + i)^{-1} \tag{2.39}$$

maps the half-plane Π_+ onto the disk $\mathcal{D}^+$. Moreover, $v(\mathbb{R}) = \Gamma_r$ and $v(\infty) = t_0$. The inverse transformation $z = v^{-1}(\zeta)$ has the form

$$z = v^{-1}(\zeta) = i(t_0 - 2z_0 + \zeta)(t_0 - \zeta)^{-1}. \tag{2.40}$$

It follows that the function

$$h(t) = \exp\{i\sigma v^{-1}(t)\} = \exp\{\sigma(t + t_0 - 2z_0)(t - t_0)^{-1}\} \tag{2.41}$$

belongs to the class $GL_\infty(\Gamma_r) \cap L_\infty^{\pm}(\Gamma_r)$, as required in the definition of the (∞, p, ϱ)-factorization.

In the next example we will describe a class of deformations of the curve Γ_r for which the functions h of the form (2.41) (more precisely, the functions that differ from (2.41) by a constant factor) continue to satisfy the conditions (2) in (2.30).

Example 2.5. Let Γ be a smooth curve which passes through the point t_0 and enjoys the following property: there exist circles γ_+ and γ_- contained in the domains $\overline{\mathcal{D}^+}$ and $\overline{\mathcal{D}^-}$, respectively, which both pass through the point t_0 and share the tangent to Γ at t_0 (see Figure 2.1). Denote by $\eta_\Gamma(t)$ the angle between the outward normal to Γ at the point t and the positively oriented real axis. Then the function

$$h(t) = \exp\{a_0(t - t_0)^{-1}\},$$

where $\arg a_0 = \eta_\Gamma(t_0)$, lies in $GL_\infty(\Gamma) \cap L_\infty^{\pm}(\Gamma)$.

The proof of this assertion rests on a simple observation concerning the behavior of a function $h(z)$ of the form (2.36) in an arbitrary strip parallel to the real line $\mathbb{R}$. Since in any such strip the function $h(z) = \exp(i\sigma z)$, $\sigma \in \mathbb{R}$, is bounded and separated away from zero, the same is true for any curve lying in such strip.

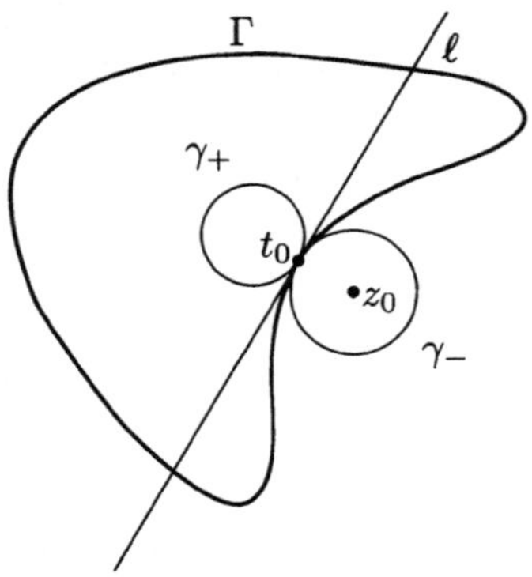

Figure 2.1

Let z_0 and r denote the center and the radius of the circle γ_-. Then $t_0 = z_0 + r\exp(i(\theta_0+\pi))$, where $\theta_0 = \eta_\Gamma(t_0)$. The fractional linear transformation

$$v(z) = r\exp(i(\theta_0+\pi))(z+i)(z-i)^{-1} + z_0,$$
$$v^{-1}(\zeta) = i\frac{\zeta - z_0 - r\exp(i\theta_0)}{\zeta - z_0 + r\exp(i\theta_0)},$$

maps Π_- (resp., Π_+) into the interior (resp., exterior) of the circle γ_-. Moreover, $v^{-1}(\gamma_-) = \mathbb{R}$, the set $v^{-1}(\gamma_+)$ is contained in Π_+ and is a line parallel to $\mathbb{R}$, and the curve $v^{-1}(\Gamma)$ lies in the strip bounded by the lines $\mathbb{R}$ and $v^{-1}(\gamma_+)$. Elementary computations show that the function $h(v(z)) = \exp\{a_0(v(z)-t_0)^{-1}\}$ is a constant times $\exp(i\sigma z)$, where $\sigma = |a_0|(2r)^{-1}$, and hence $h(v(z)) \in GL_\infty(v(\Gamma)) \cap L_\infty^+(v(\Gamma))$. It follows that $h \in GL_\infty(\Gamma) \cap L_\infty^+(\Gamma)$.

The machinery of generalized factorization with infinite index developed above is suitable for the investigation of sufficiently rich classes of Toeplitz operators, including some well-known ones. For example, let us examine the class of Toeplitz operators with symbols in the algebra $L_\infty^+(\Gamma) + C(\Gamma)$.

Proposition 2.9 *Let $a \in L_\infty^+(\Gamma) + C(\Gamma)$ and $a^{-1} \in L_\infty(\Gamma)$. Then the function a admits a generalized factorization with infinite index in $L_p(\Gamma, \varrho)$ of the form $a = bh$.*

Proof. Write $a = d + c$, where $d \in L_\infty^+(\Gamma)$, $c \in C(\Gamma)$. Pick a rational function r_ε, such that $\|c - r_\varepsilon\|_\infty < \varepsilon$, where ε is sufficiently small. Let $\{z_j\}_{j=1}^l$ be the poles of the function r_ε in the domain $\mathcal{D}_\Gamma^+$ and let $\{m_j\}_{j=1}^l$ be their multiplicities. Introduce the polynomial $u(t) = \prod_{j=1}^l (t-z_j)^{m_j}$. Then $r_\varepsilon = \widetilde{r}_\varepsilon/u$, where $\widetilde{r}_\varepsilon = ur_\varepsilon \in L_\infty^+(\Gamma)$. Therefore $a = b_\varepsilon h_\varepsilon$, where $h = h_\varepsilon = du + \widetilde{r}_\varepsilon$ and

$$b_\varepsilon = \frac{1}{u}\left[1 + \frac{c - r_\varepsilon}{d + r_\varepsilon}\right].$$

Clearly, $h_\varepsilon \in L_\infty^+(\Gamma)$ and for small values of ε,

$$h_\varepsilon^{-1} = \frac{1}{u(d + r_\varepsilon)} = \frac{1}{u(a + r_\varepsilon - c)} \in L_\infty(\Gamma).$$

Furthermore, for small ε the operator $T(ub_\varepsilon)$ is invertible in $L_p^+(\Gamma, \varrho)$ (by Proposition 2.4). Applying Theorem 2.1, we conclude that $b_\varepsilon \in \text{fact}(p, \varrho)$. □

Theorem 2.7 *Let $a \in L_\infty^+(\Gamma) + C(\Gamma)$. Then the operator $T(a)$ is normally solvable in $L_p^+(\Gamma, \varrho)$ if and only if $a^{-1} \in L_\infty(\Gamma)$. The operator $T(a)$ is a Φ-operator in $L_p^+(\Gamma, \varrho)$ if and only if $a^{-1} \in L_\infty^+(\Gamma) + C(\Gamma)$. If $a^{-1} \in L_\infty(\Gamma)$ and $a^{-1} \notin L_\infty^+(\Gamma) + C(\Gamma)$, then the operator $T(a)$ is left-invertible in $L_p^+(\Gamma, \varrho)$.*

Proof. The necessity of the condition $a^{-1} \in L_\infty(\Gamma)$ follows from Proposition 1.6. Now suppose that this condition is satisfied. Then, using the notations introduced in the proof of Proposition 2.9, we conclude that the following generalized factorization holds:

$$a(t) = b_\varepsilon(t) h_\varepsilon(t),$$

where

$$h_\varepsilon = u(d + r_\varepsilon), \quad u(t) = \prod_{j=1}^{l} (t - z_j)^{m_j}, \quad z_j \in \mathbb{D}_\Gamma^+.$$

Hence, for $T(a)$ to be a Φ-operator it is necessary and sufficient (see condition (2.31) and Theorem 2.6) that there exists a polynomial v with no zeros on the contour Γ such that $h_\varepsilon^{-1} v \in L_\infty^+(\Gamma)$. Since $L_\infty^+(\Gamma) + C(\Gamma)$ is an algebra, the product $h_\varepsilon^{-1} u$ $(= (h_\varepsilon^{-1} v)(u/v))$ lies in $L_\infty^+(\Gamma) + C(\Gamma)$. Since the latter is a closed subalgebra of $L_\infty(\Gamma)$ and

$$\lim_{\varepsilon \to 0} \|a^{-1} - h_\varepsilon^{-1} u\| = 0,$$

we conclude that $a^{-1} \in L_\infty^+(\Gamma) + C(\Gamma)$.

Now suppose that the last inclusion does not hold, but $a^{-1} \in L_\infty(\Gamma)$. Then the generalized factorization $a(t) = b_\varepsilon(t) h_\varepsilon(t)$ with infinite index satisfies condition (2.31) and, by Theorem 2.6, the operator $T(a)$ is left-invertible in $L_p^+(\Gamma, \varrho)$. □

2.9 The argument of a Blaschke product

Let us return to the case $\Gamma = \Gamma_0$ and consider an example of a more general nature. If $T(a)$ is a normally solvable operator, then, as seen from the preceding section, the entire variety of problems with infinite index (at least in $H^2(\mathbb{D}^+)$) is generated by functions h of class $H^\infty(\mathbb{D}^+)$ that are invertible in $L_\infty(\Gamma_0)$. By Theorem 2.4, such a function is representable in the form

$$h = UF,$$

where the outer function F lies in $GL^+_\infty(\Gamma_0)$ and generates an invertible operator $T(F)$, and U is an inner function. By Proposition 2.8, the function U can be approximated by Blaschke products B_ε, $\|U - B_\varepsilon\|_\infty < \varepsilon$, and so $U_1 = U\overline{B}_\varepsilon \in$ fact(p, ϱ) (see Proposition 2.5). Consequently, the function h admits a generalized factorization of the form

$$h = bB_\varepsilon, \quad \text{where} \quad b = FU\overline{B}_\varepsilon.$$

This circumstance establishes the special role that Blaschke products play in problems with infinite index.

Incidentally, let us mention here that in practice a generalized factorization of the form (2.30) with a Blaschke product in the role of the function h is not the most convenient for the study of some Toeplitz operators. For example, in the case when the function $U(t)$ has a single discontinuity at the point $t = 1$, Theorem 2.4 yields

$$U(t) = E(t)B(t),$$

where the singular inner function $E(t)$ has the form

$$E(t) = \exp\left\{\sigma\frac{t+1}{t-1}\right\}, \quad \sigma > 0,$$

and the Blaschke product

$$B(t) = \prod_{k=1}^{\infty} \frac{\overline{z}_k}{|\overline{z}_k|} \cdot \frac{z_k - t}{1 - \overline{z}_k t} \tag{2.42}$$

has the property that the set of its zeros $\{z_k\} \subset \mathbb{D}^+$ has $t = 1$ as the only limit point and

$$\sum_{k=1}^{\infty}(1 - |z_k|) < \infty.$$

Here the operator $T(U)$ has a sufficiently simple structure:

$$T(U) = T(E)T(B).$$

The direct investigation of Toeplitz operators whose symbols are singular inner function is simpler than the approach based on the approximation of such symbols by Blaschke products and will be discussed in Chapter 4. Here we will examine in more detail the properties of Blaschke functions of the form (2.42).

Let us investigate the behavior of the function $\arg B(t)$ near the point $t = 1$. To that purpose it is convenient to write $\Gamma_0 \setminus \{1\} = \overset{\frown}{(1+0, 1-0)}$, $t = \exp(i\theta)$, $\theta \in (-\pi, \pi]$. If we write $z_k = r_k \exp(i\theta_k)$, then since $z_k \to 1$ as $k \to \infty$, we have $0 < r_k < 1$, $-\pi < \theta \leq \pi$, $\lim_{k\to\infty} r_k = 1$, $\lim_{k\to\infty} \theta_k = 0$.

Theorem 2.8 *Assume that the Blaschke product* $B(t)$ *has the form* (2.42). *Then one can choose a continuous branch of the function* $\arg B(t)$ *on* $\Gamma_0 \setminus \{1\}$ *which is monotonically increasing on* $(1+\overset{\frown}{0,1}-0)$ *and satisfies*

$$\lim_{t\to 1+0} \arg B(t) = A_+ < 0, \qquad \lim_{t\to 1-0} \arg B(t) = A_- > 0.$$

Moreover, at least one of these limits is infinite and we have the formula:

$$\arg B(\exp(i\theta)) = \begin{cases} -2\Big(\sum\limits_{\theta_k\geq\theta} (\pi+\varphi_k(\theta)) + \sum\limits_{\theta_k<\theta} \varphi_k(\theta)\Big), & \text{if } \theta > 0, \\ 2\Big(\sum\limits_{\theta_k\leq\theta} (\pi-\varphi_k(\theta)) - \sum\limits_{\theta_k>\theta} \varphi_k(\theta)\Big), & \text{if } \theta < 0, \end{cases} \tag{2.43}$$

where

$$\varphi_k(\theta) = \arctan\left[\varepsilon_k \cot\left(\frac{\theta-\theta_k}{2}\right)\right], \qquad \varepsilon_k = \frac{1-r_k}{1+r_k}.$$

Proof. We will use the fact that if $0 < r < 1$, $\theta \in (-\pi, 0) \cup (0, \pi]$, then

$$\arg \frac{r-\exp(i\theta)}{1-r\exp(i\theta)} = -2\arctan\left[\frac{1-r}{1+r}\cot\frac{\theta}{2}\right]. \tag{2.44}$$

The proof of (2.44) rests on an elementary geometric argument shown for $\theta > 0$ in Figure 2.2, where $\varphi = \varphi(t) = -\arg[(r-t)/(1-rt)]$, $t = \exp(i\theta)$.

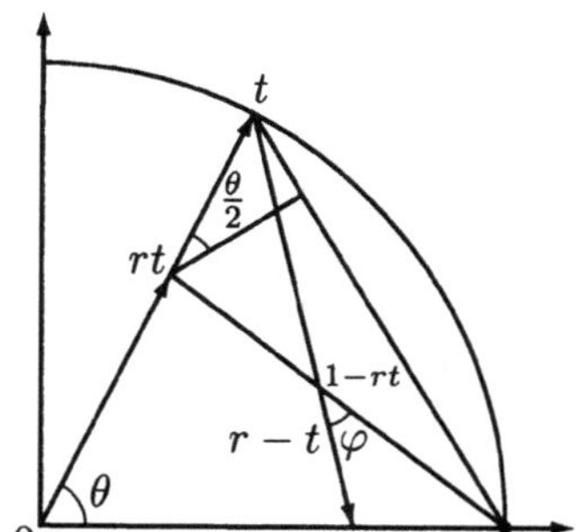

Figure 2.2

Also, $\varphi(t)$ is continuous and monotone on the set $\Gamma_0 \setminus \{1\} = (1+\overset{\frown}{0,1}-0)$ and $\lim_{t\to 1\pm 0} \varphi(t) = \mp\pi$.

Let $t_k = \exp(i\theta_k)$ and

$$B_k(t) = \frac{\overline{z}_k}{|z_k|} \cdot \frac{z_k - t}{1-\overline{z}_k t} = \frac{r_k - \exp(i(\theta-\theta_k))}{1 - r_k\exp(i(\theta-\theta_k))} = \frac{r_k - t\overline{t}_k}{1-t\overline{t}_k}.$$

We define the function $\arg B_k(t)$ in such a manner that it is continuous and monotone on the set $\Gamma_0 \setminus \{1\} = \overset{\frown}{(1+0, 1-0)}$ and satisfies

$$\lim_{t\to 1+0} \arg B_k(t) = a_{\kappa,+} < 0, \qquad \lim_{t\to 1-0} \arg B_k(t) = a_{\kappa,-} > 0.$$

In this case, $\arg B_k(t)$ has a discontinuity of the first kind at the point $t = 1$, with the jump $a_{\kappa,-} - a_{\kappa,+} = 2\pi$. Let $\theta_k \geq 0$. By formula (2.44), for $t \in \overset{\frown}{(t_k, 1-0)}$ $(\theta \in [\theta_k, \pi] \cup (-\pi, 0))$,

$$\arg B_k(\exp(i\theta)) = -2\arctan\left(\varepsilon_k \cot\frac{\theta - \theta_k}{2}\right) = -\varphi_k(\theta).$$

To ensure that $\arg B_k(t)$ is continuous at the point $\theta = \theta_k$, we put, for $t \in \overset{\frown}{(1+0, t_k)}$ $(\theta \in (0, \theta_k])$,

$$\arg B_k(\exp(i\theta)) = -2\varphi_k(\theta) - 2\pi.$$

Similarly, for $\theta_k < 0$,

$$\arg B_k(\exp(i\theta)) = \begin{cases} -2\varphi_k(\theta), & \text{if } t \in \overset{\frown}{(t_k, 1-0)}, \\ -2\varphi_k(\theta) + 2\pi, & \text{if } t \in \overset{\frown}{(1+0, t_k)}. \end{cases}$$

Observing that

$$\arg B(t) = \sum_{k=1}^{\infty} \arg B_k(t),$$

we arrive at formula (2.43), in which the convergence of the series for $\theta \neq 0$ follows from the condition $\sum_{k=1}^{\infty}(1 - r_k) < \infty$.

Now let us assume that $\theta_k \geq 0$ for infinitely many zeros z_k of the function $B(t)$. Then in the first row of (2.43) the number of terms whose modulus is larger than, say, $\pi/2$, grows unboundedly when $\theta \to +0$. It follows that $A_+ = -\infty$. Analogously, if $\theta_k < 0$ for infinitely many zeros z_k , then $A_- = +\infty$. □

The next assertion shows that the argument of a Blaschke product may have arbitrarily slow or arbitrarily fast growth when $t \to 1$, which, generally speaking, is different on different sides.

Theorem 2.9 *Suppose that the real-valued function $f(t)$ is continuous on the set $\Gamma_0 \setminus \{1\} = \overset{\frown}{(1+0, 1-0)}$, is monotonically increasing on that set, and at least one of the limits $\lim_{t\to 1\pm 0} f(t)$ is infinite. Then there exists a Blaschke product $B(t)$ of the form* (2.42) *such that*

$$|\arg B(t) - f(t)| < \text{const}, \quad t \in \Gamma_0 \setminus \{1\}. \tag{2.45}$$

Proof. Put $\psi(\theta) = f(\exp(i\theta))$ and assume that $\lim_{\theta\to+0}\psi(\theta) = -\infty$. We construct the Blaschke product $B_+(t)$ with the zeros $z_k = r_k\exp(i\theta_k)$, where $\theta_k = \psi^{-1}(-(2k-1)\pi)$, $k = 1, 2, \ldots$, and ψ^{-1} is the inverse function to ψ. We choose the numbers r_k, $0 < r_k < 1$, as follows. We denote by k_0 the index k starting with which $\cot((\theta_k - \theta_{k+1})/2) \geq 1$. We then pick arbitrary numbers r_k for $1 \leq k \leq k_0$, while for $k > k_0$ we require that

$$\max\left[\varepsilon_k \cot\frac{\theta_k - \theta_{k+1}}{2}, \varepsilon_k \cot\frac{\theta_{k-1} - \theta_k}{2}\right] = \frac{1}{2^k}, \tag{2.46}$$

where $\varepsilon_k = (1 - r_k)/(1 + r_k)$.

Let $\theta > 0$. We estimate the value of $|\arg B_+(\exp(i\theta_n)) - \psi(\theta_n)|$ for $n > k_0$. By the first row of formula (2.43) and condition (2.46),

$$\begin{aligned}
&|\arg B_+(\exp(i\theta_n)) - \psi(\theta_n)| \\
&\leq \left|-2\left(\sum_{k=1}^{n-1}(\pi + \varphi_k(\theta_n)) + \pi/2 + \sum_{k=n+1}^{\infty}\varphi_k(\theta_n)\right) - \psi(\theta_n)\right| \\
&\leq 2\sum_{k=1(k\neq n)}|\varphi_k(\theta_n)| \leq \text{const} + 2\sum_{k=k_0+1}^{\infty} 2^{-k} = \text{const},
\end{aligned}$$

because

$$\cot|(\varphi_k - \varphi_n)/2| \leq \begin{cases} \cot((\varphi_k - \varphi_{k+1})/2), & n > k, \\ \cot((\varphi_{k-1} - \varphi_k)/2), & n < k. \end{cases}$$

Since both functions $\arg B_+(\exp(i\theta))$ and $\psi(\theta)$ are monotone and

$$|\psi(\theta_n) - \psi(\theta_{n+1})| = 2\pi,$$

inequality (2.45) holds for any $\theta \in (0, \pi]$.

Now let $\theta < 0$. Then by the second row of formula (2.43) and condition (2.46),

$$\begin{aligned}
|\arg B_+(\exp(i\theta))| &= 2\left|\sum_{k=1}^{\infty}\varphi_k(\theta)\right| \\
&\leq 2\sum_{k=1}^{\infty}|\varphi_k(0)| \leq \text{const} + 2\sum_{k=k_0+1}^{\infty} 2^{-k} = \text{const},
\end{aligned}$$

since

$$\cot(\theta_k/2) \leq \cot((\theta_k - \theta_{k+1})/2), \quad k = 1, 2, \ldots.$$

Thus, if $\lim_{\theta\to-0}\psi(\theta) < \infty$, the theorem is proved. If $\lim_{\theta\to-0}\psi(\theta) = +\infty$, then proceeding in an analogous manner one can construct a function $B_-(t)$ such that $|\arg B_-(\exp(i\theta)) - \psi(\theta)| <$ const for any $\theta \in (-\pi, 0)$ and then set $B = B_+B_-$. □

Remark 2.4 In Theorem 2.8 one may require that the function $f(t)$ be monotone only in a neigborhood of the point $t = 1$.

In general, Theorem 2.9 does not allow us to construct a generalized factorization with infinite index of the function $a(t)$ given by $\arg a(t) = f(t)$, because after division by the Blaschke symbol constructed above all we can say about the remaining factor is that its argument is bounded. This is not enough for the well-known theorems on p-factorization to apply. In this connection essential help comes from the theory of factorization of u-periodic functions, developed below in Chapter 5. Let us mention, however, that in some particular cases one can manage with the classical results.

Theorem 2.10 *Suppose that $a(t) \in GL_\infty(\Gamma_0)$ is continuous on the set $\Gamma_0 \setminus \{1\}$ and has a discontinuity of second kind at $t = 1$ such that*

$$\lim_{\theta \to \pm 0} (\arg a(\exp(i\theta)) \pm \pi \log |\theta|^{-1}) = 0.$$

Then the function $a(t)$ admits a generalized factorization with infinite index in the space $L_2(\Gamma_0)$ of the form

$$a(t) = b(t)B(t),$$

where $B(t)$ is a Blaschke product with positive zeros:

$$B(t) = \prod_{k=1}^{\infty} \frac{x_k - t}{1 - x_k t}, \qquad x_k = \frac{2 - \exp(-k + 1/2)}{2 + \exp(-k + 1/2)}.$$

Proof. Let us introduce the function

$$A(\theta) = 2 \sum_{k=1}^{\infty} \arctan(\exp(-k + 1/2)\theta^{-1})$$

and estimate the quantity $\arg B(\exp(i\theta)) + A(\theta)$ for $\theta > 0$. By formula (2.43),

$$\begin{aligned} &|\arg B(\exp(i\theta)) + A(\theta)| \\ &= 2 \left| \sum_{k=1}^{\infty} \arctan(\varepsilon_k \cot(\theta/2)) - \sum_{k=1}^{\infty} \arctan(2\varepsilon_k \theta^{-1}) \right| \\ &\le 2 \sum_{k=1}^{\infty} \frac{1}{1 + \gamma_k^2} |\varepsilon_k \cot(\theta/2) - 2\varepsilon_k \theta^{-1}|, \end{aligned}$$

where $\gamma_k \in [\varepsilon_k \cot(\theta/2), 2\varepsilon_k \theta^{-1}]$, $\varepsilon_k = (1/2)\exp(-k + 1/2)$. Since $0 < \theta \le \pi$ one has $|\cot(\theta/2) - 2\theta^{-1}| < \operatorname{const} \theta$. It follows that

$$|\arg B(\exp(i\theta)) + A(\theta)| \le \operatorname{const} \theta \sum_{k=1}^{\infty} \varepsilon_k \le \operatorname{const} \theta$$

and hence

$$\lim_{\theta\to+0}(\arg B(\exp(i\theta)) + A(\theta)) = 0.$$

Now let us estimate the function $|A(\theta)-\pi\log(\theta^{-1})|$ in a sufficiently small positive half-neighborhood of zero. To that end we introduce the family of sequences $\theta_{n,\varepsilon} = \exp(-n-\varepsilon)$, $n = 1, 2, \ldots$, and the sequence of functions

$$\alpha_n(\varepsilon) = A(\theta_{n,\varepsilon}) - \pi\log(\theta_{n,\varepsilon}^{-1}), \quad \varepsilon\in[0,1].$$

When $n\to\infty$ the sequence $\{\alpha_n(\varepsilon)\}$ converges uniformly on the segment $[0,1]$, because

$$\begin{aligned}
|\alpha_{n+m}(\varepsilon)-\alpha_n(\varepsilon)| &= \Bigg|2\sum_{k=1}^{\infty}\arctan(\exp(n+m-k+(1/2)+\varepsilon)) \\
&\quad - 2\sum_{k=1}^{\infty}\arctan(\exp(n-k+(1/2)+\varepsilon)) - m\pi\Bigg| \\
&= 2\left|\sum_{k=1}^{m}(\arctan(\exp(n+m-k+(1/2)+\varepsilon)) - (\pi/2))\right| \\
&= 2\sum_{k=1}^{m}\arctan(\exp(k-n-m-(1/2)-\varepsilon)) \\
&\le 2\sum_{k=1}^{m}\exp(k-n-m-(1/2)-\varepsilon) \le \mathrm{const}\exp(-n).
\end{aligned}$$

Here we used the identity $\arctan x + \arctan x^{-1} = \pi/2$.

Denote the limit function of the sequence $\{\alpha_n(\varepsilon)\}$ by $\alpha(\varepsilon)$. We claim that $\alpha(0)-\alpha(1)=0$. Indeed, since $\alpha_{n+1}(0)=\alpha_n(1)$, we have

$$\begin{aligned}
\alpha(0)=\alpha(1) &= \lim_{n\to\infty}\left(2\sum_{k=1}^{\infty}\arctan(\exp(n-k+(1/2))) - \pi n\right) \\
&= 2\lim_{n\to\infty}\left(\sum_{k=1}^{n}\arctan(\exp(n-k+(1/2)))+\right. \\
&\quad + \sum_{k=n+1}^{2n}(\arctan(\exp(n-k+(1/2))) - \pi/2) \\
&\quad \left. + \sum_{k=2n+1}^{\infty}\arctan(\exp(n-k+(1/2)))\right)
\end{aligned}$$

$$
\begin{aligned}
&= 2\lim_{n\to\infty}\left(\sum_{k=1}^{n}\arctan(\exp(n-k+(1/2)))\right.\\
&\quad -\sum_{k=n+1}^{2n}\arctan(\exp(k-n-(1/2)))\\
&\quad \left.+\sum_{j=n+1}^{\infty}\arctan(\exp(-j+(1/2)))\right)\\
&= 2\lim_{n\to\infty}\sum_{j=n+1}^{\infty}\arctan(\exp(-j+(1/2)))=0.
\end{aligned}
$$

Similarly, $\alpha(1/2)=0$. These equalities, in conjunction with the monotonicity of the functions $A(\theta)$ and $\log(\theta^{-1})$ on the interval $(0,\pi)$ and the fact that

$$|\pi\log(\theta_{n,0}^{-1})-\pi\log(\theta_{n,\pm 1/2}^{-1})|=\pi/2$$

yield the estimate $|\alpha(\varepsilon)|<\pi/2$. This in turn means that in a sufficiently small positive neighborhood of zero $|A(\theta)-\pi\log(\theta^{-1})|<\pi/2$, and hence $|\arg B(\exp(i\theta))+\pi\log(\theta^{-1})|<\pi/2$. In a similar way one can show that there exists a negative half-neighborhood of zero in which $|\arg B(\exp(i\theta))-\pi\log|\theta^{-1}||<\pi/2$. Therefore, the function $a(t)$ is representable in the form $a(t)=b(t)B(t)$, where the function $b(t)$ has the property that, for any point $t_0\in\Gamma_0$, there exists a neighborhood $\nu(t_0)$, $\nu(t_0)\subset\Gamma_0$, such that the set $b(\nu(t_0))$ lies inside an angle of apex $z=0$ and opening less than π. Applying Theorem 2.3 we conclude that $b\in\operatorname{fact}(2,1)$. □

2.10 The argument of an outer function

Let us observe that the unboundedness of the function $\arg a(t)$ does not necessarily imply that the operator $T(a)$ ceases being a Φ-operator or even an invertible operator. To illustrate this observation we turn to outer functions.

The following result is an analogue of formula (2.43) for outer functions.

Proposition 2.10 *Let $F(t)$ be an outer function. Then for almost all θ in $(-\pi,\pi]$,*

$$\arg F(\exp(i\theta))=\frac{1}{2\pi}\int_{-\pi}^{\pi}\cot\left(\frac{\theta-\xi}{2}\right)\log|F(\exp(i\xi))|d\xi. \tag{2.47}$$

Proof. The outer function $F(z)$ is expressed through its boundary values $F(t)$ as

$$F(z)=\exp\left\{\frac{1}{2\pi i}\int_{\Gamma_0}\frac{\tau+z}{\tau-z}\cdot\frac{\log|F(\tau)|}{\tau}d\tau\right\},\quad z\in\mathbb{D}^+$$

(see Section 2.6). By the Sokhotskiĭ-Plemelj formula (1.6),

$$F(t) = \exp\left\{\frac{1}{2\pi i}\int_{\Gamma_0}\frac{\tau+t}{\tau-t}\cdot\frac{\log|F(\tau)|}{\tau}d\tau + \log|F(\tau)|\right\}.$$

The change of variables $\tau = \exp(i\xi)$, $t = \exp(i\theta)$ and the identity

$$\frac{\exp(i\xi)+\exp(i\theta)}{\exp(i\xi)-\exp(i\theta)} = \frac{1}{i}\cot\frac{\xi-\theta}{2}$$

lead to the equality

$$F(\exp(i\theta)) = |F(\exp(i\theta))|\exp\left\{\frac{i}{2\pi}\int_{-\pi}^{\pi}\cot\frac{\theta-\xi}{2}\log|F(\exp(i\xi))|d\xi\right\},$$

which in turn yields formula (2.47). □

As an example, let us consider the outer function

$$F(t) = \exp\left\{ic\log\left(2i\frac{1-t}{1+t}\right)\right\}, \quad c>0, \tag{2.48}$$

where the branch of the multi-valued function $w = \log z$ is chosen as follows: $\log z = \log|z| + i\arg z$, $-\pi/2 < \arg z < 3\pi/2$. Obviously, $F \in GH^\infty(\mathbb{D}^+)$. Moreover, it is readily verified that

$$\arg F(\exp(i\theta)) = \begin{cases} c\log\left(2\cot\frac{\theta}{2}\right), & \text{if } \theta\in(0,\pi], \\ c\log\left(2\cot\left|\frac{\theta}{2}\right| + \pi\right), & \text{if } \theta\in(-\pi,0). \end{cases}$$

Thus, despite the fact that the function $F(t)$ generates a invertible operator $T(F)$, its argument is unbounded: $\lim_{\theta\to\pm 0}\arg F(\exp(i\theta)) = +\infty$. As formula (2.47) shows, the argument of an outer function can behave in a multitude of ways. Without dwelling further upon this aspect, let us cite without proof the following result.

Proposition 2.11 *In the class of functions that are bounded on Γ_0 and continuous on $\Gamma_0\setminus\{1\}$:*

(1) *there exist functions $a(t)$ such that*

$$\lim_{\theta\to\pm 0}\arg a(\exp(i\theta)) = +\infty$$

and the operator $T(a)$ is invertible in $H^2(\mathbb{D}^+)$;

(2) *there exist functions $a(t)$ such that*

$$\lim_{\theta\to\pm 0}\arg a(\exp(i\theta)) = +\infty$$

and, for some sequence $\{\theta_k\}_{k=1}^{\infty}$,

$$\lim_{\theta_k \to -0} \arg a(\exp(\theta_k)) = -\infty,$$

and the operator $T(a)$ *is invertible in* $H^2(\mathbb{D}^+)$;

(3) *if*

$$\lim_{\theta \to +0} \arg a(\exp(i\theta)) = +\infty,$$

and if the function $\arg a(\exp(i\theta))$ *is bounded in a negative half-neighborhood of zero, then* $T(a)$ *is not a* Φ*-operator in* $H^2(\mathbb{D}^+)$;

(4) *if* $\arg a(\exp(i\theta)) = \psi(\theta) + \delta(\theta)$, *where the function* $\delta(\theta)$ *is bounded, while* $\psi(\theta)$ *is monotone on the interval* $[-\pi, \pi] \setminus \{0\}$ *and* $\lim_{\theta \to \pm 0} \psi(\theta) = +\infty$, *and if the operator* $T(a)$ *is invertible in* $H^2(\mathbb{D}^+)$, *then*

$$\arg a(\exp(i\theta)) = O(\log |\theta^{-1}|), \quad \theta \to 0.$$

The investigations carried out in this section show that the examples of symbols that generate left-invertible Toeplitz operators which are not Φ-operators should be sought among the functions whose arguments are monotone and satisfy the conditions

$$\lim_{\theta \to \pm 0} \arg a(\exp(i\theta)) = \mp\infty.$$

At the same time, the arguments of such symbols may be essentially modified via multiplication by an outer function, with preservation of the left invertibility of the Toeplitz operator. For instance, if one considers the product $a(t) = F(t)B(t)$, where $F(t)$ is given by formula (2.48) with $c > \pi$ and $B(t)$ is given by Theorem 2.10, then $\lim_{\theta \to \pm 0} \arg a(\exp(i\theta)) = +\infty$.

References and comments

2.1 Necessary and sufficient conditions on the curve Γ for the operator S_Γ to be bounded in $L_p(\Gamma)$, $1 < p < \infty$, were established by David [1]. An important role in the proof of this fundamental result was played by the works of Calderón [2], [3] and Coifman, David and Meyer [1]. See also Böttcher and Karlovich's monograph [1].

Condition (A_p) on the weight $\varrho(t)$ – a necessary and sufficient condition for the boundedness of the operator S_Γ in $L_p(\Gamma, \varrho)$ – was obtained by Hunt, Muckenhoupt and Wheeden [1].

To see how this result extends to the case of an arbitrary Carleson curve, consult Dyn′kin and Osilenker [1] or Böttcher and Karlovich [1]. Power weights were introduced and studied by Khvedelidze [1].

Formula (2.2) for S_Γ^* is borrowed from Gohberg and Krupnik [3, Chap. I].

2.2–2.3 This material is treated in more detail in Gohberg and Krupnik [3, Chaps. I, II] and Böttcher and Karlovich [1, Chap. 6], for example. Let us mention that weighted Hardy spaces were studied in many pioneering works; see, e.g., Rochberg [1], [2] and Rozenblum and Rovnyak [1].

2.4 In the context discussed here the idea of factoring continuous functions was brought forth in a series of classical works (by J. Plemelj, N. Wiener and E. Hopf, F. D. Gakhov) of the first half of the 20th century devoted to the study of various classes of singular integral equations and boundary value problems. For detailed historical information on such questions the reader should consult the monographs of Gakhov [2], Gakhov and Cherskiĭ [1], and Muskhelishvili [1].

The notion of the p-factorization of a measurable function and Theorem 2.1 originate in the works of Simonenko [2], [6] (see also Widom [3], [4], [6] and Gohberg and Krupnik [3, Chap. VIII]). For the proof of Theorem 2.1 we used an unpublished work by I. B. Simonenko.

2.5 Decomposing algebras of continuous functions were introduced in Gohberg's work [2], where one can also find Theorem 2.2 (see also Gohberg and Krupnik [3, Chap. III, §3]). Proposition 2.3 is due to Hartman and Wintner [1]. The other notions and results discussed in this section, in particular, the classes of sectorial operators, are taken from the works of Simonenko [2], [6], [7] and the monograph of Gohberg and Krupnik [3]. More on locally sectorial (matrix) functions can be found in Böttcher [2], Böttcher and Silbermann [2], [3], Böttcher, Silbermann, and Spitkovsky [1], Clancey and Gosselin [1], Clancey and Morrel [1], Douglas [2]–[4], Douglas and Sarason [1], Douglas and Widom [1], Nikol′skiĭ [1], Spitkovsky [3].

2.6 The results of this section are all from Garnett's book [1, Chap. II, Sections 5 and 6]; see also the monographs of Hoffman [1] and Koosis [1].

2.7–2.10 Theorem 2.5 was proved by Widom [1] and Devinatz [1].

The notion of the (p, ϱ)-factorization with infinite index was introduced by Grudsky in the paper [4], where one can also find Theorem 2.6.

Toeplitz operators with symbols from the algebra $L^+_\infty(\Gamma)+C(\Gamma)$ were studied by Douglas [1], [2]. Proposition 2.9 and Theorem 2.7 are close to his results (see also Böttcher and Silbermann [4, Chap. 2]).

Theorems 2.8, 2.9 and 2.10 are due to Grudsky [5].

Proposition 2.11 is due to Böttcher [1] (a full proof is also given in Böttcher and Silbermann [4, 2.26]).

2.3 [illegible] Cordes [illegible] and Kurepa [illegible] Chap. [illegible] and [illegible] for example. [illegible] [illegible] works. See [illegible] [illegible] [illegible].

2.4 [illegible] of the [illegible] of [illegible] continuous functions was [illegible] J. Plemelj, N. [illegible] half of the 20th century devoted to the [illegible] equations and boundary value [illegible] such questions the reader [illegible] and [illegible] of [illegible] and [illegible].

[illegible] function and Theorem 2.3 [illegible] see also Widom [2], [4], [5] and [illegible] and Krupnik [illegible] the proof of Theorem 2.1 [illegible]

2.5 [illegible] were introduced in Gohberg [illegible] Gohberg and Krupnik [illegible] and Mandelbrojt [illegible]. The [illegible] section, in particular, the classes [illegible] works of Simonenko [2], [3], [4] [illegible] and Krupnik [3]. More on locally sectorial [illegible] Böttcher and Silbermann [illegible] Clancey and Gosselin [1], [illegible] and Sarason [1]. [illegible]

2.6 [illegible] Chap. 2, Sections 5 [illegible] and Roch [illegible]

2.7 [illegible] and Douglas [4].

[illegible] infinite index was introduced by [illegible] Theorem 2.6.

[illegible] were studied [illegible] are close to this [illegible]

[illegible]

[illegible] also given in [illegible]

Chapter 3

Model Subspaces

In this chapter we study Toeplitz operators of the form $T(h^{\pm 1})$, where $h \in L_\infty^+(\Gamma) \cap GL_\infty(\Gamma)$. Special attention will be devoted to the subspaces $\ker T(h^{-1})$, in terms of which the kernels of the operators $T(a)$ with symbols $a \in \mathrm{fact}(\infty, p, \varrho)$ were described in Theorem 2.6. It is in this chapter that we establish a connection between model subspaces, which are standard objects of nonclassical spectral theory (see Remark 2.1), and Toeplitz operators with infinite index.

Model operators and model subspaces

3.1 Model subspaces

A function h with the properties $h \in L_\infty^+(\Gamma)$ and $h^{-1} \in L_\infty(\Gamma)$ will be called a *model function* on the curve Γ. The operator $T(h^{-1})$, acting in $L_p^+(\Gamma, \varrho)$, and its kernel $\ker T(h^{-1})$ will be referred to as the *model operator* and respectively the *model subspace* (in the space $L_p^+(\Gamma, \varrho)$) generated by the function h.

We remark that for a function h of class $L_\infty^+(\Gamma)$ the condition $h^{-1} \in L_\infty(\Gamma)$ is necessary and sufficient for the normal solvability and one-sided invertibility of the operators $T(h^{\pm 1})$.

In the space $L_p(\Gamma, \varrho)$ let us introduce the pair of complementary projectors

$$P_h = hP_\Gamma^- h^{-1} I, \quad Q_h = hP_\Gamma^+ h^{-1} I, \quad P_h + Q_h = I, \tag{3.1}$$

and the subspace

$$\mathcal{M}(h) = P_h(L_p^+(\Gamma, \varrho)). \tag{3.2}$$

Proposition 3.1 *Let h be a model function on the curve Γ. Then*

$$L_p^+(\Gamma, \varrho) = \ker T(h^{-1}) \oplus \operatorname{im} T(h), \tag{3.3}$$

where

$$\ker T(h^{-1}) = \mathcal{M}(h), \quad \operatorname{im} T(h) = Q_h(L_p^+(\Gamma, \varrho)).$$

Proof. Let $f^+ \in L_p^+(\Gamma, \varrho)$ and denote $\varphi^+ = P_h f^+$. Then $\varphi^+ = hP_\Gamma^- h^{-1} f^+ \in L_p^+(\Gamma, \varrho)$, because $P_\Gamma^- \varphi^+ = P_\Gamma^- h(I - P_\Gamma^+)h^{-1} f^+ = 0$. Moreover, $P_\Gamma^+ h^{-1} \varphi^+ = 0$, i.e., $\varphi^+ \in \ker T(h^{-1})$. Since φ^+ is the general element of the subspace $\mathcal{M}(h)$, conclude that $\mathcal{M}(h) \subset \ker T(h^{-1})$.

Conversely, if φ^+ is an arbitrary element of $\ker T(h^{-1})$, i.e., $P_\Gamma^+ h^{-1} \varphi^+ = 0$, then $h^{-1}\varphi^+ = P_\Gamma^- h^{-1} \varphi^+$. It follows that $\varphi^+ = P_h \varphi^+$, and so $\ker T(h^{-1}) \subset \mathcal{M}(h)$ as well.

Since in $L_p^+(\Gamma, \varrho)$ we have $\operatorname{im} T(h^{-1}) = \operatorname{im} P_\Gamma^+ h^{-1} I = L_p^+(\Gamma, \varrho)$, it follows that

$$\operatorname{im} Q_h = \operatorname{im}(hP_\Gamma^+ h^{-1} I) = \operatorname{im} T(h)$$

and so decomposition (3.3) holds. □

Let us introduce in the space $L_p(\Gamma, \varrho)$ the integral operator

$$(Vf)(z) = \frac{1}{2\pi i} \int_\Gamma \frac{1 - h(z)h^{-1}(\tau)}{\tau - z} f(\tau) d\tau, \quad z \in \mathcal{D}_\Gamma^+. \tag{3.4}$$

Definition (3.2) is closely connected with the representation of the elements of a model subspace by means of the operator V. Henceforth the notation $z \in \overline{\mathcal{D}_\Gamma^+}$ will mean that $z \in \mathcal{D}_\Gamma^+$ or $z \in \Gamma \setminus \mu$, where $\operatorname{meas} \mu = 0$.

Proposition 3.2 *Let h be a model function on the curve Γ. Then the following statements are equivalent:*

(1) *$f \in \mathcal{M}(h)$ in the space $L_p^+(\Gamma, \varrho)$;*

(2) *$f(z) = (Vg)(z)$, where $g \in L_p^+(\Gamma, \varrho)$, $z \in \overline{\mathcal{D}_\Gamma^+}$.*

Proof. Let $g \in L_p^+(\Gamma, \varrho)$ and

$$f(z) = (Vg)(z) = \frac{1}{2\pi i} \int_\Gamma \frac{g(\tau)}{\tau - z} d\tau - \frac{h(z)}{2\pi i} \int_\Gamma \frac{h^{-1}(\tau) g(\tau)}{\tau - z} d\tau, \quad z \in \mathcal{D}_\Gamma^+.$$

By one of the Sokhotskiĭ-Plemelj formulas, which connects the boundary values of the Cauchy-type integral

$$I(z) = \frac{1}{2\pi i} \int_\Gamma \frac{g(\tau)}{\tau - z} d\tau, \quad z \in \mathbb{C} \setminus \Gamma, \quad g \in L_p(\Gamma, \varrho),$$

with the singular integral S_Γ, namely,

$$\lim_{z \in \mathcal{D}_\Gamma^\pm,\, z \to t} I(z) = \pm\frac{1}{2} g(t) + \frac{1}{2}(S_\Gamma g)(t), \quad t \in \Gamma, \tag{3.5}$$

the nontangential $z \to t$ limit, $z \in \mathcal{D}_\Gamma^+$,

$$\begin{aligned} f(t) &= g(t) - \left(\frac{1}{2}g(t) + \frac{h(t)}{2\pi i}\int_\Gamma \frac{h^{-1}(\tau)g(\tau)}{\tau - t}d\tau\right) \\ &= \frac{1}{2\pi i}\int_\Gamma \frac{1 - h(t)h^{-1}(\tau)}{\tau - t}g(\tau)d\tau = (P_h g)(t) \end{aligned}$$

exists for almost all $t \in \Gamma$. From equality (3.2) it follows that $f \in \mathcal{M}(h)$.

Conversely, let $f \in \mathcal{M}(h)$, i.e., $f \in \ker T(h^{-1})$. Then $h^{-1}f = P^- h^{-1} f \in L_p^-(\Gamma, \varrho)$, and so

$$\begin{aligned} (Vf)(z) &= \frac{1}{2\pi i}\int_\Gamma \frac{1 - h(z)h^{-1}(\tau)}{\tau - z}f(\tau)d\tau \\ &= \frac{1}{2\pi i}\int_\Gamma \frac{f(\tau)}{\tau - z}d\tau - \frac{h(z)}{2\pi i}\int_\Gamma \frac{(P_\Gamma^- h^{-1}f)(\tau)}{\tau - z}d\tau = f(z), \quad z \in \overline{\mathcal{D}_\Gamma^+}, \end{aligned}$$

thanks to Proposition 1.1 and formula (3.5). □

We say that the model function h on the curve Γ *belongs to the class* $\mathcal{U}$ if $h^{-1} \in L_\infty^-(\Gamma)$. Obviously, any inner function u belongs to $\mathcal{U}$. Generally speaking, a function $h(z)$ of class $\mathcal{U}$ is meromorphic in $\mathcal{D}_\Gamma^-$. Moreover, when $z \to t$, $z \in \mathcal{D}_\Gamma^\pm$, the nontangential limit values $h^\pm(t)$ exist for almost all $t \in \Gamma$ and satisfy $h^+(t) = h^-(t)$. In connection with this, for model functions of class $\mathcal{U}$ the representations (3.2) and (3.4) can be somewhat simplified.

Let us introduce the integral operator

$$(V_1 g)(z) = \frac{1}{2\pi i}\int_\Gamma \frac{h(\tau) - h(z)}{\tau - z}g(\tau)d\tau, \quad z \in \mathcal{D}_\Gamma^+ \cup (\mathcal{D}_\Gamma^- \setminus \{z_j^-\}), \tag{3.6}$$

where $\{z_j^-\}$ is the set of poles of the function $h(z)$ in $\mathcal{D}_\Gamma^-$. For each $g \in L_p(\Gamma, \varrho)$, the function $(V_1 g)(z)$ is analytic in $\mathcal{D}_\Gamma^+$ and meromorphic in $\mathcal{D}_\Gamma^-$. Moreover, by virtue of the Sokhotskiĭ-Plemelj formulas (3.5), for almost any $t \in \Gamma$, $(V_1 g)(z)$ has nontangential limit values $(V_1 g)^\pm(t)$ for $z \to t$, $z \in \mathcal{D}_\Gamma^\pm$, and these values satisfy the equality

$$(V_1 g)^+(t) = (V_1 g)^-(t) = (V_1 g)(t), \quad t \in \Gamma, \tag{3.7}$$

where

$$(V_1 g)(t) = \frac{1}{2\pi i}\int_\Gamma \frac{h(\tau) - h(t)}{\tau - t}g(\tau)d\tau = (P_\Gamma^+ h P_\Gamma^- g)(t). \tag{3.8}$$

Proposition 3.3 *Let h be a model function of class $\mathcal{U}$ on the curve Γ. Then*

$$\mathcal{M}(h) = \ker T(h^{-1}) = P_\Gamma^+(h L_p^-(\Gamma, \varrho)). \tag{3.9}$$

For a function f to belong to $\mathcal{M}(h)$ it is necessary and sufficient that

$$f(z) = (V_1 g)(z), \quad z \in \overline{\mathcal{D}_\Gamma^+}, \quad g \in L_p(\Gamma, \varrho). \tag{3.10}$$

Proof. Let $f \in \mathcal{M}(h)$. Then $P_\Gamma^+ h^{-1} f = 0$, i.e., $h^{-1}f = g^- \in L_p^-(\Gamma, \varrho)$. Since $f = hg^-$, we see that $f = P_\Gamma^+ hg^-$, and so $f \in P_\Gamma^+(hL_p^-(\Gamma, \varrho))$. We have shown that $\mathcal{M}(h) \subset P_\Gamma^+(hL_p^-(\Gamma, \varrho))$. Conversely, suppose that $f = P_\Gamma^+ hg^-$, where $g^- \in L_p^-(\Gamma, \varrho)$. Then $P_\Gamma^+ h^{-1} f = P_\Gamma^+ h^{-1} P_\Gamma^+ hg^- = P_\Gamma^+ g^- = 0$. Therefore,

$$P_\Gamma^+(hL_p^-(\Gamma, \varrho)) \subset \ker T(h^{-1})(= \mathcal{M}(h)).$$

□

Remark 3.1 From the equality $h^{-1}f = g^-$, where $g^- \in L_p^-(\Gamma, \varrho)$, it follows that the function $f \in \mathcal{M}(h)$ $(h \in \mathcal{U})$, which is a priori defined in $\overline{\mathcal{D}_\Gamma^+}$, admits an analytic continuation to the domain $\mathcal{D}_\Gamma^- \setminus \{z_j^-\}$, where the points z_j^- – the zeros of the function $h^{-1}(z)$ – may be poles of $f(z)$. It is therefore natural to regard the elements of the space $\ker T(h^{-1})$ as being defined and regular everywhere in the set $\mathcal{D}_\Gamma^+ \cup \mathcal{D}_\Gamma^- \setminus \{z_j^-\}$, and to identify the interior $(z \in \mathcal{D}_\Gamma^+,\ z \to t)$ and exterior $(z \in \mathcal{D}_\Gamma^-,\ z \to t)$ nontangential limit values for almost all $t \in \Gamma$. By Proposition 3.3, V_1 can be thought of as the operator effecting the analytic continuation of a function $f(t)$ from the contour Γ to the indicated domain.

Proposition 3.4 *Let $h_j \in \mathcal{U}$, $j = 1, 2, \ldots, n$. Then $h = \prod\limits_{j=1}^{n} h_j \in \mathcal{U}$ and*

$$\mathcal{M}(h) = \mathcal{M}(h_1) \oplus h_1\mathcal{M}(h_2) \oplus \cdots \oplus \left(\prod_{j=1}^{n-1} h_j\right) \mathcal{M}(h_n). \tag{3.11}$$

Proof. We proceed by induction. The case $n = 1$ is trivial. Now suppose that decomposition (3.11) holds for $n = k$ and let $f \in \mathcal{M}(\prod_{j=1}^{k+1} h_j) = \ker T(\prod_{j=1}^{k+1} h_j^{-1})$. Since

$$T\left(\prod_{j=1}^{k+1} h_j^{-1}\right) = T(h_{k+1}^{-1})T\left(\prod_{j=1}^{k} h_j^{-1}\right),$$

we can write

$$f = \varphi + \left(\prod_{j=1}^{k} h_j\right)\psi,$$

where, by Proposition 3.1,

$$\varphi \in \mathcal{M}\left(\prod_{j=1}^{k} h_j\right) = \ker T\left(\prod_{j=1}^{k} h_j^{-1}\right)$$

and $\psi \in \ker T(h_{k+1}^{-1})$. Since

$$\varphi \in \mathcal{M}(h_1) \oplus h_1\mathcal{M}(h_2) \oplus \cdots \oplus \left(\prod_{j=1}^{k-1} h_j\right) \mathcal{M}(h_k)$$

by the inductive hypothesis, we deduce that $f \in \mathcal{M}_{k+1}$, where $\mathcal{M}_{k+1}$ denotes the right-hand side of (3.11) for $n = k+1$.

This establishes the inclusion $\mathcal{M}\left(\prod_{j=1}^{k+1} h_j\right) \subset \mathcal{M}_{k+1}$. The opposite inclusion $\mathcal{M}_{k+1} \subset \mathcal{M}\left(\prod_{j=1}^{k+1} h_j\right)$ is verified by letting the operator $T\left(\prod_{j=1}^{k+1} h_j^{-1}\right)$ act on the subspace $\mathcal{M}_{k+1}$. □

3.2 Deformation of the contour

As follows from Proposition 3.3 and Remark 3.1, in some cases the elements of a model subspace $\mathcal{M}(h)$ can be defined in a domain that is larger than $\mathcal{D}_\Gamma^+$. Hence, there are grounds to assume that the subspace $\mathcal{M}(h)$ is stable with respect to certain deformations of the contour Γ, which depend on the analytic properties of the function $h(z)$. In this section we will describe a class of admissible (in the indicated sense) deformations of the curve Γ. Let us explain this idea in more detail.

Example 3.1. Consider the classical case of the operator $T(t^{-\kappa})$, where $\kappa \in \mathbb{N}$.

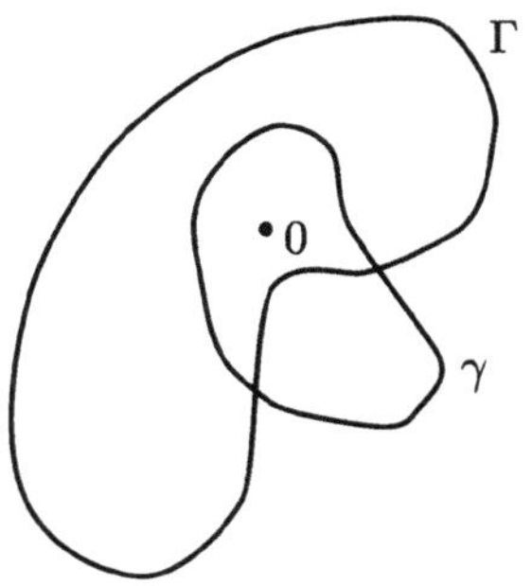

Figure 3.1

Let Γ and γ be two arbitrary smooth closed Jordan curves which encircle the origin 0 counterclockwise (Figure 3.1). Consider the operators $T_\Gamma(t^{-\kappa})$, $t \in \Gamma$, and $T_\gamma(\tau^{-\kappa})$, $\tau \in \gamma$. By Theorem 2.6,

$$\ker T_\Gamma(t^{-\kappa}) = \operatorname{span}\{t^{j-1},\ j = 1, 2, \ldots, \kappa\},$$

$$\ker T_\gamma(\tau^{-\kappa}) = \operatorname{span}\{\tau^{j-1},\ j = 1, 2, \ldots, \kappa\}.$$

Therefore, $\ker T_\Gamma(t^{-\kappa}) = \ker T_\gamma(\tau^{-\kappa})$, in the sense that both subspaces are generated by the same subspace $P_{\kappa-1}(\mathbb{C})$ of all polynomials with complex coefficients of degree at most $\kappa - 1$, but

$$\ker T_\Gamma(t^{\kappa-1}) = P_{\kappa-1}(\mathbb{C})\big|_\Gamma, \quad \ker T_\gamma(\tau^{-\kappa}) = P_{\kappa-1}(\mathbb{C})\big|_\gamma.$$

In this sense the subspace $\mathcal{M}(t^{-\kappa}) = \ker T_\Gamma(t^{-\kappa})$ is stable with respect to a broad class of perturbations of the curve Γ. Needless to say, the perturbations may be rather rough, e.g., as shown in Figure 3.1.

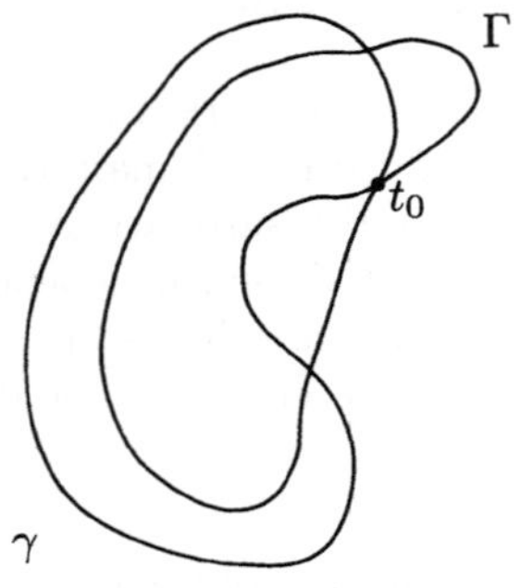

Figure 3.2

Now, in the case of Toeplitz operators with infinite index the singularities of the symbol (above it was a pole at the point 0) are of a more complicated nature, for instance, like that of an essentially singular point of an analytic function. Accordingly, the corresponding admissible deformations of the curve Γ in a neighborhood of a singular point (in Figure 3.2 this is the point $t = t_0$) must be of a subtler nature.

First let us return to the subspaces $L_p^\pm(\Gamma, \varrho)$. The analytic functions belonging to these subspaces are defined on curves γ that lie in $\overline{\mathcal{D}_\Gamma^\pm}$. Let us clarify when such functions belong to the space $L_p^\pm(\gamma, \varrho_\gamma)$, where

$$\varrho_\gamma(z) = \prod_{k=1}^m |z - t_k|^{\alpha_k}, \qquad z \in \gamma,$$

is the natural continuation of the weight $\varrho(t)$ to the curve γ. To that end let us examine first the operator

$$(S_{\Gamma,\gamma} f)(z) = \frac{1}{\pi i} \int_\Gamma \frac{f(\tau)d\tau}{\tau - z}, \quad z \in \gamma,$$

where the integral is understood as a Cauchy principal value. In the case of Carleson curves Γ and γ, the linear operator $S_{\Gamma,\gamma} : L_p(\Gamma, \varrho) \to L_p(\gamma, \varrho_\gamma)$ is bounded.

Indeed, in this case $\Gamma \cup \gamma$ is a Carleson curve, and therefore the operator S_Γ is bounded in the space $L_p(\widetilde{\Gamma}, \widetilde{\varrho})$, where $\widetilde{\Gamma} = \Gamma \cup \gamma$, $\widetilde{\varrho} = \varrho$ on Γ, $\widetilde{\varrho} = \varrho_\gamma$ on γ. Moreover, we have the representation

$$S_{\Gamma,\gamma} = 1_\gamma S_{\widetilde{\Gamma}} 1_\Gamma I,$$

where 1_M denotes the characteristic function of the set M.

Proposition 3.5 *Let Γ and γ be simple closed Carleson curves of class $\mathcal{T}$. The following assertions are true:*

(a) *if $\mathcal{D}_\gamma^\pm \subset \mathcal{D}_\Gamma^\pm$, then $L_p^\pm(\Gamma, \varrho) \subset L_p^\pm(\gamma, \varrho_\gamma)$;*

(b) *if $\mathcal{D}_\gamma^+ \subset \mathcal{D}_\Gamma^-$, then $L_p^-(\Gamma, \varrho) \subset L_p^+(\gamma, \varrho_\gamma)$.*

Proof. Suppose, for example, that $\mathcal{D}_\gamma^+ \subset \mathcal{D}_\Gamma^-$. If $f \in L_p^-(\Gamma, \varrho)$, then, by Proposition 1.1,

$$f(z) = \frac{1}{2\pi i} \int_\gamma \frac{f(\tau) d\tau}{\tau - z}, \qquad z \in \gamma.$$

Since the operator $S_{\Gamma,\gamma} : L_p(\Gamma, \varrho) \to L_p(\gamma, \varrho_\gamma)$ is bounded, $f(z) \in L_p(\gamma, \varrho_\gamma)$. Furthermore, if $f \in R^-(\Gamma)$, then $f \in L_p^+(\gamma, \varrho_\gamma)$. Since $\operatorname{clos} R^-(\Gamma) = L_p^-(\Gamma, \varrho)$, we conclude that $L_p^-(\Gamma, \varrho) \subset L_p^+(\gamma, \varrho_\gamma)$. The remaining assertions are established in analogous manner. □

The next result is in some sense a converse to Proposition 3.5.

Proposition 3.6 *Suppose the domain $\mathcal{D}_\Gamma^+$ (resp., $\mathcal{D}_\Gamma^-$) is partitioned into finitely many simply connected domains which are bounded by closed Carleson curves $\gamma_j \in \mathcal{T}$, $1 \le j \le m$. Let the functions $f_j(z)$ belong to $L_p^+(\gamma_j, \varrho_{\gamma_j})$ (resp., to $L_p^+(\gamma_j, \varrho_{\gamma_j})$ if γ_j bounds a finite domain and to $L_p^-(\gamma_j, \varrho_{\gamma_j})$ in the opposite case), and assume that for any j and k, $1 \le j, k \le m$, $f_j(z) = f_k(z)$ almost everywhere on the set $\gamma_j \cap \gamma_k$. Then the functions $f_j(z)$ are restrictions to the corresponding domains of a function $f(z)$ that is analytic in $\mathcal{D}_\Gamma^+$ (resp., in $\mathcal{D}_\Gamma^-$). Moreover, the nontangential limits, which form a function $f(t)$, exists almost everywhere on Γ and $f(t) \in L_p^+(\Gamma, \varrho)$ (resp., $f(t) \in L_p^-(\Gamma, \varrho)$).*

Proof. Let us prove the first assertion of the theorem. Denote by $f_j(t)$ the function formed by the nontangential boundary values of the function $f_j(z)$ on the set $\gamma_j \cap \Gamma$. Form the function $f(t)$ on Γ that coincides with $f_j(t)$ when $t \in \gamma_j \cap \Gamma$. We claim that $f(t) \in L_p^+(\Gamma, \varrho)$. Indeed, since $f(t) \in L_p(\Gamma, \varrho)$, the criterion provided by Proposition 1.1 yields

$$\int_\Gamma \frac{f(t)}{(t - z_0)^n} dt = \sum_{j=1}^m \int_{\gamma_j} \frac{f_j(z)}{(z - z_0)^n} dz = 0, \qquad z_0 \in \mathcal{D}_\Gamma^-.$$

Notice that this equality holds because all arcs of the form $\gamma_j \cap \gamma_k$ appear here twice and are traversed in opposite directions, while the arcs $\gamma_j \cap \Gamma$ appear only once and are traversed in positive direction, so that the domains $\mathcal{D}_\gamma^+$ and $\mathcal{D}_\Gamma^+$ remain to the left of these arcs. The second assertion of the proposition is established in much the same manner. □

In what follows we will need some sufficient conditions for membership in the space $L_p^{\pm}(\Gamma, \varrho)$.

Proposition 3.7 *Let $f^{\pm} \in L_p^{\pm}(\Gamma, \varrho)$, $1 \le p < \infty$. If g is a polynomial, $\deg g = n$, then $gf^+ \in L_p^+(\Gamma, \varrho)$ and there exists a polynomial g_1, $\deg g_1 < n$, such that $gf^- - g_1 \in L_p^-(\Gamma, \varrho)$.*

Proof. The case $n = 0$ is trivial. Let $n \ge 1$ and $r^+ \in R^+(\Gamma)$. Then $gr^+ \in R^+(\Gamma) \subset L_p^+(\Gamma, \varrho)$. Since $\operatorname{clos} R^+(\Gamma) = L_p^+(\Gamma, \varrho)$ and the multiplication operator gI is bounded on $L_p(\Gamma, \varrho)$, it follows that $gf^+ \in L_p^+(\Gamma, \varrho)$.

If $r^- \in R^-(\Gamma)$, there exists a polynomial g_1, $\deg g_1 < n$, such that the function $g(z)r^-(z) - g_1(z)$ vanishes at infinity; therefore, $gr^- - g_1 \in R^-(\Gamma) \subset L_p^-(\Gamma, \varrho)$, which in turn implies that $gr^- \in L_p^-(\Gamma, \varrho) \oplus L_n$, where $L_n = \operatorname{span}\{1, t, \dots, t^{n-1}\}$. Since the subspace $L_p^-(\Gamma, \varrho) \oplus L_n$ is closed in $L_p(\Gamma, \varrho)$, and since $R^-(\Gamma)$ is dense in $L_p^-(\Gamma, \varrho)$, we conclude that $gf^- \in L_p^-(\Gamma, \varrho) \oplus L_n$. □

Proposition 3.8 *Suppose that the function $f(t) \in L_p(\Gamma, \varrho)$, $1 < p < \infty$, admits an analytic continuation to $\mathcal{D}_\Gamma^+$ (resp., $\mathcal{D}_\Gamma^-$, including ∞), is continuous in $(\overline{\mathcal{D}_\Gamma^+}) \setminus M$ (resp., $(\overline{\mathcal{D}_\Gamma^-}) \setminus M$), where $M = \{\tau_1, \tau_2, \dots, \tau_m\} \subset \Gamma$, and the following estimates hold in interior (resp., exterior) neighborhoods of the points τ_k:*

$$|f(z)| < c_k |z - \tau_k|^{-\beta_k}, \quad z \in \mathcal{D}_\Gamma^+ \quad (resp., z \in \mathcal{D}_\Gamma^-),$$

where $c_k > 0$, $\beta_k > 0$, $k = 1, 2, \dots, m$. Then $f(t) \in L_p^+(\Gamma, \varrho)$ (resp., $f(t) \in L_p^-(\Gamma, \varrho) \oplus \mathbb{C}$).

Proof. Consider the case of the space $L_p^-(\Gamma, \varrho)$. Let $f^{\pm} = P_\Gamma^{\pm} f$. Multiply both sides of the equality $f^+ = f - f^-$ by the polynomial

$$g_1(t) = \prod_{k=1}^{m} (t - \tau_k)^{[\beta_k]+1},$$

and subtract from the two sides of the resulting equality a polynomial $g_2(t)$, $\deg g_2 \le \deg g_1$, with the property that the function $d(z) = (f(z) - f^-(z))g_1(z) - g_2(z)$ vanishes at infinity. Then

$$f^+ g_1 - g_2 = d$$

almost everywhere on Γ. The function d lies in $L_p^-(\Gamma, \varrho)$ because of the representations

$$f g_1 = f_1 + h_1, \quad f^- g_1 = f_1^- + h_2,$$

where $f_1 \in (A^-(\Gamma) \ominus \mathbb{C}) \subset L_p^-(\Gamma, \varrho)$, $f_1^- \in L_p^-(\Gamma, \varrho)$ and h_1, h_2 are polynomials such that $h_1 - h_2 = g_2$. So, $d = f^+ g_1 - g_2 \in L_p^+(\Gamma, \varrho) \cap L_p^-(\Gamma, \varrho)$, whence $f^+ g_1 - g_2 \equiv 0$. We see that $f^+ = (g_2/g_1) \in L_p^+(\Gamma, \varrho)$, and hence that $g_2 = \operatorname{const} g_1$. This means that $f^+ = \operatorname{const}$ and $f = f^- + \operatorname{const}$.

The case of the space $L_p^+(\Gamma, \varrho)$ is dealt with analogously. □

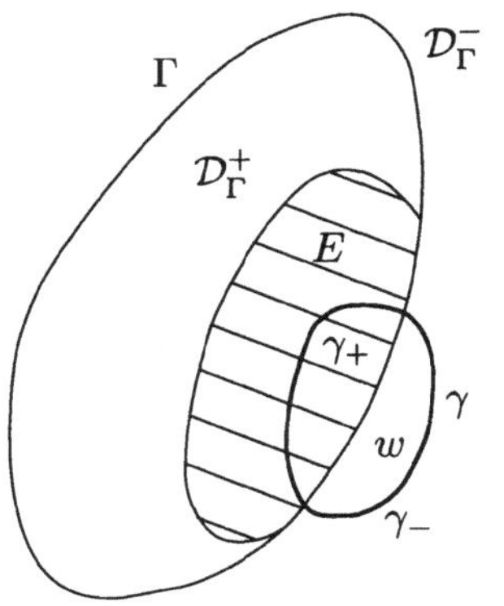

Figure 3.3

Remark 3.2 Proposition 3.8 remains true also in the case $\Gamma = \mathbb{R}$. Here in the neighborhoods of the finite points τ_k the estimates remain as above, while in a neighborhood of the point at infinity one has

$$|f(z)| < c_0|z|^{\beta_0}, \quad z \in \Pi_+ \quad (\text{resp., } z \in \Pi_-), \quad c_0 > 0, \quad \beta_0 > 0.$$

Now let us address the question of what is the domain of analyticity of the elements of $\mathcal{M}(h)$ for an arbitrary model function $h(z)$.

Proposition 3.9 *Let h be a model function on the curve Γ, and let $\{z_j^+\} \subset \mathcal{D}_\Gamma^+$ be the sequence of its zeros, with corresponding multiplicities n_j. Let $f^+(t) \in \mathcal{M}(h)(\subset L_p^+(\Gamma, \varrho))$ and*

$$f^-(t) = h^{-1}(t)f^+(t), \quad t \in \Gamma. \tag{3.12}$$

Then the function $f^-(t)$ admits an analytic continuation to the domain $\mathcal{D}_\Gamma^- \cup \mathcal{D}_\Gamma^+ \setminus \{z_j^+\}$, in the sense that there exists a function $\Phi(z)$ that is analytic in this domain, vanishes at infinity, has only poles at the points z_j^+, of orders at most n_j, and for almost all $t \in \Gamma$ has nontangential interior limits ($\lim_{z\in\mathcal{D}_\Gamma^+,\, z\to t} \Phi(z) = \Phi^+(t)$) and exterior limits ($\lim_{z\in\mathcal{D}_\Gamma^-,\, z\to t} \Phi(z) = \Phi^-(t)$) of class $L_p(\Gamma, \varrho)$, which coincide with one another and are equal to $f^-(t)$.

If, in addition, the function $h^{-1}(z)$ is bounded in some domain $E \subset \mathcal{D}_\Gamma^+$ whose boundary contains a curve $w \subset \Gamma$, then the function $f^-(t)$ admits an analytic continuation to w.

Proof. Set $\Phi^+(z) = f^+(z)/h(z)$, $\Phi^-(z) = f^-(z)$. Since $f^+ \in \ker T(h^{-1})$ in $L_p^+(\Gamma, \varrho)$, $f^- = h^{-1}f^+ \in L_p^-(\Gamma, \varrho)$. This establishes the first assertion of the proposition. To establish the second assertion, we resort to Propositions 3.4 and 3.6.

To that end, let us trace a curve γ of class $\mathcal{T}$ in such a way that the arc w divides the domain $\mathcal{D}_\gamma^+$ into two simply connected subdomains $\mathcal{D}_{w_+}^+$ and $\mathcal{D}_{w_-}^-$, bounded by the curves $w_+ = \gamma_+ \cup w$ and $w_- = \gamma_- \cup w$, where $\gamma_\pm = \gamma \cap \mathcal{D}_\Gamma^\pm$ (see Figure 3.3). Since the function $h^{-1}(z)$ is bounded in $\mathcal{D}_{w_+}^+$, $h^{-1}(z)f^+(z) \in L_p^+(w_+, \varrho_{w_+})$. Moreover, $f^-(z) \in L_p^+(w_-, \varrho_{w_+})$ and $f^-(t) = h^{-1}(t)f^+(t)$ almost everywhere on w. It follows that $f^-(z) \in L_p^+(\gamma, \varrho_\gamma)$, where by ϱ_{w_+}, ϱ_{w_-} and ϱ_γ we denote the natural continuation of the polynomial weight $\varrho(t)$ to the curves w_+, w_- and γ, respectively.

Therefore, the function $f^-(z)$ is analytic on w. □

Remark 3.3 Remark 3.1 is in fact an analogue of the first assertion of Proposition 3.9 for functions f^+ that lie in $\ker T(h^{-1})$. Proceeding in much the same way as in the proof of the second part of Proposition 3.9, one can show that for a model function $h(z)$ on Γ that admits analytic continuation to a domain larger than $\mathcal{D}_\Gamma^+$, the function $f^+(z)$ can be analytically continued through the corresponding part of the boundary Γ. More precisely, the integral representation (3.4) effects this analytic continuation via the formula

$$(Vf^+)(z) = -\frac{h(z)}{2\pi i}\int_\Gamma \frac{h^{-1}(\tau)f^+(\tau)d\tau}{\tau - z}, \quad z \in \mathcal{D}_\Gamma^-. \tag{3.13}$$

At the same time, the integral operators V and V_1 can be used to represent the function $f^-(z)$, where $f^- = h^{-1}f^+$, $f^+ \in \ker T(h^{-1})$. For example,

$$f^-(z) = \begin{cases} \dfrac{1}{2\pi i}\displaystyle\int_\Gamma \frac{h^{-1}(z) - h^{-1}(\tau)}{\tau - z} f^+(\tau)d\tau, & \text{if } z \in \mathcal{D}_\Gamma^+ \setminus \{z_j^+\}, \\ -\dfrac{1}{2\pi i}\displaystyle\int_\Gamma \frac{h^{-1}(\tau)f^+(\tau)d\tau}{\tau - z}, & \text{if } z \in \mathcal{D}_\Gamma^-. \end{cases} \tag{3.14}$$

Thus, we see that for wide classes of model functions h for a curve Γ, the functions f^+, $f^+ \in \ker T(h^{-1})$, and f^-, $f^- = h^{-1}f^+$, which are a priori defined in the domains $\mathcal{D}_\Gamma^+$ and $\mathcal{D}_\Gamma^-$, admit analytic continuation to larger domains. Whenever the intersection of these larger domains contains a curve γ of class $\mathcal{T}$ on which the function $h(z)$ is defined, the question of the relationship between the subspaces $\ker T(h^{-1})$ in the spaces $L_p^+(\Gamma, \varrho)$ and $L_p^+(\gamma, \varrho_\gamma)$, where ϱ_γ is the natural extension of the weight ϱ to γ, becomes meaningful.

Turning now to precise statements, let us introduce the notation $T_\Gamma(h^{-1})$, which indicates the curve on which the Toeplitz operator $T(h^{-1})$ is considered.

Let $\Gamma \in \mathcal{T}$ and let h be a model function on the curve Γ. We say that a curve γ of class $\mathcal{T}$ is a (h, Γ)-*admissible curve* or *contour* if

(1) the curve $\Gamma \cup \gamma$ divides the domains $\mathcal{D}_\Gamma^+$ and $\mathcal{D}_\Gamma^-$ into a finite number of simply connected components which are bounded by Carleson curves;

(2) the function $h(z)$ is defined on γ, is a model function on γ, and satisfies the condition

$$\inf_{z\in\mathcal{D}_{\Gamma,\gamma}} |h(z)| > 0, \text{ where } \mathcal{D}_{\Gamma,\gamma} = (\mathcal{D}_\Gamma^+ \cup \mathcal{D}_\gamma^+) \setminus (\mathcal{D}_\Gamma^+ \cap \mathcal{D}_\gamma^+).$$

The set of all (h,Γ)-admissible curves will be denoted by $\mathcal{T}(h,\Gamma)$. From the definition of the (h,Γ)-admissible curves it follows that if $\gamma \in \mathcal{T}(h,\Gamma)$, then

$$\mathcal{T}(h,\Gamma) = \mathcal{T}(h,\gamma).$$

Remark 3.4 The requirement $\gamma \in \mathcal{T}(h,\Gamma)$ ensures that all possible paths along which the function $h^{-1}(z)$ tends to infinity in the domains $\mathcal{D}_\Gamma^+$ and $\mathcal{D}_\gamma^+$ lie in $\mathcal{D}_\Gamma^+ \cap \mathcal{D}_\gamma^+$. In particular, the sets of zeros of the function $h(z)$ in $\mathcal{D}_\Gamma^+$ and in $\mathcal{D}_\gamma^+$ coincide.

Example 3.2. Let us return to Example 2.5. Let Γ be the contour consisting of the upper part of the circle of radius R centered at the origin and the segment $[-R,R]$ of the real line. Let $t_0 = 0$ and let γ be the circle of radius r centered at the point $\xi(0,r)$ (see Figure 3.4).

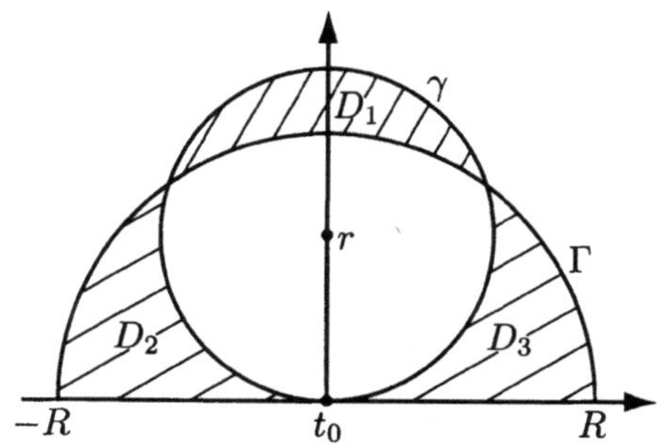

Figure 3.4

Consider the function $h(t) = \exp(-it^{-1})$. Let us show that γ is an (h,Γ)-admissible contour. According to Example 2.5, $h(t)$ is a model function for Γ as well as for γ. Hence, all that remains to be verifed is that the function $h(z)$ is separated away from zero on the set $\mathcal{D}_{\Gamma,\gamma} = D_1 \cup D_2 \cup D_3$, which is the shaded domain in Figure 3.4.

Obviously,

$$\inf_{z\in D_1} |h(z)| > 0.$$

Now let $z \in D_2 \cup D_3$. Clearly, the set $D_2 \cup D_3$ can be represented as a union of arcs of circles γ_ϱ of radius ϱ, $r < \varrho < \infty$, centered at the points $(0, \varrho)$. (Note that $\gamma = \gamma_r$.) Since the equation of the circle γ_ϱ can be written in the form

$$z = 2\varrho \sin\varphi \exp(i\varphi), \quad \varphi \in (0, \pi),$$

it follows that

$$\begin{aligned} \inf_{z \in D_2 \cup D_3} |h(z)| &= \inf_{\varrho \in (r,\infty)} \left| \exp\left(\frac{-i\exp(-i\varphi)}{2\varrho \sin\varphi} \right) \right| \\ &= \inf_{\varrho \in (r,\infty)} \left| \exp\left(\frac{-i\cot\varphi}{2\varrho} - \frac{1}{2\varrho} \right) \right| \\ &= \inf_{\varrho \in (r,\infty)} \exp\left(-\frac{1}{2\varrho} \right) = \exp\left(-\frac{1}{2r} \right) > 0. \end{aligned}$$

Theorem 3.1 (Principle of the independence of the kernel of the contour). *If* $\gamma \in \mathcal{T}(h, \Gamma)$, *then*

$$\ker T_\Gamma(h^{-1}) = \ker T_\gamma(h^{-1}).$$

Remark 3.5 Let us emphasize once more that the last equality means that the elements f^+ of the subspace $\ker T_\Gamma(h^{-1})$ in $L_p^+(\Gamma, \varrho)$ admit analytic continuation to the domain $\mathcal{D}_\gamma^+$ and have nontangential limit values f_γ^+ of class $L_p^+(\gamma, \varrho_\gamma)$ that lie in the subspace $\ker T_\gamma(h^{-1})$. The elements g^+ of the subspace $\ker T_\gamma(h^{-1})$ in $L_p^+(\Gamma, \varrho_\gamma)$ enjoy similar properties.

Proof. Let $\gamma \in \mathcal{T}(h, \Gamma)$. If $\mathcal{D}_\Gamma^+ \cap \mathcal{D}_\gamma^+ = \emptyset$, then the two operators $T_\Gamma(h^{-1})$ and $T_\gamma(h^{-1})$ are invertible, which proves the theorem in this case.

Now assume that $\mathcal{D}_\Gamma^+ \cap \mathcal{D}_\gamma^+ \neq \emptyset$ and denote

$$\Gamma_\pm = \Gamma \cap \mathcal{D}_\gamma^\pm, \quad \gamma_\pm = \gamma \cap \mathcal{D}_\Gamma^\pm.$$

If $f^+ \in \ker T_\Gamma(h^{-1})$ in $L_p^+(\Gamma, \varrho)$, then, by Proposition 3.7, the function $f^-(z) = h^{-1}(z) f^+(z)$ is analytically continuable across Γ to the domain $\mathcal{D}_\Gamma^+ \setminus \mathcal{D}_\gamma^+$, and so it is defined on γ_+. Therefore, $f^-(z)$ is defined almost everywhere on γ. But then the function $f^+(z) = h(z) f^-(z)$ is defined almost everywhere on γ. Let us show that $f^+|_\gamma \in L_p^\pm(\gamma, \varrho_\gamma)$. By Proposition 3.9 and Remark 3.3 concerning the formulas (3.4) and (3.13),

$$f^+(z) = \begin{cases} -\dfrac{h(z)}{2\pi i} \displaystyle\int_\Gamma \frac{f^+(\tau)d\tau}{\tau - z} + \frac{1}{2\pi i} \int_\Gamma \frac{h^{-1}(\tau) f^+(\tau) d\tau}{\tau - z}, & \text{if } z \in \gamma_+, \\ -\dfrac{h(z)}{2\pi i} \displaystyle\int_\Gamma \frac{h^{-1}(\tau) f^+(\tau) d\tau}{\tau - z}, & \text{if } z \in \gamma_-. \end{cases} \tag{3.15}$$

The operator $S_{\Gamma,\gamma}$, by means of which the function $f^+(z)$ is expressed, acts as a bounded operator from $L_p(\Gamma, \varrho)$ to $L_p(\gamma, \varrho_\gamma)$. Hence, $f^+(z) \in L_p(\gamma, \varrho_\gamma)$. Further,

$\Gamma \in \mathcal{T}(h,\gamma)$, and so $\mathcal{D}_\gamma^+$ is divided by the curve $\Gamma \cup \gamma$ into a finite number of simply connected domains $\mathcal{D}_j$, contained in either $\mathcal{D}_\Gamma^+$ or $\mathcal{D}_\Gamma^-$, and bounded by Carleson curves γ_j, $j = 1, 2, \dots, m$. If $\mathcal{D}_j \subset \mathcal{D}_\Gamma^+$, then, by Proposition 3.5, item (a), $f^+(z) \in L_p^+(\gamma_j, \varrho_{\gamma_j})$. If $\mathcal{D}_j \subset \mathcal{D}_\Gamma^-$, then

$$-\frac{1}{2\pi i}\int_\Gamma \frac{h^{-1}(\tau)f^+(\tau)d\tau}{\tau - z} \in L_p^-(\Gamma, \varrho)$$

and $f^+(z) \in L_p^+(\gamma_j, \varrho_{\gamma_j})$ thanks to the boundedness of the function $h(z)$ in $\mathcal{D}_j$ and the embedding $L_p^-(\Gamma, \varrho) \subset L_p^+(\gamma_j, \varrho_{\gamma_j})$ (Proposition 3.5, item (b)). Notice also that $\gamma_j \cap \gamma_k \subset \Gamma$ for $j \neq k$, and that, by the integral representation (3.15), the corresponding limit values of the functions $f^+|_{\mathcal{D}_j}$ and $f^+|_{\mathcal{D}_k}$ coincide almost everywhere on $\gamma_j \cap \gamma_k$. Therefore, we are under the conditions of Proposition 3.6, which tells us that $f^+(z) \in L_p^+(\gamma, \varrho_\gamma)$.

Using in a similar manner formulas (3.14), one verifies that $f^- = h^{-1}f^+ \in L_p^-(\gamma, \varrho_\gamma)$. This shows that $\ker T_\Gamma(h^{-1}) \subset \ker T_\gamma(h^{-1})$. But $\Gamma \in \mathcal{T}(h,\gamma)$. Whence the opposite inclusion follows. □

Theorem 3.1 will play an important role in the sequel, because when we want to describe the subspace $\ker T(h^{-1})$ it will enable us to choose the most convenient admissible curves. In what follows, for these purposes we will often use circles (or lines).

3.3 Model subspaces on $\mathbb{R}$

The definitions of the model function and the model subspace as well as the properties discussed in Sections 3.1 and 3.2 carry over almost verbatim to the case of the real line $\mathbb{R}$. At the same time, this particular case will play an essential role in the constructions that follow. For that reason, in this section we will give a number of results specific for the case of the real line $\mathbb{R}$.

Let h be a model function for $\mathbb{R}$ and let $a \in \mathbb{R}$ be a fixed number. We say that the function h *admits the complex shift* I_a if the line $\mathbb{R}_a = \{z : z = x - ia, x \in \mathbb{R}\}$ is an $(h, \mathbb{R})$-admissible contour. In order that a model function h on $\mathbb{R}$ admit the complex shift I_a it is necessary and sufficient that it have an analytic continuation to the half-plane $\Pi_a = \{z : \operatorname{Im} z > -a\}$ and be bounded there (when $a < 0$ this requirement is satisfied automatically), and that the function h^{-1} be bounded in the strip $\{z : -a < \operatorname{Im} z < 0\}$ when $a > 0$ or the strip $\{z : 0 < \operatorname{Im} z < -a\}$ when $a < 0$. Clearly, if the model function h on $\mathbb{R}$ admits the complex shift I_a, then $h(x - ia) = (I_a h)(x)$ is a model function on $\mathbb{R}$.

Example 3.3 The simplest example of a model function on $\mathbb{R}$ that admits a complex shift is $h(x) = \exp(i\sigma x)$, $\sigma > 0$. Indeed, for any $\alpha \in \mathbb{R}$ the function

$$h(x - i\alpha) = \exp(i\sigma(x - i\alpha)) = h(x)\exp(\sigma\alpha)$$

differs from the original model function only by a constant factor.

Other examples of model functions on $\mathbb{R}$ that admit a complex shift will be considered in Section 3.9.

Proposition 3.10 *Let h be a model function on $\mathbb{R}$ that admits the complex shift I_a. Then the transformation I_a, acting on functions by the rule $(I_a f)(x) = f(x - ia)$, is an isomorphism of the subspace $\mathcal{M}(h)$ of $L_p^+(\mathbb{R}, \varrho)$ onto the subspace $\mathcal{M}(I_a h)$ of $L_p^+(\mathbb{R}, \varrho)$.*

Proof. By Proposition 3.1, $\mathcal{M}(h) = \ker T_{\mathbb{R}}(h^{-1})$ in $L_p^+(\mathbb{R}, \varrho)$. By Theorem 3.1, $\ker T_{\mathbb{R}}(h^{-1}) = \ker T_{\mathbb{R}_a}(h^{-1})$, that is, $f(x) \in \mathcal{M}(h)$ if and only if $f(x) \in L_p^+(\mathbb{R}_a, \varrho_{\mathbb{R}_a})$ and $P_{\mathbb{R}_a}^+(h^{-1}f) = 0$. The last two conditions are equivalent to the requirements

$$f(t) = \frac{1}{\pi i} \int_{\mathbb{R}_a} \frac{f(\tau)d\tau}{\tau - t}$$

and

$$\frac{1}{2}h^{-1}(t)f(t) + \frac{1}{2\pi i} \int_{\mathbb{R}_a} \frac{h^{-1}(\tau)f(\tau)d\tau}{\tau - t} = 0, \quad t = x - ia, \quad x \in \mathbb{R}.$$

Making the change of variables $\tau = \xi - ia$ in the integrals, we conclude that for $x \in \mathbb{R}$,

$$f(x - ia) = \frac{1}{\pi i} \int_{\mathbb{R}} \frac{f(\xi - ia)d\xi}{\xi - x}$$

and

$$\frac{1}{2}h^{-1}(x - ia)f(x - ia) + \frac{1}{2\pi i} \int_{\mathbb{R}} \frac{h^{-1}(\xi - ia)f(\xi - ia)}{\xi - x} d\xi = 0,$$

which is equivalent to the conditions

$$I_a f \in L_p^+(\mathbb{R}, \varrho) \quad \text{and} \quad P_{\mathbb{R}}^+(I_a h^{-1} I_a f) = 0.$$

The latter are in turn equivalent to

$$I_a f \in \mathcal{M}(I_a h) \quad \text{in} \quad L_p^+(\mathbb{R}, \varrho). \qquad \square$$

The technique of fractional linear isomorphisms $\mathcal{K}^{\pm}$ of the form (2.5), (2.4) enables us to establish a connection between the model subspaces on $\mathbb{R}$ and those on any circle (or line) Γ_v in the complex plane.

Proposition 3.11 *Let $h(x)$ be a model function on the real line $\mathbb{R}$ and let Γ_v be a circle (or a line) which is taken into $\mathbb{R}$ by a fractional linear transformation $x = v(t)$ of the from (2.4). Then $h_v(t) = h(v(t))$ is a model function on Γ_v and*

$$\mathcal{M}(h_v) = \mathcal{K}(\mathcal{M}(h)),$$

where the operator $\mathcal{K}$ has the form (2.5).

Proof. This assertion is a direct consequence of Propositions 2.1 and 3.1:

$$\mathcal{K}(\mathcal{M}(h)) = \mathcal{K}(hP_{\mathbb{R}}^{-}h^{-1}(L_p^+(\mathbb{R},\varrho))) = h_v P_{\Gamma_v}^{-} h_v^{-1}(L_p^+(\Gamma_v, \varrho_{\Gamma_v})) = \mathcal{M}(h_v). \quad \square$$

Let $\sigma > 0$. Denote by $\mathcal{B}_\sigma$ and $\mathcal{L}_{p,\varrho}^\sigma$, $1 < p < \infty$, the sets of functions of exponential type at most σ that belong to $L_\infty(\mathbb{R})$ and $L_p(\mathbb{R},\varrho)$, respectively. Since functions of class $\mathcal{L}_{p,\varrho}^\sigma$ are analytic on $\mathbb{R}$, they cannot have singularities at the nodes of the weight $\varrho(x)$. Consequently, $\mathcal{L}_{p,\varrho}^\sigma = \mathcal{L}_{p,\varrho_0}^\sigma$, where $\varrho_0(x) = |x+i|^\alpha$.

The class $\mathcal{B}_\sigma$, which was introduced and first studied by S. N. Bernstein, enjoys a number of remarkable properties. In particular, for any p, $1 < p < \infty$, $\mathcal{L}_p^\sigma \subset \mathcal{B}_\sigma$, where we put $\mathcal{L}_p^\sigma = \mathcal{L}_{p,1}^\sigma$. Furthermore, if $f \in \mathcal{B}_\sigma$, then

$$|f(z)| \le c \cdot \exp(\sigma|y|)$$

for all $z = x + iy$ (see Akhiezer [1], p. 184).

We will use the notations

$$\mathcal{B}_\sigma^\pm = \mathcal{B}_\sigma \cap H^\infty(\Pi_\pm), \quad \mathcal{L}_{p,\varrho}^{\sigma\pm} = \mathcal{L}_{p,\varrho}^\sigma \cap H^{p,\varrho}(\Pi_\pm).$$

Functions in the classes $\mathcal{B}_\sigma$ and $\mathcal{B}_\sigma^\pm$ also admit an integral representation connected with the Fourier transformation.

Proposition 3.12 *The following assertions are true:*

(a) $f \in \mathcal{B}_\sigma \Longrightarrow f(z) = f(0) + z \int_{-\sigma}^{\sigma} g(s)\exp(izs)ds, \quad g \in L_2(-\sigma,\sigma);$

(b) $z \in \mathcal{B}_\sigma^+ \Longrightarrow f(z) = f(0) + z \int_{0}^{\sigma} g(s)\exp(izs)ds, \quad g \in L_2(0,\sigma);$

(c) $f \in \mathcal{B}_\sigma^- \Longrightarrow f(z) = f(0) + z \int_{-\sigma}^{0} g(s)\exp(izs)ds, \quad g \in L_2(-\sigma,0).$

Here $z \in \mathbb{C}$.

Proof. This proposition is a consequence of Theorem 1.4 and the fact that

$$f \in \mathcal{B}_\sigma \implies \frac{f(x)-f(0)}{x} \in \mathcal{L}_2^\sigma. \quad \square$$

Proposition 3.12 allows us to obtain some "crude" estimates for functions of class $\mathcal{L}_{p,\varrho}^\sigma$.

Proposition 3.13 *Any function* $f \in \mathcal{L}_{p,\varrho}^\sigma$ *obeys the estimate*

$$|f(z)| \le c_1|z|\exp(\sigma|y|) + c_2.$$

If $f \in \mathcal{L}_{p,\varrho}^{\sigma\pm}$, *then*

$$|f(z)| \le c_1|z|\exp(\mp\sigma y) + c_2.$$

Here $z = x + iy \in \mathbb{C}$ *and* $c_{1,2}$ *are some constants that are independent of* z.

Proof. As observed above, $\mathcal{L}^\sigma_{p,\varrho} = \mathcal{L}^\sigma_{p,\varrho_0}$, where $\varrho_0(x) = |x+i|^\alpha$, $-1 < \alpha < p-1$. Therefore, $f \in \mathcal{L}^\sigma_{p,\varrho}$ means that $f \in \mathcal{L}^\sigma_{p,\varrho_0}$, and so

$$u(x) = \frac{f(x) - f(0)}{x} \in \mathcal{L}^\sigma_p.$$

But $\mathcal{L}^\sigma_p \subset \mathcal{B}_\sigma$. Hence,

$$|u(z)| = \left|\frac{f(z) - f(0)}{z}\right| \le c_1 \exp(\sigma|y|).$$

This yields the first assertion. The remaining assertions of the proposition are established in an analogous manner. □

The estimates of the functions of class $\mathcal{L}^\sigma_{p,\varrho}$ provided by Proposition 3.13 are far from sharp. Sharp estimates will be given below.

The importance of the classes $\mathcal{L}^\sigma_{p,\varrho}$ and $\mathcal{L}^{\sigma\pm}_{p,\varrho}$ for Toeplitz operators with infinite index is shown by the equality $\mathcal{L}^{\sigma+}_{p,\varrho} = \mathcal{M}(\exp(i\sigma x))$. Turning next to the task of establishing this equality, let us recall that the case of the space $L_2(\mathbb{R})$ was already considered in Section 1.6.

Proposition 3.14 *Assume that $\sigma > 0$ and that the Toeplitz operator $T(\exp(-i\sigma x))$ acts in $L^+_p(\mathbb{R}, \varrho)$. Then*

(a) $\mathcal{L}^{\sigma+}_{p,\varrho} = \ker T(\exp(-i\sigma x))$;

(b) $\exp(\mp i\sigma x)(\mathcal{L}^{\sigma\pm}_{p,\varrho}) = \mathcal{L}^{\sigma\mp}_{p,\varrho}$;

(c) $\mathcal{L}^\sigma_{p,\varrho} = \mathcal{L}^{\sigma+}_{p,\varrho} \oplus \mathcal{L}^{\sigma-}_{p,\varrho}$;

(d) $S(\mathcal{L}^\sigma_{p,\varrho}) = \mathcal{L}^\sigma_{p,\varrho}$.

Consequently, the spaces $\mathcal{L}^\sigma_{p,\varrho}$ and $\mathcal{L}^{\sigma\pm}_{p,\varrho}$ are closed in $L_p(\mathbb{R}, \varrho)$.

Proof. Let $f \in \mathcal{L}^{\sigma+}_{p,\varrho}$. Then, by Proposition 3.13,

$$|\exp(-i\sigma z)f(z)| \le c_1|z|, \quad z \in \overline{\Pi}_-.$$

Since $\exp(-i\sigma x)f(x) \in L_p(\mathbb{R}, \varrho)$, Proposition 3.8 and Remark 3.2 show that $\exp(-i\sigma x)f(x) \in L^-_p(\mathbb{R}, \varrho)$. This proves that $\mathcal{L}^{\sigma+}_{p,\varrho} \subset \ker T(\exp(-i\sigma x))$. Moreover, by Proposition 3.13,

$$|\exp(-i\sigma z)f(z)| \le c_1|z| + c_2\exp(\sigma y), \quad z \in \Pi_+,$$

i.e., $\exp(-i\sigma x)f(x) \in \mathcal{L}^{\sigma-}_{p,\varrho}$. This shows that $\exp(-i\sigma x)\mathcal{L}^{\sigma+}_{p,\varrho} \subset \mathcal{L}^{\sigma-}_{p,\varrho}$.

In a similar manner one verifies that $\exp(i\sigma x)\mathcal{L}^{\sigma-}_{p,\varrho} \subset \mathcal{L}^{\sigma+}_{p,\varrho}$. This proves assertion (b).

Let us show that $\ker T(\exp(-i\sigma x)) \subset \mathcal{L}^{\sigma+}_{p\varrho}$. By Proposition 3.3, any function f belonging to $\ker T(\exp(-i\sigma x))$ has the integral representation (3.6):

$$f(z) = \frac{1}{2\pi i}\int_{\mathbb{R}} \frac{\exp(i\sigma\tau) - \exp(i\sigma z)}{\tau - z} g(\tau)d\tau, \quad g \in L_p(\mathbb{R}, \varrho), \quad z \in \mathbb{C}.$$

Therefore, f is an entire function (see Remark 3.3). Applying the Hölder inequality, we have

$$|f(z)| \le c_1\|g\|_{p,\varrho}\left(\int_{\mathbb{R}}\left|\frac{\exp(i\sigma\tau) - \exp(i\sigma z)}{\tau - z}\right|^q \varrho^{-q/p}(\tau)d\tau\right)^{1/q},$$

where the weight $\varrho(\tau)$ has the form (1.1). The function under the integral sign is locally integrable, and when $|\tau| \to \infty$ it behaves like $O(\tau^{-(1+\varepsilon)})$, where $\varepsilon > 0$. Therefore, the integral exists for all $z \in \mathbb{C}$. Let us estimate the integral, dividing it into two integrals, over the sets $M_1 = \{\tau : |\tau - \operatorname{Re} z| < 1\}$ and $M_2 = \mathbb{R} \setminus M_1$, respectively:

$$\begin{aligned}
|f(z)| \le\ & c_2(1 + |\exp(i\sigma z)|)\left(\int_{M_1} |\tau + i|^{-\frac{\alpha_0 q}{p}} \prod_{k=1}^{m} |\tau - x_k|^{-\frac{\alpha_k q}{p}} d\tau\right.\\
& \left. + \int_{M_2}\left|\frac{\tau + i}{\tau - i}\right|^q |\tau + i|^{-(\frac{\alpha_0}{p}+1)q} \prod_{k=1}^{m} |\tau - x_k|^{-\frac{\alpha_k q}{p}} d\tau\right)^{1/q}\\
\le\ & c_3(1 + |\exp(i\sigma z)|)\left(\sup_{\tau\in M_1} |\tau + i|^{-\frac{\alpha q}{p}} \int_{M_1} \prod_{k=1}^{m} \left|\frac{\tau - x_k}{\tau + i}\right|^{-\frac{\alpha_k q}{p}} d\tau\right.\\
& \left. + \int_{\mathbb{R}} |\tau + i|^{-(\frac{\alpha}{p}+1)q} \prod_{k=1}^{m} |1 - \tau^{-1}x_k|^{-\frac{\alpha_k q}{p}} d\tau\right)^{1/q}\\
\le\ & c_4(1 + \exp(-\sigma \operatorname{Im} z))(|z|^{-\frac{\alpha}{p}} + 1),
\end{aligned}$$

since the first integral is uniformly bounded in z, while the second converges because $(\alpha/p + 1)q > 1$. Thus, $f(z)$ is an entire function of exponential type at most σ, and since $f \in L^+_p(\mathbb{R}, \varrho)$, we conclude that $f \in \mathcal{L}^{\sigma+}_{p,\varrho}$. This establishes assertion (a).

Now let $f \in \mathcal{L}^{\sigma}_{p,\varrho}$. By Proposition 3.13, $\exp(-i\sigma x)f \in L^-_p(\mathbb{R}, \varrho)$. But $f = f^+ + f^-$, where $f^\pm = P^\pm f \in L^\pm_p(\mathbb{R}, \varrho)$. Therefore,

$$\exp(-i\sigma x)f^+ = \exp(-i\sigma x)f - \exp(-i\sigma x)f^- \in L^-_p(\mathbb{R}, \varrho),$$

and so

$$f^+ \in \ker T(\exp(-i\sigma x)) = \mathcal{L}^{\sigma+}_{p,\varrho}.$$

In much the same way one can verify that $f^- \in \mathcal{L}^{\sigma-}_{p,\varrho}$. This proves assertions (c) and (d). □

3.4 Boundary behavior

In the cases in which the class $\mathcal{T}(h,\Gamma)$ contains circles or lines, Theorem 3.1 and Proposition 3.11 reduce the description of the model subspace $\mathcal{M}(h)$ to the simplest case, that of model subspaces for the real line $\mathbb{R}$. Here we give another important application of Theorem 3.1, which provides a description of a model space $\mathcal{M}(h)$ in terms of the boundary behavior of its elements.

Throughout this section it will be assumed that Γ is a smooth curve and that the model function h satisfies the following two conditions:

(a) $h(t)$ is continuous on the set $\overline{\mathcal{D}_\Gamma^+} \setminus \{t_k\}_{k=1}^n$, where $t_k \in \Gamma$;

(b) there exists a pair of smooth curves $\gamma_\pm$, $\gamma_\pm \in \mathcal{T}(h,\Gamma)$ such that $\gamma_\pm \subset \overline{\mathcal{D}_\Gamma^\pm}$ and in sufficiently small neighborhoods of the points t_k, $1 \le k \le n$, the estimates

$$d(t,\gamma_\pm) \ge c|t-t_k|^{1+\lambda_k}, \quad t\in\Gamma, \quad \lambda_k>0, \tag{3.16}$$

hold, where $d(t,\gamma)$ denotes the distance from the point t to the curve γ and the constant c does not depend on t.

Before formulating the main result of this section, let us give the following auxiliary result.

Proposition 3.15 *Let γ be a simple smooth curve emanating from the point t_0. Then for any $q>1$ and $-1<s<q-1$,*

$$\int_\gamma \frac{|\tau-t_0|^s}{|\tau-z|^q}|d\tau| \le c|z-t_0|^s d^{1-q}(z,\gamma), \quad z\in\mathbb{C}\setminus\gamma, \tag{3.17}$$

where the constant c does not depend on z.

Proof. Let us introduce a new coordinate system XOY, with the origin in the point t_0, such that the axes OX and OY are directed along the tangent and the normal to the curve γ at t_0, respectively. In the new coordinate system the curve γ can be given by the equation $\tau(x)=x+i\varphi(x)$, where $x\in[0,x_0]$ and $\varphi(x)$ is a real-valued and continuously differentiable function on $[0,x_0]$. Clearly, $\varphi(x)/x\to 0$ when $x\to 0$, $x\in[0,x_0]$.

Let (x_r,y_r) be the coordinates of the point z in the new coordinate system, $r=|z-t_0|$ and $s>0$. Then the function x^s is monotonically increasing on $[0,x_0]$ and consequently

$$\begin{aligned} I(z) &= \int_\Gamma \frac{|\tau-t_0|^s}{|\tau-z|^q}|d\tau| \le c_1\int_0^{x_0}\frac{(x^2+\varphi^2(x))^{s/2}dx}{((x-x_r)^2+(\varphi(x)-y_r)^2)^{q/2}} \\ &\le c_2\int_0^{x_0}\frac{x^s dx}{((x-x_r)^2+(\varphi(x)-y_r)^2)^{q/2}}. \end{aligned}$$

Assuming that $x_r > 0$ and r is sufficiently small, let us break the last integral into three parts:

$$I(z) \le c_2 \left(\int_0^{x_r - d(z,\gamma)} + \int_{x_r - d(z,\gamma)}^{x_r + d(z,\gamma)} + \int_{x_r + d(z,\gamma)}^{x_0} \right)$$
$$\le c_3 \left(x_r^s \int_0^{x_r - d(z,\gamma)} (x_r - x)^{-q} dx + (x_r + d(z,\gamma))^s d^{-q}(z,\gamma) \int_{x_r - d(z,\gamma)}^{x_r + d(z,\gamma)} dx \right.$$
$$\left. + \left(\frac{x_r + d(z,\gamma)}{d(z,\gamma)} \right)^s \int_{x_r + d(z,\gamma)}^{x_0} (x - x_r)^{s-q} dx \right).$$

In the third term we used the inequality

$$x(x - x_r)^{-1} \le (x_r - d(z,\gamma)) d^{-1}(z,\gamma),$$

which holds because the function $x(x - x_r)^{-1}$ is monotonically decreasing on the segment $[x_r + d(z,\gamma), x_0]$. Further, since $d(z,\gamma) \le r$, $x_r \le r$, we have

$$I(z) \le c_4 \left(x_r^s (d^{1-q}(z,\gamma) - x_r^{1-q}) + (x_r + d(z,\gamma))^s d^{1-q}(z,\gamma) \right.$$
$$\left. + \left(\frac{x_r + d(z,\gamma)}{d(z,\gamma)} \right)^s (d^{s-q+1}(z,\gamma) - (x_0 - x_r)^{s-q+1}) \right) \le c r^s d^{1-q}(z,\gamma).$$

If $x_r \le 0$, then taking into account that $-x_r \le d(z,\gamma)$ and proceeding in an analogous manner, we deduce that

$$I(z) \le c_2 \left(\int_0^{x_r + d(z,\gamma)} + \int_{x_r + d(z,\gamma)}^{x_0} \right)$$
$$\le c_5 \left(\frac{(x_r + d(z,\gamma))^s}{d(z,\gamma)^q} \int_0^{x_r + d(z,\gamma)} dx + \left(\frac{x_0}{x_0 - x_r} \right)^s \int_{x_r + d(z,\gamma)}^{x_0} (x - x_r)^{s-q} dx \right)$$
$$\le c_6 (d^{s-q+1}(z,\gamma) + (d^{s-q+1}(z,\gamma) - (x_0 - x_r)^{s-q+1})) \le c r^s d^{1-q}(z,\gamma).$$

Here, to estimate the last integral we used the fact that the function $x(x - x_r)^{-1}$ ($x_r < 0$!) is monotonically decreasing on the segment $[x_r + d(z,\gamma), x_0]$.

The case $s \le 0$ is dealt with in much the same manner. □

Theorem 3.2 *Let h be a model function on a smooth curve Γ which satisfies conditions (a) and (b). Let $\{z_j^+\} \subset \mathcal{D}_\Gamma^+$ be its zero set and let $\{n_j\}$ denote the multiplicity of z_j^+. Also, let the discontinuity points t_k, $t_k \in \Gamma$, of the function $h(t)$ coincide with the first n ($n \le m$) nodes of a weight $\varrho(t)$ of the form (2.1) with exponents α_k, which may be equal to zero.*

For a function $f^+(t)$ to lie in the subspace $\mathcal{M}(h)$ in $L_p^+(\Gamma, \varrho)$ it is necessary and sufficient that the function $f^-(t) = h^{-1}(t)f^+(t)$ admit an analytic continuation to the domain $\mathbb{C} \setminus (\{t_k\}_{k=1}^n \cup \{z_j^+\})$ that vanishes at infinity and has the property that the points z_j^+ are its regular points or poles of order at most n_j, while the points t_k are regular, isolated or nonisolated singular points, in the vicinity of which the following estimates hold:

$$|f^-(z)| \leq o_k(z)|h(z)|^{-1}|z - t_k|^{-\frac{\alpha_k}{p}} d^{-\frac{1}{p}}(z, \gamma_-), \quad z \in \overline{\mathcal{D}_\Gamma^+}, \tag{3.18}$$

$$|f^-(z)| \leq o_k(z)|z - t_k|^{-\frac{\alpha_k}{p}} d^{-\frac{1}{p}}(z, \gamma_+), \quad z \in \overline{\mathcal{D}_\Gamma^-}, \tag{3.19}$$

where $\gamma_\pm$ satisfy conditions (b) and (3.16) and $\lim_{z \to t_k} o_k(z) = 0$.

Proof. Suppose that the conditions of the theorem are satisfied. Then, by Proposition 3.1 and Theorem 3.1, $\mathcal{M}(h) = \ker T_\Gamma(h^{-1}) = \ker T_{\gamma_-}(h^{-1})$ in $L_p^+(\gamma_-, \varrho_{\gamma_-})$.

The case $\mathcal{D}_\Gamma^+ \subset \mathcal{D}_{\gamma_-}^-$ is not interesting, since the condition $\gamma_- \in \mathcal{T}(h, \Gamma)$ ensures that the operators $T_\Gamma(h^{-1})$ and $T_{\gamma_-}(h^{-1})$ are invertible, and consequently $\mathcal{M}(h) = \{0\}$.

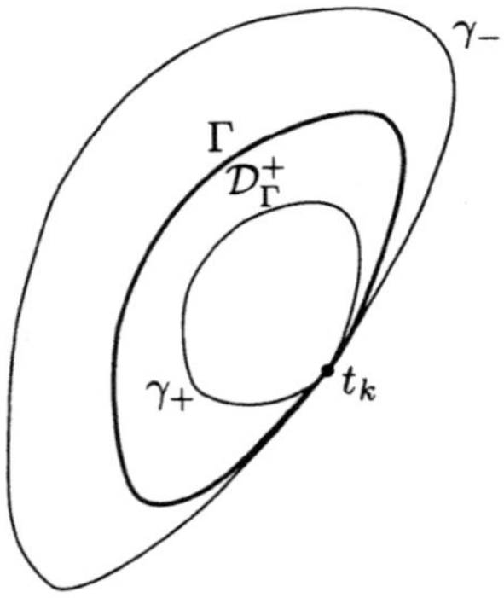

Figure 3.5

Thus, let $\mathcal{D}_\Gamma^+ \subset \mathcal{D}_{\gamma_-}^+$ (see Figure 3.5). If $f^+ \in \mathcal{M}(h)$, then, by Proposition 3.9, the function $f^-(z)$ admits an analytic continuation that enjoys the analyticity properties described in the theorem. Hence, it only remains to verify the estimates (3.18) and (3.19).

Let $z \in \overline{\mathcal{D}_\Gamma^+}$. By formula (3.14),

$$f^-(z) = \frac{h^{-1}(z)}{2\pi i} \int_{\gamma_-} \frac{1 - h(z)h^{-1}(\tau)}{\tau - z} g(\tau) d\tau,$$

where $g(\tau) \in L_p(\gamma_-, \varrho_{\gamma_-})$. Suppose that the point z lies in a sufficiently small neighborhood of the point t_k.

Observing that the function $h(z)$ is bounded in $\overline{\mathcal{D}_\Gamma^+}$, we have

$$|h(z)||f^-(z)| \le c_1 \int_{\gamma_-} \frac{|1-h(z)h^{-1}(\tau)|}{|\tau - z|}|g(\tau)||d\tau| \le c_2 \int_{\gamma_-} \frac{|g(\tau)|}{|\tau - z|}|d\tau|.$$

The function $|g(\tau)|^p \varrho(\tau)$ is Lebesgue integrable on γ_-. Hence, for any $\varepsilon > 0$ there exists a neighborhood γ_ε of the point t_k, $\gamma_\varepsilon \subset \gamma_-$, such that

$$\left(\int_{\gamma_\varepsilon} |g(\tau)|^p \varrho(\tau)|d\tau| \right)^{1/p} \le \varepsilon.$$

Fixing the curve γ_ε and applying Hölder's inequality, we conclude that

$$\begin{aligned}
|h(z)||f^-(z)| &\le c_2 \left(\int_{\gamma_\varepsilon} \frac{|g(\tau)|}{|\tau - z|}|d\tau| + \int_{\gamma_- \setminus \gamma_\varepsilon} \frac{|g(\tau)|}{|\tau - z|}|d\tau| \right) \\
&\le c_2 \left(\int_{\gamma_\varepsilon} \frac{|g(\tau)|\varrho^{1/p}(\tau)}{|\tau - z|\varrho^{1/p}(\tau)}|d\tau| + \frac{1}{d(z, \gamma_- \setminus \gamma_\varepsilon)} \int_{\gamma_- \setminus \gamma_\varepsilon} |g(\tau)||d\tau| \right) \\
&\le c_3 \left(\left(\int_{\gamma_\varepsilon} |g(\tau)|^p \varrho(\tau)|d\tau| \right)^{1/p} \left(\int_{\gamma_\varepsilon} \varrho^{-\frac{p}{q}}(\tau)|\tau - z|^{-q}|d\tau| \right)^{1/q} \right. \\
&\quad \left. + \frac{\|g\|_{p,\varrho}}{d(z, \gamma_- \setminus \gamma_\varepsilon)} \left(\int_{\gamma_-} \varrho^{-\frac{q}{p}}(\tau)|d\tau| \right)^{1/q} \right) \\
&\le c_4 \left(\varepsilon \left(\int_{\gamma_\varepsilon} |\tau - t_k|^{-\frac{q\alpha_k}{p}} |\tau - z|^{-q}|d\tau| \right)^{1/q} \right. \\
&\quad \left. + \frac{\|g\|_{p,\varrho}}{d(z, \gamma_- \setminus \gamma_\varepsilon)} \left(\int_{\gamma_-} \prod_{j=1}^m |\tau - t_j|^{-\frac{q\alpha_j}{p}} |d\tau| \right)^{1/q} \right).
\end{aligned}$$

Since $q = p/(p-1)$, $-1 < \alpha_j < p - 1$, the last two integrals converge and all the assumptions of Proposition 3.15 are satisfied. Hence, by inequalities (3.17) and (3.16),

$$\begin{aligned}
|h(z)||f(z)| &\le c_5 \left(\varepsilon |z - t_k|^{-\frac{\alpha_k}{p}} d^{-\frac{1}{p}}(z, \gamma_-) + d^{-1}(z, \gamma_- \setminus \gamma_\varepsilon) \right) \\
&\le c_5 |z - t_k|^{-\frac{\alpha_k}{p}} d^{-\frac{1}{p}}(z, \gamma_-) \left(\varepsilon + |z - t_k|^{\frac{\alpha_k}{p}} d^{\frac{1}{p}}(z, \gamma_-) d^{-1}(z, \gamma_- \setminus \gamma_\varepsilon) \right).
\end{aligned}$$

Since the curve γ_ε is fixed,

$$|z - t_k|^{\frac{\alpha_k}{p}} d^{\frac{1}{p}}(z, \gamma_-) d^{-1}(z, \gamma_- \setminus \gamma_\varepsilon) < \varepsilon$$

for sufficiently small $|z - t_k|$. Therefore, the necessity of the estimate (3.18) is established.

If now $z \in \overline{\mathcal{D}_\Gamma^-}$, we use the equalities

$$\mathcal{M}(h) = \ker T_\Gamma(h^{-1}) = \ker T_{\gamma_+}(h^{-1}) \quad \text{in} \quad L_p^+(\gamma_+, \varrho_{\gamma_+}).$$

Resorting again to formula (3.14), we obtain

$$f^-(z) = -\frac{1}{2\pi i}\int_{\gamma_+} \frac{h^{-1}(\tau)g(\tau)}{\tau - z}\,d\tau, \quad \text{where} \quad g \in L_p(\gamma_+, \varrho_{\gamma_+}).$$

Repeating the technical construction used in the proof of estimate (3.18), but employing now inequality (3.16) for the curve γ_+, one verifies that estimate (3.19) holds.

Now let us prove the sufficiency of the conditions of the theorem. To that end we will show that under those conditions $f^+(t) \in L_p^+(\Gamma, \varrho)$, and $f^-(t) = h^{-1}(t)f^+(t) \in L_p^-(\Gamma, \varrho)$. Indeed, the function $f^-(z)$ is analytic in the domain $\overline{\mathcal{D}_\Gamma^-}\setminus\{t_k\}_{k=1}^n$ and vanishes at infinity. Furthermore, by (3.19) and (3.16), in exterior neighborhoods of the points t_k ($z \in \mathcal{D}_\Gamma^-$) it obeys the estimates

$$|f^-(z)| \le c_3|z - t_k|^{-\frac{\alpha_k}{p}} d^{-\frac{1}{p}}(z, \gamma_+) \le c_4|z - t_k|^{-\frac{1+\alpha_k+\lambda_k}{p}}.$$

Therefore, by Proposition 3.8, $f^-(t) \in L_p^-(\Gamma, \varrho)$. Turning now to the function $f^+(z)$, we observe that it is analytic in the domain $\mathcal{D}_\Gamma^+$, continuous in $\overline{\mathcal{D}_\Gamma^+}\setminus\{t_k\}_{k=1}^n$, and, according to (3.18) and (3.16), satisfies the following estimates in interior neighborhoods of the points t_k ($z \in \mathcal{D}_\Gamma^+$):

$$|f^+(z)| \le c_5|z - t_k|^{-\frac{1+\alpha_k+\lambda_k}{p}}.$$

Applying again Proposition 3.8, we conclude that $f^+(t) \in L_p^+(\Gamma, \varrho)$. □

Remark 3.6 In the space $L_p^\pm(\Gamma, \varrho)$ there exist functions that grow arbitrarily fast when $z \to t$ ($z \in \mathcal{D}_\Gamma^\pm$, $t \in \Gamma$) (of course, on a set of sufficiently small measure). The constraints (3.18) and (3.19) regulate the growth of the functions $f^+(t) \in \ker T(h^{-1})$ in the vicinity of the points t_k, namely, they provide an upper bound for the growth. At the same time, among the elements of the null space $\ker T(h^{-1})$ there exist functions that are bounded or even decreasing in the vicinity of the points t_k.

Remark 3.7 Requiring the existence of a curve γ_- that satisfies condition (3.16) reduces somewhat the generality of the constraints on the model functions h considered in this section, because we are asking that h admit analytic continuation across the boundary of the domain $\mathcal{D}_\Gamma^+$. The assertions of the theorem remain valid if we merely require that there exist an (h, Γ)-admissible curve γ_+, $\gamma_+ \subset \overline{\mathcal{D}_\Gamma^+}$ which satisfies condition (3.16). One can show that in this case there exists an

(h, Γ)-admissible curve γ_1 which is smooth in the vicinity of the points $\{t_k\}$ and is "equidistant" from the curves Γ and γ_+, in the sense that

$$d(t, \Gamma \cup \gamma_+) \geq c|t - t_k|^{1+\lambda_k}, \quad \lambda_k > 0, \quad t \in \gamma_1.$$

Here the estimates (3.18) and (3.19) of Theorem 3.2 are replaced by

$$|f(z)| \leq o_k(z)|h(z)|^{-1}|z - t_k|^{-r_k}, \quad z \in \overline{\mathcal{D}^+_{\gamma_1}},$$

and

$$|f(z)| \leq o_k(z)|z - t_k|^{-r_k}, \quad r_k > 0, \quad z \in \overline{\mathcal{D}^-_{\gamma_1}}.$$

Let us mention, though, that in what follows Theorem 3.2 will be used only in the form (3.18), (3.19).

Bases and interpolation in model subspaces

In this section we describe various types of bases in the model subspaces of the space $L_p(\mathbb{R}, \varrho)$, $1 < p < \infty$, generated by the inner functions $B(x)$ and $\exp(ix\sigma)$. The technique developed in the preceding section allows us to extend these results to model subspaces of a more general form. Here we solve the interpolation problem, which enables us to expand an arbitrary element of the model subspace into a Lagrange interpolation series. The special attention devoted to this problem is motivated by the important role it plays in the nonelliptic theory of singular integral equations (see Chapter 6).

3.5 Bases

A system of elements $\{e_k\}$, $k \in \mathbb{Z}$, of a complex Banach space X is called a *basis* of X if any element $x \in X$ can be uniquely represented as a series

$$x = \sum_{k \in \mathbb{Z}} c_k e_k, \quad \text{where } c_k \in \mathbb{C}, \tag{3.20}$$

which converges in the norm of X. A system of elements $\{e_k\} \subset X$ is said to be *uniformly minimal* if

$$\operatorname{dist}\left(\frac{e_n}{\|e_n\|}, \mathcal{L}_n\right) \geq \Delta > 0 \quad \text{for all } n \in \mathbb{Z}, \tag{3.21}$$

where $\mathcal{L}_n = \operatorname{clos\,span}\{e_k, k \in \mathbb{Z}, k \neq n\}$. Every basis is a uniformly minimal system. For each uniformly minimal system there exists a biorthogonal system of functionals $\{f_n\}$ in the dual space X^*: $f_n(e_k) = \delta_{nk}$, $k, n \in \mathbb{Z}$. In the decomposition (3.20), $c_k = f_k(x)$. This biorthogonal system of functionals is often called the *dual* or *conjugate system* to $\{e_k\}$. In a reflexive space X, a system $\{e_k\}$ is a basis if and only if the dual system $\{f_n\}$ is a basis in the dual space.

If X_1 and X_2 are Banach spaces, $T : X_1 \to X_2$ is an isomorphism and $\{e_k\}$ is a basis in X_1, then $\{u_k\} = \{Te_k\}$ is a basis in X_2. Using Banach's theorem on the inverse operator one readily obtains the converse statement: if $T : X_1 \to X_2$ is a bounded linear operator and $\{e_k\}$ and $\{Te_k\}$ are bases in X_1 and X_2, respectively, then T is an invertible operator.

Proposition 3.16 *Let $\{e_k\}$ be a system of elements of the normed space X which possesses a dual system of functionals $\{f_n\}$, $f_n \in X^*$. Then*

$$\operatorname{dist}\left(\frac{e_n}{\|e_n\|}, \mathcal{L}_n\right) = \|e_n\|^{-1}\|f_n\|^{-1}, \quad n \in \mathbb{Z}. \tag{3.22}$$

Proof. The proof of formula (3.22) is carried out by a standard argument along the lines of the Hahn-Banach theorem (see, e.g., Dunford and Schwartz [1], II.3.12–II.3.16). □

We will need one more auxiliary result.

Proposition 3.17 *Let X_1, X_2 be Banach spaces, $T : X_1 \to X_2$ a bounded linear operator, and $\{e_k\}$ a system of elements of X_1. If the system $\{Te_k\}$ is uniformly minimal in X_2 and $\inf_k(\|Te_k\|) \geq \Delta > 0$, then the system $\{e_k\}$ is uniformly minimal in X_1.*

3.6 The Carleson condition and interpolation in $H^{p,\varrho}(\Pi_\pm)$

Let $\{z_k\}_{k\in\mathbb{Z}}$ be a set of points in one of the half-planes Π_+ or Π_-. We say that $\{z_k\}_{k\in\mathbb{Z}}$ satisfies the *Carleson condition* if

$$\inf_{k\in\mathbb{Z}} \prod_{n\neq k} \left|\frac{z_k - z_n}{z_k - \overline{z}_n}\right| \geq \Delta > 0, \quad \text{for all } n \in \mathbb{Z}. \tag{3.23}$$

If the set $\{z_k\}$ lies in the strip $\{z : 0 < c_1 < \operatorname{Im} z < c_2 < \infty\}$, the Carleson condition (3.23) is equivalent to the separation condition (Garnett [1], p. 287)

$$\inf_{k\neq n} |z_k - z_n| \geq d > 0. \tag{3.24}$$

Condition (3.23) plays an important role in the theory of bases and complex interpolation. In particular, the following interpolation theorem holds.

Theorem 3.3 *Let $f \in H^{p,\varrho}(\Pi_\pm)$, $1 < p < \infty$, let the function ϱ have the form* (1.1), *and let the set $\{z_k\}_{k\in\mathbb{Z}}$ lie in $\Pi_\pm$ and satisfy condition* (3.23). *Then*

$$\pm \sum_{k\in\mathbb{Z}} \operatorname{Im} z_k \, d_k^\pm |f(z_k)|^p \leq c_p(\Delta) \|f\|_{p,\varrho}^p, \tag{3.25}$$

where $c_p(\Delta) > 0$ *is a constant that does not depend on* f *and*

$$d_k^{\pm} = (|z_k| + 1)^{\alpha_0} \prod_{j=1}^{m} |z_k - x_j|^{\alpha_j}.$$

Proof. In the case $\varrho(x) \equiv 1$, $d_k^{+} = 1$, $k \in \mathbb{Z}$, our Teorem 3.3 is just the well-known Carleson-Shapiro-Shields theorem (see Shapiro and Shields [1]) on interpolation in $H^p(\Pi_+)$, $1 < p < \infty$. Theorem 3.3 as stated can be reduced to this particular case with the help of Proposition 1.2: $f \in H^{p,\varrho}(\Pi_\pm)$ if and only if $\varrho_p^{\pm} f \in H^p(\Pi_\pm)$, where the function $\varrho_p^{\pm}$ has the form (1.6). Note that the estimate (3.25) does not depend on the choice of branches of the multi-valued functions in formula (1.16), because all the branches have the same modulus. □

3.7 Sine-type functions

An entire function $s(z)$ of exponential type σ is called a *sine-type function* if it satisfies the following conditions:

(1) all its zeros lie in some strip $|y| \leq h < \infty, \quad z = x + iy$;

(2) outside any strip $|y| \leq h + \varepsilon$, $\varepsilon > 0$, it obeys the estimate

$$m_1 \leq |s(z)| \exp(-\sigma |y|) \leq m_2$$

where $m_1 = m_1(\varepsilon)$ and $m_2 = m_2(\varepsilon)$ are positive finite numbers.

From the last inequality it follows, via an application of the Phragmén–Lindelöf principle, that $s \in \mathcal{B}_\sigma$. The function $\widetilde{s}^{\pm} = \exp(\pm i\sigma z)s(z)$ has exactly the same zeros as the function $s(z)$, is an entire function of exponential type 2σ, and outside the strip $|y| \leq h + \varepsilon$ satisfies the estimates

$$m_1 \exp(\sigma(|y| \mp y)) \leq |\widetilde{s}^{\pm}(z)| \leq m_2 \exp(\sigma(|y| \mp y)).$$

Consequently, $\widetilde{s}^{\pm} \in \mathcal{B}_{2\sigma}^{\pm}$. Here are several properties of the sequence of zeros $\{z_k\}_{k\in\mathbb{Z}}$ of sine-type functions.

(i) If the set $\{z_k\}$ is *separated*, i.e., $|z_k - z_j| \geq d > 0$, $k \neq j$, then outside the disks $c_k(r) = \{z : |z - z_k| < r\}$, $0 < r \leq d/2$, the function $s(z)$ obeys the estimate

$$|s(z)| \geq m_r \exp(\sigma |y|), \tag{3.26}$$

where $m_r > 0$ does not depend on z.

(ii) If the set $\{z_k\}$ is separated, then

$$|s'(z_k)| \geq a > 0, \quad k \in \mathbb{Z}. \tag{3.27}$$

(iii) Under the additional constraint $s(0) = 1$, the function $s(z)$ admits the representation

$$s(z) = \prod_{k\in\mathbb{Z}} \left(1 - \frac{z}{z_k}\right) = \lim_{n\to\infty} \prod_{k=-n}^{n} \left(1 - \frac{z}{z_k}\right), \tag{3.28}$$

where the right-hand side converges uniformly on every compact subset.

(iv) A necessary and sufficient condition for a function that satisfies a condition of type (3.28) to be a sine-type function is that $\sup_k |\operatorname{Im} z_k| < \infty$ and that the function

$$\prod_{k\in\mathbb{Z}} \left(1 - \frac{z}{\operatorname{Re} z_k}\right)$$

be a sine-type function.

(v) If $\{\psi_k\} \in l_p(\mathbb{Z})$, $1 < p < \infty$, and s is a sine-type function of type σ and form (3.28), then

$$\widetilde{s}(z) = \prod_{k\in\mathbb{Z}} \left(1 - \frac{z}{z_k + \psi_k}\right)$$

is a sine-type function of type σ.

3.8 Bases of entire functions

We will first consider the subspace $\mathcal{L}^{\sigma+}_{p,\varrho}$. Let $s(z)$ be a sine-type function of type $\sigma/2$ whose set of zeros $\{z_k\}_{k\in\mathbb{Z}}$ lies in the strip $|\operatorname{Im} z| \le h < \infty$ and satisfies the separation condition (3.24), and let

$$s^+(z) = \exp(iz\sigma/2)s(z).$$

Let us introduce the system of functions

$$u_k(x) = \frac{s^+(x)}{(s^+)'(z_k)(x - z_k)}, \quad k \in \mathbb{Z}. \tag{3.29}$$

Recall that $\mathcal{L}^{\sigma}_{p,\varrho} = \mathcal{L}^{\sigma}_{p,\varrho_0}$, where the weight $\varrho(x)$ has the form (1.1) and $\varrho_0(x) = |x + i|^\alpha$. Consider the discrete weight

$$\delta = \{\delta_k\} = \{|z_k + (h+1)i|^\alpha\} \tag{3.30}$$

and let $l_{p,\delta}(\mathbb{Z})$ be the space of all sequences $\xi = \{\xi_k\}$ with finite norm

$$\|\xi\| = \left(\sum_{k\in\mathbb{Z}} |\xi_k|^p \delta_k\right)^{1/p}.$$

Theorem 3.4 *Let $\{f_k\} \in l_{p,\delta}(\mathbb{Z})$, where the weight δ is defined by (3.30). Then the series*

$$f(x) = \sum_{k \in \mathbb{Z}} f_k u_k(x) \tag{3.31}$$

(1) *converges in the norm of the space $L_p(\mathbb{R}, \varrho)$;*

(2) *represents a function of class $\mathcal{L}^{\sigma+}_{p,\varrho}$;*

(3) *effects an isomorphism ($\{f_k\} \mapsto f(x)$) of the space $l_{p,\delta}(\mathbb{Z})$ onto the space $\mathcal{L}^{\sigma+}_{p,\varrho}$;*

(4) *solves the interpolation problem $f(z_k) = f_k$, $k \in \mathbb{Z}$, in $\mathcal{L}^{\sigma+}_{p,\varrho}$.*

Proof. Let us consider a segment of the series (3.31):

$$F_{l,m}(x) = \sum_{k=l}^{m} f_k u_k(x).$$

Since $\exp(-ix\sigma)s^+(x) \in L^-_\infty(\mathbb{R})$ implies that $\exp(-ix\sigma)u_k(x) \in L^-_p(\mathbb{R}, \varrho)$ and hence that $u_k \in \mathcal{L}^{\sigma+}_{p,\varrho}$, we see that $F_{l,m} \in \mathcal{L}^{\sigma+}_{p,\varrho}$. Let $A(x) = \exp(ix\sigma)$, $\alpha = h + d/2$. Since the model function $A(x)$ admits the transformation $A(x) \mapsto A(x - i\alpha)$, it follows from Proposition 3.10 that

$$F_{l,m}(x - i\alpha) \in P^+(\exp(i(x - i\alpha)\sigma)L^-_p(\mathbb{R}, \varrho)) = \mathcal{L}^{\sigma+}_{p,\varrho}$$

and

$$\|F_{l,m}(x)\| \le c_1 \|F_{l,m}(x - i\alpha)\| = c_1 \sup_g \left| \int_{\mathbb{R}} F_{l,m}(x - i\alpha)\overline{g(x)}dx \right|$$

with $g \in L^+_q(\mathbb{R}, \varrho^{1-q})$, $\|g\| = 1$, $q = p/(p-1)$, and

$$F_{l,m}(x - i\alpha) = \sum_{k=l}^{m} f_k \frac{s^+(x - i\alpha)}{(s^+)'(z_k)(x - \xi_k)},$$

where the numbers $\xi_k = z_k + i\alpha$, being the roots of the function $s^+(z - i\alpha)$, lie in the strip

$$0 < d/2 \le \operatorname{Im} z \le 2h + d/2$$

and satisfy the separation condition (3.24). But then the set $\{\xi_k\}_{k \in \mathbb{Z}}$ satisfies the Carleson condition (3.23) and, by Theorem 3.3,

$$\sum_{k \in \mathbb{Z}} \varrho_+(\xi_k)|f(\xi_k)|^p \le c_p(\Delta)\|f\|^p \tag{3.32}$$

whenever $f \in L_p^+(\mathbb{R}, \varrho)$, $1 < p < \infty$. Consequently,

$$\begin{aligned} \|F_{l,m}(x - i\alpha)\| &= \sup_g \left| \sum_{k=l}^{m} \frac{f_k}{(s^+)'(z_k)} \int_{\mathbb{R}} \frac{\overline{g(x)} s^+(x)}{x - \xi_k} dx \right| \\ &= 2\pi \sup_g \left| \sum_{k=l}^{m} \frac{f_k V(\xi_k)}{(s^+)'(z_k)} \right|, \end{aligned}$$

where $V = P^+(\overline{g} s^+) \in L_q^+(\mathbb{R}, \varrho^{1-q})$ and $\|V\| \le c_2 \|g\|$.

Using estimate (3.27) for the function $(s^+)'(z)$ and inequality (3.32) for the function $V(z)$, we conclude that

$$\begin{aligned} \|F_{l,m}(x - i\alpha)\| &\le \frac{2\pi}{a} \sup_g \left[\left(\sum_{k=l}^{m} |f_k|^p \delta_k \right)^{1/p} \left(\sum_{k=l}^{m} |V(\xi_k)|^q \delta_k^{1-q} \right)^{1/q} \right] \\ &\le \frac{2\pi}{a} c_q^{1/q}(\Delta) \left(\sum_{k=l}^{m} |f_k|^p \delta_k \right)^{1/p}. \end{aligned}$$

We have thus established that $\{F_{l,m}\}$ is a fundamental sequence in $\mathcal{L}_{p,\varrho}^{\sigma+}$ and that the series (3.31) converges.

Now let $f(x) \in \mathcal{L}_{p,\varrho}^{\sigma+}$. Then $f(x - i\alpha) \in \mathcal{L}_{p,\varrho}^{\sigma+}$ and, by Theorem 3.3, $\{f(\xi_k - i\alpha)\} = \{f(z_k)\} \in l_{p,\delta}(\mathbb{Z})$. If $g = \sum_{k \in \mathbb{Z}} f(z_k) u_k$, then $h = f - g \in \mathcal{L}_{p,\varrho}^{\sigma+}$ and $h(z_k) = 0$, $k \in \mathbb{Z}$. Let us show that $h(z) \equiv 0$. By Proposition 3.13, for any $z = x + iy$, $z \in \mathbb{C}$,

$$|h(z)| \le a_1 |z| \exp(-\sigma y) + a_2, \quad a_1, a_2 > 0.$$

Considering the function $\psi(z) = h(z)/s^+(z)$, we remark that inequality (3.26) holds everywhere outside of the disjoint disks $c_k(z)$. It follows that for any z in the complement of the set $M_r = \bigcup_{k \in \mathbb{Z}} c_k(r)$ one has the estimate

$$|\psi(z)| \le \frac{a_1 |z| \exp(-\sigma y) + a_2}{m_r \exp((\sigma/2)(|y| - y)} \le a_3 |z| + a_4,$$

where $a_3, a_4 > 0$. In the complement of the set $M_r \cup \widetilde{c}(r)$, where $\widetilde{c}(r) = \{z : |z| < r\}$,

$$|\psi(z)| \le a_5 |z|, \quad a_5 > 0.$$

Consequently, for those values of z,

$$|\psi(z)/z| \le a_5.$$

Applying the maximum principle for analytic functions and a generalization of the Liouville principle, we conclude that $\psi(z)$ must be a linear function. But $\psi s^+ \in L_p(\mathbb{R}, \varrho)$, and so, by estimate (3.26), $\psi \equiv 0$, whence $f \equiv g$. □

Setting $s^+(x) = \exp(ix\sigma) - 1$, we deduce from Theorem 3.4 that a system of functions of the form (1.30) is a biorthogonal basis in the space $\mathcal{L}^{\sigma+}_{p,\varrho}$.

Remark 3.8 Let $a \in \mathbb{R}$, $\mathbb{R}_a = ai + \mathbb{R}$, $\varrho(x) = |x+i|^\alpha$ and $\varrho_a(x) = |x + i(|a|+1)|^\alpha$, $-1 < \alpha < p-1$. Then a series of the form (3.31) converges in the space $L_p(\mathbb{R}_a, \varrho_a)$ and represents a function of class $L_p^+(\mathbb{R}_a, \varrho_a)$. This follows upon observing that

$$\left\| \sum_{k=l}^{m} c_k u_k(z) \right\|_{L_p(\mathbb{R}_a, \varrho_a)} \le \left\| \sum_{k=l}^{m} c_k u_k(x+ia) \right\|_{L_p(\mathbb{R}, \varrho)}$$

and $u_k(x+ia) \in \mathcal{L}^{\sigma+}_{p,\varrho}$, $k \in \mathbb{Z}$, $a \in \mathbb{R}$, and applying Proposition 3.10. But then the series (3.31) converges also in the space $L_p(\Gamma, \varrho_a)$, where Γ is an arbitrary smooth curve that lies in the half-plane $\Pi_a = \{\operatorname{Im} z > a\}$ and approaches the line $\mathbb{R}_a$ asymptotically at infinity. Indeed this is a consequence of the boundedness of the operator $S_{\mathbb{R}_a,\Gamma}$ and the estimate

$$\begin{aligned} \left\| \sum_{k=l}^{m} c_k u_k(z) \right\|_{L_p(\Gamma, \varrho_a)} &= \left\| S_{\mathbb{R}_a,\Gamma} \left(\sum_{k=l}^{m} c_k u_k(x+ia) \right) \right\|_{L_p(\Gamma, \varrho_a)} \\ &\le \text{const} \left\| \sum_{k=l}^{m} c_k u_k \right\|_{L_p(\mathbb{R}_a, \varrho_a)}. \end{aligned}$$

By Proposition 3.5, the sum of the series (3.31) belongs to $L_p^+(\Gamma, \varrho_a)$.

3.9 Bases of meromorphic functions

Next we turn our attention to the space $\mathcal{M}(B)$ generated by a Blaschke product B. Let us introduce the following systems of functions:

$$\omega_k(x) = \frac{B(x)}{B'(z_k)(x - z_k)}, \quad k \in \mathbb{Z}, \tag{3.33}$$

$$e_k(x) = \frac{i}{2\pi} \cdot \frac{1}{x - \overline{z}_k}, \quad k \in \mathbb{Z}, \tag{3.34}$$

$$B(z) = \prod_{k \in \mathbb{Z}} \nu_k \frac{z - z_k}{z - \overline{z}_k}, \quad \nu_k = \frac{|1 + z_k^2|}{1 + z_k^2}, \quad z_k \in \Pi_+. \tag{3.35}$$

The zero set $\{z_k\}$ of the function $B(z)$ a priori satisfies the Blaschke condition (1.19). Our immediate goal is to prove that the Carleson condition (3.23), when imposed to this set, is necessary and sufficient for the systems of functions (3.33) and (3.34) to be bases in $\mathcal{M}(B)$.

Let us remark that the systems of functions (3.33) and (3.34) are conjugate to one another. Indeed,

$$\langle e_k, \omega_j \rangle = \int_{\mathbb{R}} \overline{e_k(x)} \omega_j(x) dx = \frac{1}{2\pi} \int_{\mathbb{R}} \frac{B(x) dx}{B'(z_j)(x - z_j)(x - z_k)} = 0$$

whenever $k \neq j$, by Cauchy's theorem, and

$$\langle e_k, \omega_k \rangle = \frac{1}{2\pi i} \cdot \frac{1}{B'(z_k)} \int_{\mathbb{R}} \frac{B(x)dx}{(x - z_k)^2} = \frac{1}{B'(z_k)} \left(\frac{B(x)}{x - z_k} \right) \Big|_{x = z_k} = 1.$$

We will also need the following auxiliary assertion.

Proposition 3.18 *Assume that $f \in L_p^+(\mathbb{R}, \varrho)$, $1 < p < \infty$, and that the Carleson condition* (3.23) *is satisfied. Let*

$$\varphi_k(f) = \int_{\mathbb{R}} \overline{f(x)} \omega_k(x) dx, \quad k \in \mathbb{Z},$$

where the function ω_k is given by (3.33). *Then*

$$\sum_{k \in \mathbb{R}} |\varphi_k(f)|^p \varrho_+(z_k) (\operatorname{Im} z_k)^{1-p} \leq a_p(\Delta) \|f\|^p, \tag{3.36}$$

where $a_p(\Delta)$ is a constant that does not depend on f, and $\varrho_+(z)$ has the form (1.6).

Proof. Since $z_k \in \Pi_+$, setting $g = P^+(\overline{f}B)$ we obtain

$$\varphi_k(f) = \frac{1}{B'(z_k)} \int_{\mathbb{R}} \frac{\overline{f(x)} B(x) dx}{x - z_k} = \frac{1}{B'(z_k)} \int_{\mathbb{R}} \frac{g(x) dx}{x - z_k} = \frac{2\pi i g(z_k)}{B'(z_k)}.$$

We can write $B(x) = \frac{x - z_k}{x - \overline{z}_k} B_k(x)$, where $B_k(x) = B(x) \frac{x - \overline{z}_k}{x - z_k}$. Hence

$$|B'(z_k)| = \left| \frac{B_k(z_k)}{2 \operatorname{Im} z_k} \right| \geq \frac{\Delta}{2 \operatorname{Im} z_k}.$$

By Theorem 3.3,

$$\begin{aligned}
&\sum_{k \in \mathbb{Z}} |\varphi_k(f)|^p \varrho_+(z_k) (\operatorname{Im} z_k)^{1-p} \\
&\quad = (2\pi)^p \sum_{k \in \mathbb{Z}} \left| \frac{g(z_k)}{B'(z_k)} \right|^p \varrho_+(z_k) (\operatorname{Im} z_k)^{1-p} \\
&\quad \leq \left(\frac{4\pi}{\Delta} \right)^p \sum_{k \in \mathbb{Z}} |g(z_k)|^p \varrho_+(z_k) \operatorname{Im} z_k \\
&\quad \leq \left(\frac{4\pi}{\Delta} \right)^p c_p(\Delta) \|g\|^p \leq a_p(\Delta) \|f\|^p.
\end{aligned}$$

□

Notice also that when $f \in L_q^+(\mathbb{R}, \varrho^{1-q})$, where $q = p/(p-1)$, inequality (3.36) becomes

$$\sum_{k \in \mathbb{Z}} |\varphi_k(f)|^q (\varrho_+(z_k))^{1-q} (\operatorname{Im} z_k)^{1-q} \leq a_q(\Delta) \|f\|^q. \tag{3.37}$$

From the set of zeros $\{z_k\}$ of the function $B(z)$ one constructs the discrete weight $\delta = \{\delta_k\}$, where

$$\delta_k = \operatorname{Im} z_k \cdot \varrho_+(z_k) = \operatorname{Im} z_k |z_k + i|^{\alpha_0} \sum_{j=1}^{m} |z_k - x_j|^{\alpha_j}. \tag{3.38}$$

Now we are ready to prove the analogue of Theorem 3.4 for the subspace $\mathcal{M}(B)$ and the set of functions $\{\omega_k\}$ of the form (3.33).

Theorem 3.5 *Suppose that the set of zeros $\{z_k\}$ of a function $B(z)$ of the form (3.35) satisfies the Carleson condition (3.23) and that $\{f_k\} \in l_{p,\delta}(\mathbb{Z})$, where the weight $\delta = \{\delta_k\}_{k\in\mathbb{Z}}$ is given by formula (3.38). Then the series*

$$f(x) = \sum_{k\in\mathbb{Z}} f_k \omega_k(x) \tag{3.39}$$

(1) *converges in the norm of $L_p(\mathbb{R}, \varrho)$,*

(2) *represents a function of class $\mathcal{M}(B)$,*

(3) *effects an isomorphism ($\{f_k\} \mapsto f(x)$) of the space $l_{p,\delta}(\mathbb{Z})$ onto $\mathcal{M}(B)$,*

(4) *solves the interpolation problem $f(z_k) = f_k$, $k \in \mathbb{Z}$, in $L_p^+(\mathbb{R}, \varrho)$.*

Proof. First we repeat the proof of Theorem 3.4, using Proposition 3.18 to show that the sequence $\{F_{l,m}\}$, where $F_{l,m} = \sum_{k=l}^{m} f_k\omega_k$, is fundamental:

$$\begin{aligned}
\|F_{k,m}\| &= \sup_g \left| \sum_{k=l}^{m} f_k \int_{\mathbb{R}} \frac{\overline{g(x)} B(x)dx}{B'(z_k)(x-z_k)} \right| = \sup_g \left| \sum_{k=l}^{m} f_k \varphi_k(g) \right| \\
&\le \sup_g \left[\left(\sum_{k=l}^{m} |f_k|^p \delta_k \right)^{1/p} \left(\sum_{k=l}^{m} |\varphi_k(g)|^q \delta_k^{1-q} \right)^{1/q} \right] \\
&\le a_q^{1/q}(\Delta) \left(\sum_{k=p}^{m} |f_k|^p \delta_k \right)^{1/p}
\end{aligned}$$

thanks to inequality (3.37). This establishes assertions (1) and (2).

To verify assertion (3), let us show that any function $f \in \mathcal{M}(B)$ admits a series expansion of the form (3.39). Let $f \in \mathcal{M}(B)$. By Theorem 3.3, $\{f(z_k)\} \in l_{p,\delta}(\mathbb{Z})$. Denote the sum of the series $\sum_{k\in\mathbb{Z}} f(z_k)\omega_k$ by g. Clearly, $g \in \mathcal{M}(B)$, and so $h = f - g \in \mathcal{M}(B) \subset L_p(\mathbb{R}, \varrho)$. Since $h(z_k) = 0$, $k \in \mathbb{Z}$, Theorem 2.4 for the real line $\mathbb{R}$ shows that $h/B \in L_p^+(\mathbb{R}, \varrho)$. At the same time, by the definition of the subspace $\mathcal{M}(B)$, $h/B \in L_p^-(\mathbb{R}, \varrho)$. We see that $h = 0$ a.e., i.e., $f = g$ a.e.

Moreover, the series (3.39) is the Lagrange series for its sum,

$$f(x) = \sum_{k\in\mathbb{Z}} f(z_k)\omega_k(x),$$

and therefore it solves the interpolation problem of finding a function of class $L_p^+(\mathbb{R}, \varrho)$ (as a matter of fact, that function belongs to the subspace $\mathcal{M}(B)$) whose values on the set $\{z_k\}$ form a prescribed element of $l_{p,\delta}(\mathbb{Z})$. □

If one follows a more general approach to the problems considered above, one can strengthen the main result formulated in Theorem 3.5. First of all, one has the following theorem, which includes the assertion that a system of rational fractions of the form (3.34) is a basis in the subspace $\mathcal{M}(B)$.

Theorem 3.6 *Let the function $B(x)$ have the form* (3.35). *Then the following assertions are equivalent.*

(1) *The system of functions $\{\omega_k\}$ given by* (3.33) *is a basis in the subspace $\mathcal{M}(B)$ of $L_p(\mathbb{R}, \varrho)$.*

(2) *The system of functions $\{e_k\}$ given by* (3.34) *is a basis in the subspace $\mathcal{M}(B)$ of $L_p(\mathbb{R}, \varrho)$.*

(3) *The set of zeros $\{z_k\}$ of the function $B(z)$ satisfies the Carleson condition* (3.23).

Proof. (1) $\Longleftrightarrow$ (2). Recall that $\mathcal{M}(B) = \ker T(\overline{B})$. Denote by $\mathcal{V}$ the isomorphism of the space $\mathcal{M}(B)$ onto itself acting as

$$\mathcal{V}f = B\overline{f} \quad (\mathcal{V}^2 = I).$$

Letting $\nu_k(x) = 1/(x - \overline{z}_k)$ and $\widetilde{\omega}_k(x) = B(x)/(x - z_k)$, we observe that $\nu_k, \widetilde{\omega}_k \in \mathcal{M}(B)$ and $\widetilde{\omega}_k = \mathcal{V}\nu_k$, $k \in \mathbb{Z}$. Therefore, the systems of functions $\{\nu_k\}$ and $\{\widetilde{\omega}_k\}$ are (or are not) bases simultaneously in $\mathcal{M}(B)$. The same is obviously true for the systems $\{\omega_k\}$ and $\{e_k\}$.

The implication (3) $\Longrightarrow$ (1) is a consequence of Theorem 3.5. It remains to verify that (2) $\Longrightarrow$ (3). Suppose that (2) holds. Since any basis is a uniformly minimal system, inequality (3.21) and Proposition 3.16 yield

$$\|e_n\|_{p,\varrho}^{-1}\|\omega_n\|_{q,\varrho^{1-q}}^{-1} = \operatorname{dist}\left(\frac{e_n}{\|e_n\|_{p,\varrho}}, \mathcal{L}_n\right) \geq \Delta > 0,$$

where $n \in \mathbb{Z}$ and $\mathcal{L}_n = \operatorname{clos\,span}\{e_k, k \in \mathbb{Z}, k \neq n\}$ in $L_{p,\varrho}(\mathbb{R})$. Since

$$\|\omega_n\|_{q,\varrho^{1-q}} = \frac{2\pi\|e_n\|_{q,\varrho^{1-q}}}{|B'(z_n)|},$$

we have

$$\frac{|B'(z_n)|}{2\pi}\left(\|e_n\|_{p,\varrho}\|e_n\|_{q,\varrho^{1-q}}\right)^{-1}\geq\Delta>0.$$

By Hölder's inequality,

$$\begin{aligned}(2\pi)^2\|e_n\|_{p,\varrho}\|e_n\|_{q,\varrho^{1-q}} &\geq (2\pi)^2\|e_n\|_2^2\\ &=\int_{-\infty}^{\infty}\frac{dx}{(x-z_n)(x-\overline{z}_n)}=\frac{2\pi i}{z_n-\overline{z}_n}=\frac{\pi}{\operatorname{Im} z_n}.\end{aligned}$$

Observing that

$$|B'(z_n)|=\frac{1}{2\operatorname{Im} z_n}\prod_{j\neq n}\left|\frac{z_n-z_j}{z_n-\overline{z}_j}\right|,$$

we finally conclude that

$$\sup_n\prod_{j\neq n}\left|\frac{z_n-z_j}{z_n-\overline{z}_j}\right|\geq\Delta. \qquad \square$$

The proof of the equivalence (1) $\Longleftrightarrow$ (2) of Theorem 3.6 given above is based on a simple but productive idea which enables us, on the one hand, to consider a generalization of the problem on rational bases and, on the other, to establish yet another connection between this problem and the theory of Toeplitz operators.

Let ν be an arbitrary inner function of class $H^\infty(\Pi_+)$ and let $\{z_k\}$ be the set of zeros of a Blaschke product B of the form (3.35). Let us introduce the projector $P_\nu=\nu P^-\overline{\nu}P^+$ of the space $L_p(\mathbb{R},\varrho)$ onto the model subspace $\mathcal{M}(\nu)$. Then

$$\begin{aligned}P_\nu e_n &= \frac{i}{2\pi}\left(\frac{1}{2}\cdot\frac{1}{x-\overline{z}_n}-\frac{\nu(x)}{2\pi i}\int_{\mathbb{R}}\frac{\overline{\nu(t)}dt}{(t-\overline{z}_n)(t-x)}\right)\\ &= \frac{i}{2\pi}\cdot\frac{1-\overline{\nu(z_n)}\nu(x)}{x-\overline{z}_n}\in\mathcal{M}(\nu).\end{aligned}$$

The function

$$\nu_n(x)=\frac{i}{2\pi}\cdot\frac{1-\overline{\nu(z_n)}\nu(x)}{x-\overline{z}_n} \tag{3.40}$$

is called the *reproducing kernel* of the subspace $\mathcal{M}(\nu)$, because of its reproducing property: for every function $f\in\mathcal{M}(\nu)$,

$$\begin{aligned}\langle f,\nu_z\rangle &= -\frac{i}{2\pi}\int_{\mathbb{R}}f(x)\left(\frac{1-\nu(z)\overline{\nu(x)}}{x-z}\right)dx\\ &= f(z)-\frac{\nu(z)}{2\pi i}\int_{\mathbb{R}}\frac{\overline{\nu(x)}f(x)}{x-z}dx=f(z), \quad z\in\Pi_+,\end{aligned}$$

because $\overline{\nu}f\in L_p^-(\mathbb{R},\varrho)$. The following question arises: when does a system of functions $\{\nu_n\}$, $n\in\mathbb{Z}$, of the form (3.40) constitute a basis in $\mathcal{M}(\nu)$? Notice that if $\nu = B$, then $\nu_n = e_n$ and so the problem of bases of reproducing kernels is a generalization of the problem of bases of rational fractions in the subspace $\mathcal{M}(B)$.

The following result holds.

Theorem 3.7 *Let ν be an inner function of class $H^\infty(\Pi_+)$. Let the Blaschke product B have the form* (3.35) *and let its zeros $\{z_k\}$ satisfy the condition*

$$\sup_n |\nu(z_n)| < 1. \tag{3.41}$$

Then the following assertions are equivalent:

(1) *The system of functions $\{\nu_n\}$ given by* (3.40) *is a basis in the subspace $\mathcal{M}(\nu)$ of $L_p(\mathbb{R}, \varrho)$.*

(2) *The set $\{z_n\}$ satisfies the Carleson condition* (3.23) *and the Toeplitz operator $T(\overline{B}\nu)$ is invertible in $L_p^+(\mathbb{R}, \varrho)$.*

Proof. First of all, let us show that the invertibility of the operator $T(\overline{B}\nu)$ in $L_p^+(\mathbb{R}, \varrho)$ is equivalent to the invertibility of the operator $P_\nu : \mathcal{M}(B) \to \mathcal{M}(\nu)$. Indeed, the operator $\nu P^- \overline{\nu} I = P^- + P_\nu$ is a projector of $L_p(\mathbb{R}, \varrho)$ onto the subspace $L_p^-(\mathbb{R}, \varrho) + \mathcal{M}(\nu)$. Since the projectors P^- and P_ν are orthogonal, the invertibility of the operator $P_\nu : \mathcal{M}(B) \to \mathcal{M}(\nu)$ is equivalent to that of the operator $\nu P^- \overline{\nu} I : L_p^-(\mathbb{R}, \varrho) + \mathcal{M}(B) \to L_p^-(\mathbb{R}, \varrho) + \mathcal{M}(\nu)$. But $L_p^-(\mathbb{R}, \varrho) + \mathcal{M}(B) = BL_p^-(\mathbb{R}, \varrho)$, $L_p^-(\mathbb{R}, \varrho) + \mathcal{M}(\nu) = \nu L_p^-(\mathbb{R}, \varrho)$, and the invertibility of the operator $\nu P \overline{\nu} I : BL_p^-(\mathbb{R}, \varrho) \to L_p^-(\mathbb{R}, \varrho)$ is equivalent to that of the operator $P^- \overline{\nu} BI : L_p^-(\mathbb{R}, \varrho) \to L_p^-(\mathbb{R}, \varrho)$. Introducing the isomorphism $\mathcal{Y} : L_p^-(\mathbb{R}, \varrho) \to L_p^+(\mathbb{R}, \varrho)$, $\mathcal{Y} f = \overline{f}$, we see that

$$P^- \overline{\nu} BI|_{L_p^-} = \mathcal{Y} P^+ \overline{B} \nu \mathcal{Y}|_{L_p^-} = \mathcal{Y} T(\overline{B}\nu) \mathcal{Y}|_{L_p^-},$$

which proves the validity of our assertion.

(1) $\Longrightarrow$ (2). If $\{\nu_n\}$ is a basis in $\mathcal{M}(\nu)$ of the form (3.40), then $\{\nu_n\}$ is a uniformly minimal system and, by Proposition 3.17, $\{e_n\}$ is also uniformly minimal, because $\nu_n = P_\nu e_n$ and

$$\inf_n \frac{\|\nu_n\|}{\|e_n\|} \geq \frac{1}{2\pi} \inf_n (1 - |\nu(z_n)|) \geq \frac{1}{2\pi}(1 - \sup_n |\nu(z_n)|) > 0,$$

thanks to condition (3.41). It follows from the proof of Theorem 3.6 that in this case the set $\{z_n\}$ satisfies the Carleson condition (3.23). It remains to show that $P_\nu : \mathcal{M}(B) \to \mathcal{M}(\nu)$ is an isomorphism. By Theorem 3.6, $\{e_n\}$ is a basis in $\mathcal{M}(B)$, and so the bounded operator P_ν maps a basis of the space $\mathcal{M}(B)$ into a basis of the space $\mathcal{M}(\nu)$. But then P_ν establishes an algebraic isomorphism between $\mathcal{M}(B)$ and $\mathcal{M}(\nu)$ and, by Banach's theorem on the inverse operator, P_ν is an isomorphism of Banach spaces.

(2) $\Longrightarrow$ (1). By Theorem 3.6, $\{e_n\}$ is a basis in $\mathcal{M}(B)$. Since $P_\nu : \mathcal{M}(B) \to \mathcal{M}(\nu)$ is an isomorphism, $\{\nu_n\} = \{P_\nu e_n\}$ is a basis in $\mathcal{M}(\nu)$. □

3.10 Boundary interpolation

Theorem 3.4 is more general than Theorem 3.5 in the sense that it allows interpolation on sets that lie in an arbitrary strip $|\operatorname{Im} z| < h$, and in particular, on subsets of $\mathbb{R}$, which is of special importance in the nonelliptic theory of singular integral equations. At the same time, the zero sets of sine-type functions, which lie on $\mathbb{R}$, are relatively poor. For example, they do not contain all separated sets (Levin and Ostrovskiĭ [1]). However, if $B(z)$ is a Blaschke product with separated set of zeros lying on the line $\operatorname{Im} z = a > 0$, then, first, $B(z)$ admits all complex shifts $B(x) \mapsto B(x - i\alpha)$ with $-\infty < \alpha < a$ and, secondly, the condition that its set of zeros be separated is equivalent to the Carleson condition (3.23). This allows us to use such functions for interpolation on $\mathbb{R}$.

Before stating the corresponding result, let us give a number of estimates for Blaschke products.

Proposition 3.19 *Let $B(z)$ be a Blaschke products with set of zeros $\{z_k\} \subset \Pi_+$. Then*

$$-\log|B(z)|^2 \geq \sum_k \frac{4yy_k}{|z - \overline{z}_k|^2}, \qquad z = x + iy, \quad z_k = x_k + iy_k. \tag{3.42}$$

Conversely, if

$$d = \inf_k \left| \frac{z - z_k}{z - \overline{z}_k} \right| > 0,$$

then, for $y \geq 0$,

$$-\log|B(z)|^2 \leq (1 + 2\log d^{-1}) \sum_k \frac{4yy_k}{|z - \overline{z}_k|^2}. \tag{3.43}$$

A proof can be found in Garnett [1], Ch. VII, 1.

Proposition 3.20 *Suppose that the set of zeros $\{z_j\} = \{x_j + iy_j\}$ of a Blaschke product $B(z)$ of the form* (3.35) *satisfies the conditions*

$$|x_j - x_{j+1}| \geq d > 0, \quad 0 < h_1 \leq y_j \leq h_2 < \infty, \quad j \in \mathbb{Z},$$

and let

$$c_\varepsilon(z_j) = \{z : |z - z_j| \leq \varepsilon\}, \quad 0 < \varepsilon < \min\{d/2, h_1\}.$$

Then the following estimates hold:

$$|B(z)| \leq M_1 \varepsilon^{-1}, \quad z \in \mathbb{C} \setminus \bigcup_{j \in \mathbb{Z}} c_\varepsilon(\overline{z}_j), \tag{3.44}$$

$$|B(z)| \leq M_2 |z - \overline{z}_j|^{-1}, \quad z \in c_\varepsilon(\overline{z}_j), \tag{3.45}$$

$$|B(z)| \geq M_3 \varepsilon, \quad z \in \mathbb{C} \setminus \bigcup_{j \in \mathbb{Z}} c_\varepsilon(z_j), \tag{3.46}$$

where the constants M_1, M_2, M_3 do not dependent on z, ε and j.

Proof. It suffices to verify estimates (3.44) and (3.45) for points $z \in \Pi_-$. Let $z = x + iy$, $y < 0$, and let z_n be the zero that minimizes the distance $|x - x_n|$. Let us estimate the Blaschke product

$$B_n(z) = \frac{z - \overline{z}_n}{z - z_n} \cdot B(z). \tag{3.47}$$

By inequality (3.42),

$$\begin{aligned} |B_n(z)| &\leq \exp\left\{2\sum_{j\neq n} \frac{-yy_j}{|z - \overline{z}_j|^2}\right\} = \exp\left\{2\sum_{j\neq n} \frac{-yy_j}{(x - x_j)^2 + (y + y_j)^2}\right\} \\ &\leq \exp\left\{2\sum_{j\neq n} \frac{-yy_j}{\frac{d^2(n-j)^2}{4} + (y + y_j)^2}\right\}. \end{aligned}$$

If $-2h_2 \leq y < 0$, then

$$|B_n(z)| \leq \exp\left\{\frac{16h_2^2}{d^2}\sum_{j\neq n} \frac{1}{(n-j)^2}\right\} \leq \exp\left\{\frac{32h_2^2}{d^2}\sum_{j=1}^{\infty} \frac{1}{j^2}\right\} < \text{const},$$

uniformly in $n \in \mathbb{Z}$. If, however, $y < -2h_2$, then

$$\begin{aligned} &\sum_{j\neq n} \frac{-2yy_j}{\frac{d^2(n-j)^2}{4} + (y + y_j)^2} \leq 16h_2 \sum_{j=1}^{\infty} \frac{-y}{d^2j^2 + 4(y + h_2)^2} \\ &\leq 16h_2 \int_0^{\infty} \frac{-y\,dt}{d^2t^2 + 4(y + h_2)^2} = \frac{8h_2|y|}{d(y + h_2)} \int_0^{\infty} \frac{d\tau}{1 + \tau^2} < \text{const}, \end{aligned}$$

which also yields an estimate $|B_n(z)| < \text{const}$ that is uniform for all $n \in \mathbb{Z}$. Now inequalities (3.44) and (3.45) follow from (3.47). To prove inequality (3.46) we can use estimate (3.43) and repeat the preceding arguments. □

Let $h > 0$. Consider a Blaschke product $B(z)$ of the form (2.28) whose set of zeros is $\{z_k\} = \{x_k + i(2h)\}$, $x_k \in \mathbb{R}$, and satisfies the separation condition (3.24):

$$B(z) = \prod_{k\in\mathbb{Z}} \frac{\overline{z}_k}{z_k} \cdot \frac{z - z_k}{z - \overline{z}_k}, \quad z_k = x_k + i(2h). \tag{3.48}$$

Let us also introduce the Blaschke product

$$A(z) = \prod_{k\in\mathbb{Z}} \frac{\overline{\xi}_k}{\xi_k} \cdot \frac{z - \xi_k}{z - \overline{\xi}_k}, \quad \xi_k = x_k + ih, \tag{3.49}$$

of class $H^\infty(\Pi_+)$. Consider the function

$$a(z) = A(z + ih) = \prod_{k\in\mathbb{Z}} \frac{\overline{\xi}_k}{\xi_k} \cdot \frac{z - x_k}{z - \overline{z}_k}, \tag{3.50}$$

which is analytic in the half-plane $\operatorname{Im} z > -h$ and bounded in the half-plane $\operatorname{Im} z \geq -h$, $|a(z)| \leq 1$.

Next, let us introduce the system of meromorphic functions

$$a_k(z) = \frac{a(z)}{a'(x_k)(z - x_k)}, \qquad k \in \mathbb{Z}, \tag{3.51}$$

and show that analogues of Theorems 3.4 and 3.5 hold for it.

Theorem 3.8 *Let a function $B(z)$ of the form* (3.48) *have a separated set of zeros. If $f \in \mathcal{M}(B)$ in $L_p^+(\mathbb{R}, \varrho)$, then*

(1) *$\{f(x_k)\} \in l_{p,\delta}(\mathbb{Z})$, where the weight $\delta = \{\delta_k\}$ has the form* (3.38)*;*

(2) *$f(x) = \sum_{k\in\mathbb{Z}} f(x_k)a_k(x)$, where the function $a_k(x)$ is defined by equality* (3.51) *and the series converges in $L_p(\mathbb{R}, \varrho)$;*

(3) *the mapping defined in* (1) *and* (2) *establishes an isomorphism of the subspace $\mathcal{M}(B)$ onto the subspace $l_{p,\delta}(\mathbb{Z})$.*

Proof. The proof of these assertions is based on Theorems 3.5 and 3.6. First let us show that the model subspace $\mathcal{M}(B)$ generated by a Blaschke product of the form (3.48) depends in a certain sense only on its poles. This subspace is invariant under a certain class of perturbations of the zeros of $B(z)$, generated by complex-shift operators that keep its poles fixed. Using appropriate complex shifts, the main conclusions of the present theorem follow from the fact that isomorphisms take bases into bases.

The set of zeros $\{z_k\} = \{x_k + i(2h)\}$ of the function $B(z)$ satisfies the Carleson condition (3.23). Hence, by Theorem 3.6, the system of functions

$$e_k(x) = \frac{1}{x - x_k + i(2h)}, \qquad k \in \mathbb{Z},$$

is a basis in the subspace $\mathcal{M}(B)$ of $L_p^+(\mathbb{R}, \varrho)$. By Proposition 3.19, the functions $B(z)$ and $B^{-1}(z)$ are bounded in the strip $-h \leq \operatorname{Im} z \leq 0$. Consequently, the model function $B(x)$ admits the complex shift I_h, and $B(x - ih)$ is a model function in $L_p^+(\mathbb{R}, \varrho)$. By Proposition 3.10, the operator I_h, given by $(I_h f)(x) = f(x - ih)$, effects an isomorphism of the subspace $\mathcal{M}(B)$ onto the subspace $\mathcal{M}(I_h B)$.

It follows that the system of functions

$$e_k(x - ih) = \frac{1}{x - x_k + ih}, \qquad k \in \mathbb{Z},$$

is a basis in the space $\mathcal{M}(I_h B)$. But, by Theorem 3.6, the model subspace $\mathcal{M}(A)$, where the function A is given by (3.49), has exactly the same basis. We conclude that in $L_p^+(\mathbb{R}, \varrho)$ the model subspaces $\mathcal{M}(A)$ and $\mathcal{M}(I_h B)$ coincide.

Thus, the operator I_h establishes an isomorphism $\mathcal{M}(B) \to \mathcal{M}(A)$. But then the operator I_{-h}, the inverse of I_h, establishes an isomorphism $\mathcal{M}(A) \to \mathcal{M}(B)$. By Theorem 3.5, $g(x) \in \mathcal{M}(A)$ if and only if

$$g(x) = \sum_{k\in\mathbb{Z}} g(\xi_k)\omega_k(x) \quad \text{and} \quad \|g\| \asymp \|\{g(\xi_k)\}\|.$$

(Here, as usual, $\|g\| \asymp \|\{g(\xi_k\}\|$ means that there exist positive constants c_1 and c_2 such that $c_1\|g\| \le \|\{g(\xi_k)\}\| \le c_2\|g\|$.) Applying the operator I_{-h}, we conclude that $f(x) = g(x + ih) \in \mathcal{M}(B)$ if and only if

$$f(x) = \sum_{k\in\mathbb{Z}} f(\xi_k - ih)\omega_k(x + ih) \quad \text{and} \quad \|f\| \asymp \|\{f(\xi_k - ih)\}\|.$$

It remains to observe that $f(\xi_k - ih) = f(x_k)$, $\omega_k(x + ih) = a_k(x)$. □

References and comments

3.1–3.4 According to Beurling's theorem, in the case where $\Gamma = \Gamma_0$ and $h(t)$ is an inner function, the model subspaces are the shift-invariant subspaces. An extensive literature is devoted to shift-invariant subspaces; see, e.g., Nikol′skiĭ [1], Nakazi [2], Aleman, Richter and Sundberg [1].

In the theory of singular integral equations with infinite index, model subspaces appeared for the first time in 1976–77 in the works of Dybin [4] and Grudsky and Dybin [1]. From that moment on the successive development of results, as presented in the sections of this chapter, was achieved in a series of papers of the authors (see, e.g., Grudsky [4], [5] and Dybin [6], [13]).

Theorem 3.1 is due to Grudsky [4].

Concerning the classes $\mathcal{L}_p^\sigma$, see Akhiezer [1], Dzhrbashyan and Rafaelyan [1], Grozev and Rahman [1].

Proposition 3.14 is due to Dybin [4], [13]. Proposition 3.15 is due to Grudsky (see Grudsky and Dybin [2]). Theorem 3.2 belongs to Grudsky [4].

3.5–3.10 Proposition 3.17 is borrowed from the work of Hrushchev, Nikol′skiĭ and Pavlov [1] (see p. 228).

Sine-type functions were introduced by Levin [1] (see also Levin and Ostrovskiĭ [1]). Theorem 3.4 was proved by Levin [1] for the case $\varrho \equiv 1$, by Dzhrbashyan and Rafaelyan [1] for the case $\varrho(x) = |x + i|^\alpha$, and by Dybin [13] for the case $\varrho \in (A_p)$. Proposition 3.18 and Theorem 3.5 for $\varrho \equiv 1$ were proved by Dzhrbashyan [1], [2] (see also Tumarkin [1]). The case of power weights was considered by Martirosyan [1], [2]. The case of weights $\varrho \in (A_p)$ was studied by Dybin [14].

Theorem 3.6 was proved by Dybin [14] and was inspired by the paper of Khrushchev, Nikol′skiĭ and Pavlov [1]. Theorem 3.7 is essentially contained in the second part of that paper.

Proposition 3.19 is taken from the book of Garnett [1], Chap. VII, 1. Proposition 3.20 is due to Grudsky, while Theorem 3.8 was obtained by Dybin [14].

Chapter 4

Toeplitz Operators with Oscillating Symbols

In this chapter we consider Toeplitz operators whose symbols possess the best known kinds of discontinuities of oscillatory type: almost periodic (a.p.) discontinuities, semi-almost periodic discontinuities, and whirl points of power type. For each of these cases the central problem is the construction of a generalized factorization (2.30) of the symbol, which enables us to apply Theorem 2.6.

Almost periodic discontinuities

As already mentioned in the Introduction, for us the case of almost periodic discontinuities is the standard model for studying Toeplitz operators with infinite index. The simplest model operator for this type of discontinuities is the operator $T(\exp(-i\sigma x))$, $\sigma > 0$, on the real line $\mathbb{R}$, and the model subspace that it generates coincides with $\mathcal{L}^{\sigma+}_{p,\varrho}$. These objects were studied in Section 3.3. Here, following the plan outlined above, we first consider Toeplitz operators on $\mathbb{R}$ with uniformly almost periodic symbols. After that we describe the model subspaces on bounded curves corresponding to this type of discontinuities. Further, we define the class of standard almost periodic symbols. The Toeplitz operators whose symbols have this type of discontinuity are investigated in the maximal depth allowed by the results of Chapters 2 and 3. In addition, for Toeplitz equations of this type we give versions of well-posed problems, in which the existence and uniqueness of the solution together with continuous dependence on the data are guaranteed. In conclusion, we consider Toeplitz operators whose symbols have general a.p. discontinuities.

4.1 Uniformly almost periodic functions

A function $f(x)$, defined and continuous on the real line $\mathbb{R}$, is called a *uniformly almost periodic function* if for every $\varepsilon > 0$ there exists a number $l = l(\varepsilon)$ such that in every interval of length l there is at least one number τ for which

$$|f(x+\tau) - f(x)| < \varepsilon$$

for all $x \in \mathbb{R}$. The set of all uniformly almost periodic functions will be denoted by $\Pi_c(\mathbb{R})$. The class $\Pi_c(\mathbb{R})$ admits the following alternative description. Recall that a function of the form

$$\alpha(x) = \sum_{j=1}^{n} c_j \exp(i\sigma_j x),$$

where $\sigma_j \in \mathbb{R}$ and $c_j \in \mathbb{C}$, is called an *almost periodic polynomial.*

Theorem 4.1 *The closure of the set of all almost periodic polynomials in the norm of the space $L_\infty(\mathbb{R})$ coincides with $\Pi_c(\mathbb{R})$. The class $\Pi_c(\mathbb{R})$ is an inverse closed subalgebra of the algebra $L_\infty(\mathbb{R})$.*

Let us denote by $\Pi_w(\mathbb{R})$ the subalgebra of $\Pi_c(\mathbb{R})$ consisting of all functions of the form

$$\alpha(x) = \sum_{j=1}^{\infty} c_j \exp(i\sigma_j x), \quad \text{where } \sigma_j \in \mathbb{R} \quad \text{and} \quad \sum_{j=1}^{\infty} |c_j| < \infty. \tag{4.1}$$

In addition to Theorem 4.1, we will need the following results, also well known in the theory of almost periodic functions.

Theorem 4.2 *Suppose that $a(x) \in \Pi_c(\mathbb{R})$ and*

$$\inf_{x\in\mathbb{R}} |a(x)| > 0. \tag{4.2}$$

Then the function $\arg a(x)$ can be defined so that

$$\arg a(x) = \sigma x + \psi(x), \tag{4.3}$$

where $\sigma \in \mathbb{R}$ and $\psi(x) \in \Pi_c(\mathbb{R})$.

Theorem 4.3 *Let $a(x) \in \Pi_w(\mathbb{R})$ and let the function $F(z)$ be analytic on the closure of the range of the function $a(x)$. Then $F(a(x)) \in \Pi_w(\mathbb{R})$.*

Let $a(x) \in \Pi_c(\mathbb{R})$ and let condition (4.2) be satisfied. The *almost periodic index* of the function $a(x)$ is defined to be the real number

$$\sigma(a) = \lim_{l\to\infty} \frac{1}{2l} \arg a(x)\Big|_{-l}^{l}. \tag{4.4}$$

By Theorem 4.2, this limit exists and is equal to the number σ that appears in formula (4.3). We note that the almost periodic index has the same properties as the usual p-factorization index. Namely, it is stable under small perturbations of the function $a(x)$ in the metric of $L_\infty(\mathbb{R})$ and whenever a_1 and a_2 in $\Pi_c(\mathbb{R})$ satisfy the condition (4.2), one has

$$\sigma(a_1 a_2) = \sigma(a_1) + \sigma(a_2). \tag{4.5}$$

Let $\Pi_w^+(\mathbb{R})$ (resp., $\Pi_w^-(\mathbb{R})$) denote the subalgebra of $\Pi_w(\mathbb{R})$ consisting of the functions of the form (4.1) for which all numbers σ_j are nonnegative (resp., nonpositive). Notice that $\Pi_w^+(\mathbb{R}) \subset L_\infty^+(\mathbb{R})$ (resp., $\Pi_w^-(\mathbb{R}) \subset L_\infty^-(\mathbb{R})$).

We now turn to the construction of a generalized factorization for functions of class $\Pi_c(\mathbb{R})$. As a preliminary step we will establish the following important auxiliary result.

Theorem 4.4 *Suppose that $a \in \Pi_w(\mathbb{R})$, condition (4.2) is satisfied, and $\sigma(a) = 0$. Then $\log a(x) \in \Pi_w(\mathbb{R})$.*

Proof. Let us observe at the outset that from Theorem 4.3 it follows that $|a| = (a\bar{a})^{1/2} \in \Pi_w(\mathbb{R})$. Further, equality (4.3) yields

$$a = |a| \exp(i\psi).$$

The function ψ lies in $\Pi_c(\mathbb{R})$ and hence, by Theorem 4.1, it can be written as $\psi = \psi_1 + \psi_2$, where ψ_1 is an almost periodic polynomial and $\psi_2 \in \Pi_c(\mathbb{R})$ has a sufficiently small norm. Applying again Theorem 4.3, we have $|a|^{-1} \in \Pi_w(\mathbb{R})$ and $\exp(-i\psi_1) \in \Pi_w(\mathbb{R})$. But then

$$\exp(i\psi_2) = \frac{a}{|a|} \exp(-i\psi_1) \in \Pi_w(\mathbb{R})$$

and so, by the same Theorem 4.3, $\log(\exp(i\psi_2)) = i\psi_2 \in \Pi_w(\mathbb{R})$. Therefore $\psi \in \Pi_w(\mathbb{R})$, whence

$$\log a = \log |a| + i\psi \in \Pi_w(\mathbb{R}).$$

□

Any function $a \in \Pi_w(\mathbb{R})$ that satisfies condition (4.2) admits a generalized factorization in the space $L_p(\mathbb{R}, \varrho)$ with infinite index with the function $\exp(i\sigma x)$ as the factor h (see Section 2.7).

Theorem 4.5 *Let $a(x) \in \Pi_w(\mathbb{R})$ and let condition (4.2) be satisfied. Then*

$$a(x) = a^-(x) \exp(i\sigma x) a^+(x), \tag{4.6}$$

where $\sigma = \sigma(a)$, $(a^+)^{\pm 1} \in \Pi_w^+(\mathbb{R})$, $(a^-)^{\pm 1} \in \Pi_w^-(\mathbb{R})$.

Proof. Let us introduce the function $b(x) = \exp(-i\sigma x)a(x)$. By (4.5), $\sigma(b) = 0$. But then, by Theorem 4.4, $\log b \in \Pi_{\mathrm{w}}(\mathbb{R})$, whence

$$\log b(x) = \sum_{j=1}^{\infty} d_j \exp(i\gamma_j x), \quad \gamma_j \in \mathbb{R}, \quad \sum_{j=1}^{\infty} |d_j| < \infty.$$

Let

$$\psi^+(x) = \sum_{\gamma_j \geq 0} d_j \exp(i\gamma_j x), \quad \psi^-(x) = \sum_{\gamma_j < 0} d_j \exp(i\gamma_j x).$$

Denoting

$$a^{\pm}(x) = \exp(\psi^{\pm}(x)),$$

we obtain the factorization (4.6). Indeed, the functions $\psi^{\pm}(z)$ are analytic and bounded in the respective half-planes $\Pi_{\pm}$. By Theorem 4.3, $a^{\pm} \in \Pi_{\mathrm{w}}(\mathbb{R})$. But the functions $a^{\pm}(z)$ are also analytic and bounded in $\Pi_{\pm}$, respectively, and therefore their series expansions of the form (4.1) contain no terms with negative, respectively positive exponents σ_j, i.e., $a^{\pm} \in \Pi_{\mathrm{w}}^{\pm}(\mathbb{R})$. Similarly, $(a^{\pm})^{-1} = \exp(-\psi^{\pm}) \in \Pi_{\mathrm{w}}^{\pm}(\mathbb{R})$. Equality (4.6) is obvious. □

Already at this point we could apply Theorem 2.6 to construct a theory of one-sided invertibility of the operator $T(a)$ in the case when $a \in \Pi_{\mathrm{w}}(\mathbb{R})$. However, we shall do that directly for functions of class $\Pi_{\mathrm{c}}(\mathbb{R})$.

We say that a function $d \in L_\infty(\mathbb{R})$ *is a multiplier that does not affect factorizability* (MNAF, for short) if $a \in \mathrm{fact}(p, \varrho)$ implies $da \in \mathrm{fact}(p, \varrho)$. Clearly, the functions $d^{\pm}$ satisfying conditions (1.13) are MNAFs.

Theorem 4.6 *Let $d \in L_\infty(\mathbb{R})$. Then the following assertions are equivalent:*

(1) *d is an* MNAF*;*

(2) *$d^{-1} \in L_\infty(\mathbb{R})$ and there exists a sequence of functions $\{d_j\}_{j=1}^{\infty}$ admitting a standard factorization, such that*

$$\lim_{j\to\infty} \|d_j - d\|_\infty = 0;$$

(3) *$d \in \mathrm{fact}(p, \varrho)$ for all p, $1 < p < \infty$, and all weights ϱ of the form* (1.1) *with the same factorization index.*

Let us state the following corollary of Theorems 2.1 and 4.6 that will be needed below.

Proposition 4.1 *If $d \in GC(\mathbb{R})$, then d is an* MNAF.

Theorem 4.7 *Let $a \in \Pi_c(\mathbb{R})$ and let condition (4.2) be satisfied. Then the function a admits a generalized factorization with infinite index in each space $L_p(\mathbb{R}, \varrho)$, $1 < p < \infty$. This factorization has the form*

$$a(x) = b^-(x)\exp(i\sigma x)b^+(x), \quad \sigma = \sigma(a), \tag{4.7}$$

where $b = a\exp(-i\sigma x) \in \operatorname{fact}(p, \varrho)$ and $b = b^+b^-$ is a (p, ϱ)-factorization of the function b with index $\kappa = 0$.

Moreover, if $\sigma = 0$, then the operator $T(a)$ is invertible $L_p^+(\mathbb{R}, \varrho)$ and

$$[T(a)]^{-1} = \frac{1}{b^+}P^+\frac{1}{b^-}I.$$

If $\sigma > 0$, then $T(a)$ is left invertible and $\dim\operatorname{coker} T(a) = \infty$, one of its left inverses has the form

$$[T(a)]^{-1} = P^+\frac{1}{b^+\exp(i\sigma x)}P^+\frac{1}{b^-}I,$$

and a function f belongs to $\operatorname{im} T(a)$ if and only if it satisfies the orthogonality condition

$$\int_{\mathbb{R}} \frac{1}{b^-(x)}\left(P^-\frac{1}{b^+(\tau)}\cdot\frac{\exp(-i\sigma\tau)}{(\tau+i)^n}\right)(x)f(x)dx = 0, \quad n \in \mathbb{N}. \tag{4.8}$$

If $\sigma < 0$, then $T(a)$ is right invertible, one of its right inverses has the form

$$[T(a)]^{-1} = \frac{1}{b^+}P^+\frac{1}{\exp(i\sigma x)b^-}I,$$

and a function f belongs to $\ker T(a)$ if and only if

$$f = \frac{1}{b^+}P^+\frac{\psi}{b^-}, \quad \textit{where } \psi \in \mathcal{L}^{|\sigma|+}_{p,\varrho}.$$

Proof. Let us show that $b = a\exp(-i\sigma x) \in \operatorname{fact}(p, \varrho)$, $1 < p < \infty$. To this end, let us pick a sequence $\{b_n\} \subset \Pi_w(\mathbb{R})$ such that, for all n, $\sigma(b_n) = 0$,

$$\inf_{x\in\mathbb{R}} |b_n(x)| > \frac{1}{2}\inf_{x\in\mathbb{R}} |b(x)| \quad \text{and} \quad \lim_{n\to\infty} \|b(x) - b_n(x)\|_\infty = 0.$$

By Theorem 4.5, each of the functions b_n admits a standard factorization with index zero. Hence, by Theorem 4.7, b is an MNAF, that is, $b \in \operatorname{fact}(p, \varrho)$, $1 < p < \infty$. This establishes representation (4.7).

All the assertions concerning the invertibility of the operator $T(a)$ are consequences of Theorem 2.6. The description of the subspace $\operatorname{im} T(a)$ in the case $\sigma > 0$ is a consequence of Proposition 1.1. Indeed, $f \in \operatorname{im} T(a)$ if and only if $f = P^+b^+b^-\exp(i\sigma x)\varphi$, where $\varphi \in L_p^+(\Gamma, \varrho)$. This is equivalent to

$$\varphi = \frac{1}{\exp(i\sigma x)b^+}P^+\frac{f}{b^-} \in L_p^+(\mathbb{R}, \varrho),$$

i.e., to (4.8).

The description of the subspace $\ker T(a)$ in the case $\sigma < 0$ is a corollary of Proposition 3.14. Indeed, $f \in \ker T(a)$ if and only if $P^+b^+b^-\exp(i\sigma x)f = 0$ or if and only if $P^+\exp(i\sigma x)P^+b^-b^+f = 0$. The latter means that $P^+b^-b^+f \in \ker T(\exp(i\sigma x))$, i.e.,

$$f = \frac{1}{b^+}P^+\frac{\psi}{b^-}, \qquad \text{where} \qquad \psi \in \mathcal{L}_{p,\varrho}^{|\sigma|+}. \qquad \square$$

The results of the preceding chapter concerning the description of bases and biorthogonal systems in $\mathcal{L}_{p,\varrho}^{|\sigma|}$ (see Section 3.8) allow us to give a more precise description of the subspace $\ker T(a)$ when $\sigma < 0$ and simpler (compared to (4.8)) conditions for membership in the subspace $\operatorname{im} T(a)$ for $\sigma > 0$. We will return to this question later, after the study of the analogues of the subspace $\mathcal{L}_{p,\varrho}^{|\sigma|+}$ on bounded curves.

4.2 Model subspaces on bounded smooth curves

If we apply a fractional linear transformation of the form (2.4) to the model function $\exp(i\sigma x)$, $\sigma \in \mathbb{R}$, we arrive at the function

$$h_0(t) = \exp(a_0(t - t_0)^{-1}), \quad a_0 \in \mathbb{C} \setminus \{0\}, \tag{4.9}$$

which will be considered on a bounded smooth closed contour Γ ($t_0 \in \Gamma$). First, following Sections 3.1 and 3.2, we will describe the curves Γ on which $h_0(t)$ is a model function, and then we will describe the class $\mathcal{T}(h_0, \Gamma)$ of admissible deformations of Γ.

Let $l(t)$ denote the tangent to the curve Γ at the point t and $\eta_\Gamma(t)$ denote the magnitude of the angle made by the outward normal to Γ at t and the positively oriented real axis ($0 \le \eta_\Gamma(t) < 2\pi$). Further, denote by $O_+(t)$ and $O_-(t)$ the families of circles passing through the point t and centered at points on the inward (O_+) and respectively outward (O_-) normal to Γ at t.

Proposition 4.2 *Let Γ be a simple closed smooth contour and let $t_0 \in \Gamma$. Then the following assertions are equivalent:*

(1) *the function $h_0(t)$ given by (4.9) is a model function on Γ;*

(2) $\arg a_0 = \eta_\Gamma(t_0)$ *and*

$$|\eta_\Gamma(t) - \eta_\Gamma(t_0)| \le \text{const}\,|t - t_0|, \quad t \in \Gamma; \tag{4.10}$$

(3) $\arg a_0 = \eta_\Gamma(t_0)$ *and there exists a pair of circles $\gamma_\pm \in O_\pm(t_0)$ with the property that $\mathcal{D}_{\gamma_\pm}^+ \subset \mathcal{D}_\Gamma^\pm$;*

(4) $\arg a_0 = \eta_\Gamma(t_0)$ *and*

$$d(t, l(t_0)) \le \text{const}\,|t - t_0|^2, \quad t \in \Gamma. \tag{4.11}$$

Proof. Assertion (1) is equivalent to the statement that function $h_0(z)$ is bounded on the set $\overline{\mathcal{D}}_\Gamma^+$ and bounded away from zero on Γ. Let $z - t_0 = r(z)\exp(i\theta(z))$, where $r(z) = |z - t_0|$, $\theta(z) = \arg(z - t_0)$. Then

$$\begin{aligned}|h_0(z)| &= \left|\exp\left(\frac{|a_0|}{r(z)}\exp(i(\arg a_0 - \theta(z)))\right)\right| \\ &= \exp\left(\frac{|a_0|}{r(z)}\cos(\arg a_0 - \theta(z))\right).\end{aligned}$$

The function $h_0(z)$ is continuous on $\mathbb{C}\setminus\{t_0\}$. Hence, if it belongs to $L_\infty^+(\Gamma)$, then $|h_0(z)|$ is bounded on any ray emanating from t_0 and lying in the half-plane that is bounded by the tangent $l(t_0)$ and contains the inward normal to Γ at t_0. The latter is equivalent to the inequality $\cos(\arg a_0 - \theta(z)) \le 0$ being satisfied for $\pi/2 \le \eta_\Gamma(t_0) - \theta(z) \le 3\pi/2$. But that is possible only when $\arg a_0 = \eta_\Gamma(t_0)$.

Once this condition is satisfied, we have $h_0 \in L_\infty^+(\Gamma)$, and the fact that $h_0(t)$ is bounded away from zero on Γ is equivalent to the inequality

$$\left|\frac{|a_0|}{r(t)}\cos(\eta_\Gamma(t_0) - \theta(t))\right| \le \text{const}, \quad t \in \Gamma,$$

or to the inequality

$$|\cos(\eta_\Gamma(t_0) - \theta(t))| \le \text{const}\,|t - t_0|, \quad t \in \Gamma. \tag{4.12}$$

Observing that $d(t, l(t_0)) = r(t)|\cos(\eta_\Gamma(t_0) - \theta(t))|$ and[1]

$$\begin{aligned}|\eta_\Gamma(t_0) - \eta_\Gamma(t)| &\asymp |\sin(\eta_\Gamma(t_0) - \eta_\Gamma(t))| \asymp |\cos(\pi/2 + \eta_\Gamma(t) - \eta_\Gamma(t_0))| \\ &\asymp |\cos(\theta(t) - \eta_\Gamma(t_0))|,\end{aligned}$$

we deduce that (4.12) $\Longleftrightarrow$ (4.10) $\Longleftrightarrow$ (4.11).

The fractional linear transformation

$$z = \nu(\zeta) = \frac{\exp(i\eta_\Gamma(t_0))}{i(\zeta - t_0)}$$

takes the point t_0 to ∞ and takes the model function $h_0(t)$ on the curve Γ into the model function $h(z) = \exp(i|a_0|z)$ on the curve $\gamma = \nu(\Gamma)$. But the functions $\exp(\pm i|a_0|z)$ are bounded on γ if and only if γ lies in a strip parallel to $\mathbb{R}$. This shows that (1) $\Longleftrightarrow$ (3). □

Remark 4.1 From the proof of Proposition 4.2 it follows that if a function h_0 of the form (4.9) is a model function on the curve Γ, then $h_0^{-1} \in L_\infty^-(\Gamma)$, i.e., $h_0 \in \mathcal{U}$.

[1] Here the symbol $\asymp$ ($f(t) \asymp g(t)$) denotes that there exist positive constants c_1 and c_2 such that $c_1|f(t)| \le |g(t)| \le c_2|f(t)|$, for any $t \in \Gamma$.

Proposition 4.3 *Under the assumptions of Proposition* 4.2, *let the function* $h_0(t)$ *given by* (4.9) *be a model function on* Γ. *Then* $\gamma \in \mathcal{T}(h_0, \Gamma)$ *if and only if the function* $h_0(z)$ *is a model function on* γ.

Proof. This proposition is an obvious consequence of the definition of the class $\mathcal{T}(h_0, \Gamma)$ (see Section 3.2). □

Proposition 4.4 *Under the assumptions of Proposition* 4.2, *let the function* $h_0(t)$ *given by* (4.9) *be a model function on* Γ. *Then*

(1) $O_+(t_0) \subset \mathcal{T}(h_0, \Gamma)$;

(2) *in the complex plane* $\mathbb{C}$ *there exist a neighborhood* v *of the point* t_0 *and a pair of curves* $\gamma_\pm$, *such that* $\gamma_\pm \in \mathcal{T}(h_0, \Gamma)$, $(v \cap \gamma_\pm) \subset (v \cap \overline{\mathcal{D}_\Gamma^\pm})$ *and the following inequalities hold:*

$$d(t, \gamma_\pm) \geq \text{const}\, |t - t_0|^2, \quad t \in v \cap \Gamma. \tag{4.13}$$

Proof. This proposition is an elementary corollary of assertion (3) of Proposition 4.2.

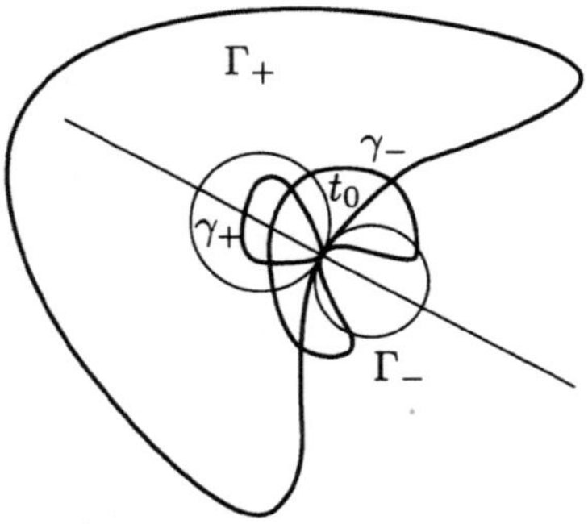

Figure 4.1

First, according to that assertion, choose circles $\Gamma_\pm \in O_\pm(t_0)$ such that $\mathcal{D}_{\Gamma_\pm}^+ \subset \mathcal{D}_\Gamma^\pm$. Next, choose curves $\gamma_\pm$, $\gamma_\pm \in \mathcal{T}(h_0, \Gamma)$, and a neighborhood v, $t_0 \in v$, such that $\gamma_+ \subset \mathcal{D}_{\Gamma_+}^+$ and $d(t, \gamma_+) \geq \text{const}\, |t - t_0|^2$, $t \in v \cap \Gamma$, while $\gamma_- \subset v \cap \mathcal{D}_{\Gamma_-}^+$ and $d(t, \gamma_-) \geq \text{const}\, |t - t_0|^2$, $t \in v \cap \Gamma$ (see Figure 4.1.)

For example, $\gamma_\pm$ can be selected not only as shown in the figure, but also so that $\gamma_+ = \Gamma_+$ and γ_- coincides with Γ_- in some neighborhood of t_0. Clearly, the curves $\gamma_\pm$ satisfy inequality (4.13). □

Remark 4.2 The presence of circles among the (h_0, Γ)-admissible curves allows us to use Theorem 3.1 to describe the model subspace $\mathcal{M}(h_0)$, and then also the apparatus developed in Sections 3.8 and 3.9, which allows us to represent the elements of $\mathcal{M}(h_0)$ in terms of their decomposition with respect of the corresponding analytic bases.

The existence of curves $\gamma_\pm$ satisfying condition (4.13) allows us to apply Theorem 3.2 and to describe the subspace $\mathcal{M}(h_0)$ in terms of the boundary behavior of its elements.

Now let Γ be a simple closed smooth contour, $\{t_1, \dots, t_n\} \subset \Gamma$, and let the following conditions be satisfied:

$$|\eta_\Gamma(t) - \eta_\Gamma(t_j)| \leq \text{const}\,|t - t_j|, \quad t \in \Gamma, \quad j = 1, 2, \dots, n. \tag{4.14}$$

Further, suppose that on Γ we are given the function

$$h(t) = \prod_{j=1}^{n} h_j(t), \tag{4.15}$$

where

$$h_j(t) = \exp(a_j(t - t_j)^{-1}), \tag{4.16}$$

and

$$\arg a_j = \eta_\Gamma(t_j), \quad j = 1, 2, \dots, n. \tag{4.17}$$

On Γ, consider the space $L_p(\Gamma, \varrho)$ with a weight ϱ of the form (2.1), where $m \geq n$.

The next theorem is the main result of this section and provides a description of the model subspace $\mathcal{M}(h)$.

Theorem 4.8 *If on a simple smooth closed contour Γ, containing the points $t_1, \dots, t_n$, conditions* (4.14) *and* (4.17) *are satisfied, then every function h of the form* (4.15), (4.16) *is a model function for Γ. Moreover, the following assertions are equivalent.*

(1) $f \in \mathcal{M}(h)$ *in the space* $L_p^+(\Gamma, \varrho)$.

(2) $f \in \ker T(h^{-1})$ *in the space* $L_p^+(\Gamma, \varrho)$.

(3) *The function f has the form*

$$f = \prod_{j=1}^{n} \left(\prod_{s=1}^{j-1} h_s \right) \sum_{k \in \mathbb{Z}} c_{jk} e_{jk} \qquad \left(\prod_{s=1}^{0} = 1 \right), \tag{4.18}$$

where

$$e_{jk}(t) = b_{jk} \frac{\exp(|a_j|\tau_j^{-1} + a_j(t - t_j)^{-1}) - 1}{t - \tau_{jk}}, \tag{4.19}$$

$b_{jk} = -|a_j|(i|a_j| + 2\pi k r_j)^{-1}$, $\tau_{jk} = t_j - ia_j(i|a_j|r_j^{-1} + 2\pi k)^{-1}$, r_j *is some fixed positive number, and*

$$\{c_{jk}\} \in l_{p,\delta_j}(\mathbb{Z}), \quad \text{where } \delta_j = \{(|k|+1)^{p-\alpha_j-2}\}_{k\in\mathbb{Z}}. \tag{4.20}$$

(4) *The function f has the form*

$$f = \prod_{j=1}^{n} f_j, \tag{4.21}$$

where $f_j \in L_p^+(\Gamma, \varrho)$, and f_j is either a constant or admits an analytic continuation to the whole extended complex plane, except for the essentially singular point t_j, in the neighborhood of which one has the estimates

$$|f_j(z)| \leq O_j(z)|z - t_j|^{-\frac{2+\alpha_j}{p}}, \quad z \in \overline{\mathcal{D}_\Gamma^+}, \tag{4.22}$$

$$|f_j(z)| \leq O_j(z)|z - t_j|^{-\frac{2+\alpha_j}{p}} \exp\left\{|a_j||z - t_j|^{-1}\cos\left[\eta_\Gamma(t_j) - \arg(z - t_j)\right]\right\}, \tag{4.23}$$

$z \in \overline{\mathcal{D}_\Gamma^-}$, $\lim_{z\to t_j} O_j(z) = 0$. Also, $f(\infty) = 0$.

Proof. Under the hypotheses of the theorem, each of the functions h_j of the form (4.16) is, thanks to Proposition 4.2, a model function on Γ and belongs to the class $\mathcal{U}$ (see Remark 4.1). By Proposition 3.4, the function h given by (4.15) is also a model function on Γ and belongs to the class $\mathcal{U}$. The equivalence (1) $\Longleftrightarrow$ (2) is a consequence of Proposition 3.1.

Now let us show that (1) $\Longleftrightarrow$ (3). By equality (3.11),

$$f \in \mathcal{M}(h) \quad \Longleftrightarrow \quad f = \sum_{j=1}^{n}\left(\prod_{s=1}^{j-1} h_s\right) f_j, \quad \text{where } f_j \in \mathcal{M}(h_j).$$

Let us denote by γ_j a circle of radius $r_j/2$ such that $\gamma_j \in O_+(t_j)$, and then consider the fractional linear transformation

$$x = v_j(t) = -i(1 + r_j \exp[i\eta_\Gamma(t_j)](t - t_j)^{-1}),$$

satisfying the conditions

$$v_j(t_j) = \infty, \quad v_j(\gamma_j) = \mathbb{R}, \quad v_j(\mathcal{D}_{\gamma_j}^+) = \Pi_+,$$

as well as the inverse transformation

$$t = v_j^{-1}(x) = t_j - ir_j \exp[i\eta_\Gamma(t_j)](x + i)^{-1}.$$

By Proposition 2.1, the isomorphism $\mathcal{K}_j$ defined by

$$(\mathcal{K}_j f)(x) = \frac{ir_j \exp[i\eta_\Gamma(t_j)] f(v_j^{-1}(x))}{x+i}, \quad (\mathcal{K}_j^{-1} g)(t) = \frac{g(v_j(t))}{t-t_j}$$

maps $L_p^+(\gamma_j, \varrho)$ onto $L_p^+(\mathbb{R}, \varrho_j)$, where $\varrho_j(x) = |x+i|^{p-2}\varrho(v_j^{-1}(x))$. Decreasing, if necessary, the radius r_j so that $\gamma_j \subset \mathcal{D}_\Gamma^+$, we ensure that $\varrho_j(x) \asymp |x+i|^{p-\alpha_j-2}$, and hence that $L_p^+(\mathbb{R}, \varrho_j) = L_p^+(\mathbb{R}, |x+i|^{p-\alpha_j-2})$. Applying Theorem 3.4 (for $s^+(x) = \exp(ix\sigma) - 1$) and Proposition 3.11, and then Proposition 3.4, we obtain equality (4.18). It remains to verify the convergence of the series in this equality on Γ and the fact that f lies in $L_p^+(\Gamma, \varrho)$.

By Proposition 4.4, there exists a circle $\gamma_- \in O_-(t_j)$ which lies inside the domain $\mathcal{D}_\Gamma^-$. Orienting γ_- so that the domain $\mathcal{D}_\Gamma^+$ will lie to its left and applying the fractional linear transformation v_j and the isomorphism $\mathcal{K}_j$, we place ourselves in the situation considered in Section 3.8. This establishes the $L_p^+(\Gamma, \varrho)$ convergence of the series $\sum_{k\in\mathbb{Z}} c_{jk}e_{jk}$ and hence of the series in (4.18).

Turning now to the proof of the equivalence (1) $\iff$ (4), we observe that, by Theorem 3.2, $f \in \mathcal{M}(h)$ if and only if $f(t) = h(t)f^-(t)$, where the function f^- belongs to $L_p^-(\Gamma, \varrho)$ and obeys the estimates (3.18) and (3.19). This proves the analyticity of the function $f(z)$ in $\mathbb{C} \setminus \{t_1, \dots, t_n\}$, the equality $f(\infty) = 0$, and, resorting to Proposition 4.4 and the equality

$$|h_j(z)| = \exp(|a_j||z-t_j|^{-1} \cos[\eta_\Gamma(t_j) - \arg(z-t_j)]),$$

also the estimates (4.22) and (4.23) for f. It remains to verify that f admits the representation (4.21).

To this end, let us surround each point t_j, $1 \le j \le n$, by a simple closed smooth contour γ_j that avoids the zeros of the function $f(z)$ and such that $\overline{\mathcal{D}}_{\gamma_j}^+ \cap \overline{\mathcal{D}}_{\gamma_k}^+ = \emptyset$ whenever $j \neq k$. Since the function $f(z)$ is continuous and does not vanish on γ_j, Theorem 2.1 guarantees that it admits a standard factorization

$$f(z) = \varphi_j^+(z)(z-z_j)^{\kappa_j}\varphi_j^-(z), \quad z \in \gamma_j,$$

where $\kappa_j \in \mathbb{Z}$, $z_j \in \mathcal{D}_{\gamma_j}^+$, $z_j \neq t_j$, and the functions $\varphi_j^\pm$ are analytic in the respective domains $\mathcal{D}_{\gamma_j}^\pm$ and different from zero therein. But then the function $f(z)/((z-z_j)^{\kappa_j}\varphi_j^-(z))$ is analytic and bounded in $\overline{\mathcal{D}}_{\gamma_j}^+$, and the function

$$\frac{f(z)}{\prod_{j=1}^{n} (z-z_j)^{\kappa_j}\varphi_j^-(z)}$$

is analytic in the whole complex plane, except at infinity, where it may have a pole of finite order. By the generalized Liouville theorem,

$$f(z) = R(z)\prod_{j=1}^{n} \varphi_j^-(z), \quad \text{where } R(z) = P(z)\prod_{j=1}^{n}(z-z_j)^{\kappa_j} \tag{4.24}$$

and $P(z)$ is a polynomial. Since $f(\infty) = 0$ and $\varphi_j^-(\infty) \neq 0$, $1 \leq j \leq n$, the fraction $R(z)$ is proper.

Let $z_{j_1}, z_{j_2}, \ldots, z_{j_k}$ be the poles of $R(z)$, with corresponding orders $\delta_1, \delta_2, \ldots, \delta_k$. Decompose $R(z)$ into a product of fractions such that

$$R(z) = \prod_{q=1}^{k} R_q(z), \qquad \text{where } R_q(z) = \frac{P_q(z)}{(z - z_{j_q})^{\delta_q}}$$

and P_q are polynomials, $\deg P_q \leq \delta_q$. Further, set

$$f_j(z) = \begin{cases} \varphi_j^-(z) R_j(z), & \text{if } j \in \{j_1, j_2, \ldots, j_k\}, \\ \varphi_j^-(z), & \text{if } j \notin \{j_1, j_2, \ldots, j_k\}. \end{cases}$$

Then clearly $f = \prod_{j=1}^n f_j$. Since $\varphi_i^-(z_j) \neq 0$, $i \neq j$, it follows from equality (4.24) that at the point z_j, $j \in \{j_1, j_2, \ldots, j_k\}$, the function φ_j^- has a zero of order at least δ_j, and consequently the function $f_j(z)$, $1 \leq j \leq n$, is analytic in the whole complex plane except, possibly, at the point t_j. The same equality, written for the function f_j, yields the estimates (4.22) and (4.23). Finally, combining inequality (4.22) and Proposition 3.8 one concludes that $f_j \in L_p^+(\Gamma, \varrho)$. □

Let us address the question of whether the estimates (4.22) and (4.23) are sharp. As it turns out, they cannot be improved, in the sense that nothing can be said about the rate of convergence to zero of the functions $O_j(z)$ as $z \to t_j$ *simultaneously for all functions from* $\mathcal{M}(h)$. Let us state the corresponding result for the case of a single discontinuity point on Γ_0.

Proposition 4.5 *Let the model subspace $\mathcal{M}(h)$ in $L_p(\Gamma_0, |t-1|^\alpha)$, $-1 < \alpha < p-1$, be generated by the model function*

$$h(t) = \exp\left(\sigma \frac{t+1}{t-1}\right), \qquad \sigma > 0.$$

Then for any positive function $O(t)$ that tends to zero as $t \to 1$, there exist a function $f \in \mathcal{M}(h)$ and a sequence of points $\{\tau_n\} \subset \Gamma_0$ converging to 1 such that the following inequality holds for all τ_n:

$$|f(\tau_n)| \geq O(\tau_n) |\tau_n - 1|^{-\frac{2+\alpha}{p}}. \tag{4.25}$$

Proof. The function h differs from a function h_0 of the form (4.9) only by a constant factor, and so it lies in the class $\mathcal{U}$. Therefore, in order to construct the model subspace $\mathcal{M}(h)$ one can use Proposition 3.3 and the integral representation (3.8):

$$f(t) = \frac{1}{2\pi i} \int_{\Gamma_0} \frac{h(\tau) - h(t)}{\tau - t} g(\tau) d\tau, \qquad g \in L_p(\Gamma_0, |t-1|^\alpha).$$

Observing that the equation $h(t) = 1$ has infinitely many solutions on the circle Γ_0, we take the piecewise constant function $g(t)$ as follows:

$$g(t) = \begin{cases} |\tau_n - 1|^{-\frac{2+\alpha}{p}}, & t \in (\tau_n', \tau_n), \qquad n = 1, 2, \ldots, \\ 0, & t \in \Gamma_0 \setminus \bigcup_{n=1}^{\infty}(\tau_n', \tau_n), \end{cases}$$

where $\{\tau_n\}$ is a sequence of roots of the equations $h(t) = 1$ lying in Π_+ ($\operatorname{Im}\tau_n > 0$) with the property that $|\tau_{n+1} - 1|/|\tau_n - 1| < \delta < 1$ and $\lim_{n\to\infty} O(\tau_n)n^2 = 0$. The points $\tau_n' \in \Gamma_0$ are chosen so that $|\tau_n' - 1| < |\tau_n - 1|$ and $|\tau_n' - \tau_n| = |\tau_n - 1|^2 n^{-2}$.

First of all, we claim that $g \in L_p(\Gamma_0, |t-1|^\alpha)$. Indeed,

$$\begin{aligned} \|g\|_{p,\varrho}^p &= \int_{\Gamma_0} |g(t)|^p |t-1|^\alpha |dt| = \sum_{n=1}^{\infty} \int_{\tau_n'}^{\tau_n} |\tau_n - 1|^{-2-\alpha} |t-1|^\alpha |dt| \\ &\le c_1 \sum_{n=1}^{\infty} |\tau_n - 1|^{-2-\alpha} |\tau - 1|^\alpha |\tau_n - \tau_n'| \le c_2 \sum_{n=1}^{\infty} n^{-2} < \infty. \end{aligned}$$

Now let us prove inequality (4.25). Denote $l_n = (\tau_n', \tau_n)$. Then

$$\begin{aligned} |f(\tau_n)| &= \frac{1}{2\pi} \left| \int_{\Gamma_0} \frac{h(\tau) - 1}{\tau - \tau_n} g(\tau) d\tau \right| \\ &\ge \frac{1}{2\pi} \left| \int_{l_n} \frac{h(\tau) - 1}{\tau - \tau_n} |\tau_n - 1|^{-\frac{2+\alpha}{p}} d\tau \right| - \frac{1}{2\pi} \left| \int_{\Gamma_0 \setminus l_n} \frac{h(\tau) - 1}{\tau - \tau_n} g(\tau) d\tau \right| \\ &\overset{\text{def}}{=} I_1(\tau_n) - I_2(\tau_n). \end{aligned}$$

Let us estimate $I_1(\tau_n)$ from below. Let $\tau = \exp(i\theta(\tau)) \in l_n$, $\tau_n = \exp(i\theta_n)$. Then $d\tau = i\exp(i\theta(\tau))|d\tau|$ and $\arg(\tau - \tau_n) = -\pi/2 + (\theta(\tau) + \theta_n)/2$. Consequently,

$$\begin{aligned} 2\pi |\tau_n - 1|^{\frac{2+\alpha}{p}} I_1(\tau_n) &= \left| \int_{l_n} \frac{h(\tau) - 1}{|\tau - \tau_n|} \exp\left(i \frac{\theta(\tau) - \theta_n}{2} \right) |d\tau| \right| \\ &\ge \left| \int_{l_n} \frac{h(\tau) - 1}{|\tau - \tau_n|} \cos \frac{\theta(\tau) - \theta_n}{2} |d\tau| \right| \\ &\quad - \left| \int_{l_n} \frac{h(\tau) - 1}{|\tau - \tau_n|} \sin \frac{\theta(\tau) - \theta_n}{2} |d\tau| \right| \\ &\overset{\text{def}}{=} I_{1,1}(\tau_n) - I_{1,2}(\tau_n). \end{aligned}$$

Next, for $I_{1,2}(\tau_n)$ we have the upper bound

$$I_{1,2}(\tau_n) \le \int_{l_n} |h(\tau) - 1||d\tau| \le \text{const}\, |\tau_n - 1|^2 n^{-2}.$$

Now let us estimate $I_{1,1}(\tau_n)$ from below. To that end, we observe that $i\sigma(1+\tau)/(1-\tau) \in \mathbb{R}$ for all $\tau \in \Gamma_0$. It follows that

$$h(\tau) - 1 = \cos\left(i\sigma \frac{1+\tau}{1-\tau} \right) - 1 + i \sin\left(i\sigma \frac{1+\tau}{1-\tau} \right),$$

with

$$\cos\left(i\sigma\frac{1+\tau_n}{1-\tau_n}\right)=1, \quad \sin\left(i\sigma\frac{1+\tau_n}{1-\tau_n}\right)=0.$$

By the Lagrange theorem, for any $\tau \in l_n$ we have

$$\left|\sin\left(i\sigma\frac{1+\tau}{1-\tau}\right)\right|=\left|\cos\left(i\sigma\frac{1+\tau'}{1-\tau'}\right)\right|\left|i\sigma\left(\frac{1+\tau}{1-\tau}-\frac{1+\tau_n}{1-\tau_n}\right)\right|,$$

where $\tau' \in l_n$. Using the fact that, for sufficiently large n and any $\tau \in l_n$, $\cos((\theta(\tau)-\theta_n)/2) > 1/2$, we obtain

$$\begin{aligned} I_{1,1}(\tau_n) &\geq c_1\int_{l_n}\frac{|h(\tau)-1|}{|\tau-\tau_n|}|d\tau| \geq c_2\int_{l_n}\frac{|\sin(i\sigma\frac{1+\tau}{1-\tau})|}{|\tau-\tau_n|}|d\tau| \\ &= c_2|\tau_n-1|^{-1}\int_{l_n}\left|\sin\left(i\sigma\frac{1+\tau}{1-\tau}\right)\right|\left|i\sigma\left(\frac{1+\tau}{1-\tau}-\frac{1+\tau_n}{1-\tau_n}\right)\right|^{-1}\frac{|d\tau|}{|1-\tau|} \\ &= c_3|\tau_n-1|^{-1}\int_{l_n}\left|\cos\left(i\sigma\frac{1+\tau'}{1-\tau'}\right)\right|\frac{|d\tau|}{|1-\tau|} \geq c_4|\tau_n-1|^{-2}|\tau_n-\tau_n'|, \end{aligned}$$

because for n large enough

$$\begin{aligned} \left|\cos\left(i\sigma\frac{1+\tau'}{1-\tau'}\right)\right| &\geq 1-\left|\cos\left(i\sigma\frac{1+\tau_n}{1-\tau_n}\right)-\cos\left(i\sigma\frac{1+\tau'}{1-\tau'}\right)\right| \\ &\geq 1-c_5\left|\frac{1+\tau_n}{1-\tau_n}-\frac{1+\tau'}{1-\tau'}\right| \geq 1-c_5\frac{|\tau_n-\tau_n'|}{|1-\tau_n||1-\tau_n'|} \\ &\geq 1-c_5n^{-2} > \frac{1}{2}. \end{aligned}$$

Clearly,

$$I_{1,1}(\tau_n) \geq c_5 n^{-2}.$$

Comparing the estimates for $I_{1,1}(\tau_n)$ and $I_{1,2}(\tau_n)$, we conclude that

$$I_1(\tau_n) \geq c_6|\tau_n-1|^{-\frac{2+\alpha}{p}}n^{-2}.$$

Let us estimate the quantity $I_2(\tau_n)$ from above. To that end, we notice that

$$\operatorname{supp} g \subset \bigcup_{k=1}^{\infty} l_k \subset l_n \cup \gamma_1 \cup \gamma_2,$$

where

$$\gamma_1=(\tau_{n-1}',-1), \quad \gamma_2=(1,\tau_{n+1})$$

(here Γ_0 is positively oriented).

Applying Hölder's inequality, we get

$$\begin{aligned} I_2(\tau_n) &\leq c_7 \left(\int_{\gamma_1 \cup \gamma_2} |g(\tau)|^p |\tau - 1|^\alpha |d\tau| \right)^{\frac{1}{p}} \\ &\quad \times \left(\int_{\gamma_1 \cup \gamma_2} |\tau - \tau_n|^{-q} |\tau - 1|^{-\frac{\alpha q}{p}} |d\tau| \right)^{\frac{1}{q}} \\ &\leq c_8 \left[\left(\int_{\gamma_1 \cup (-1,1)} |\tau - \tau_n|^{-q} |\tau - 1|^{-\frac{\alpha q}{p}} |d\tau| \right)^{\frac{1}{q}} \right. \\ &\quad \left. + \left(\int_{\gamma_2} |\tau - \tau_n|^{-q} |\tau - 1|^{-\frac{\alpha q}{p}} |d\tau| \right)^{\frac{1}{q}} \right]. \end{aligned}$$

To estimate the last integrals, we apply inequality (3.17). Setting $z = \tau_n$, $t_0 = 1$, and observing that $d(\tau_n, \gamma_1 \cup (-1,1)) = |\tau_n - \tau'_{n-1}|$, $d(\tau_n, \gamma_2) = |\tau_n - \tau_{n+1}|$, we have

$$\begin{aligned} I_2(\tau_n) &\leq c_9 \left[d^{\frac{1-q}{q}}(\tau_n, \gamma_1 \cup (-1,1)) + d^{\frac{1-q}{q}}(\tau_n, \gamma_2) \right] |\tau_n - 1|^{-\frac{\alpha}{p}} \\ &\leq c_9 \left(|\tau_n - \tau'_{n-1}|^{-\frac{1}{p}} + |\tau_n - \tau_{n+1}|^{-\frac{1}{p}} \right) |\tau_n - 1|^{-\frac{\alpha}{p}}. \end{aligned}$$

But, by the choice of the sequence $\{\tau_n\}$, for n large enough we have

$$|\tau_n - \tau_{n+1}| \asymp |\tau_n - 1|, \quad |\tau_n - \tau'_{n-1}| \asymp |\tau_{n-1} - 1|.$$

It follows that

$$I_2(\tau_n) \leq c_{10} |\tau_n - 1|^{-\frac{1+\alpha}{p}}.$$

Consequently,

$$|f(\tau_n)| \geq c_{11} |\tau_n - 1|^{-\frac{2+\alpha}{p}} n^{-2} \left(1 - c_{10} |\tau - 1|^{\frac{1}{p}} n^2 \right).$$

When $n \to \infty$, the quantity $|\tau_n - 1|$ decays no slower than δ^n, and so $\lim_{n\to\infty} |\tau_n - 1|^{1/p} n^2 = 0$. Therefore,

$$|f(\tau_n)| \geq c_{12} |\tau_n - 1|^{-\frac{2+\alpha}{p}} n^{-2} = O(\tau_n) |\tau_n - 1|^{-\frac{2+\alpha}{p}} \frac{c_{12}}{O(\tau_n) n^2}.$$

This establishes (4.25), because for n large enough, $O(\tau_n) n^2 < c_{12}$. □

Now let us return to the real line $\mathbb{R}$ and give the analogue of Theorem 4.8 for the model function $h(x) = \exp(i\sigma x)$.

Theorem 4.9 *Let $h(x) = \exp(i\sigma x)$, $\sigma > 0$, and let the weight $\varrho(x)$ have the form (1.1) and hence behave like $\varrho(x) \asymp |x|^\alpha$ for large values of $|x|$, where $-1 < \alpha < p - 1$, $\alpha = \sum_{j=0}^{n} \alpha_j$. Then, in the space $L_p^+(\mathbb{R}, \varrho)$,*

$$\ker T(h^{-1}) = \mathcal{M}(h) = \mathcal{L}_{p,\varrho}^{\sigma+}.$$

Moreover, the following conditions are equivalent:

(1) $f \in \mathcal{L}^{\sigma+}_{p,\varrho}$;

(2) *the function* f *has the form*

$$f(x) = \sum_{k\in\mathbb{Z}} c_k \frac{\exp(i\sigma x) - 1}{x - 2\pi k\sigma^{-1}},$$

where $\{c_k\} \in l_{p,\delta}(\mathbb{Z})$ *and* $\delta = \{(|k|+1)^\alpha\}_{k\in\mathbb{Z}}$;

(3) $f(z)$ *is an entire function of exponential type at most* σ *which obeys the estimates*

$$|f(z)| \le O(z)|z+i|^{-\frac{\alpha}{p}}, \quad \operatorname{Im} z \ge 0, \tag{4.26}$$

$$|f(z)| \le O(z)|z-i|^{-\frac{\alpha}{p}} \exp(-\sigma \operatorname{Im} z), \quad \operatorname{Im} z \le 0, \tag{4.27}$$

where

$$\lim_{|z|\to\infty} O(z) = 0. \tag{4.28}$$

The estimates (4.26) and (4.27) are sharp in the sense that for any positive function $O(z)$ *that satisfies condition (4.28) one can find a function* $f(z) \in \mathcal{L}^{\sigma+}_{p,\varrho}$ *and a real sequence* $\{x_n\}$ *that converges to infinity, such that*

$$|f(x_n)| \ge O(x_n)|x_n + i|^{-\frac{\alpha}{p}}. \tag{4.29}$$

Proof. Define $\mathcal{K}$ and its inverse by

$$(\mathcal{K}f)(t) = \frac{1}{1-t} f\left(i\frac{1+t}{1-t}\right), \quad (\mathcal{K}^{-1}g)(x) = \frac{2i}{x+i} g\left(\frac{x-i}{x+i}\right). \tag{4.30}$$

Clearly, the mapping $\mathcal{K}$ effects an isomorphism of the space $L^+_p(\mathbb{R}, |x+i|^\alpha)$ onto the space $L^+_p(\Gamma_0, |t-1|^{p-2-\alpha})$ and also of the subspace $\mathcal{L}^{\sigma+}_{p,|x+i|^\alpha} = \mathcal{M}(\exp(i\sigma x))$ onto the subspace $\mathcal{M}(\exp(\sigma(t+1)/(t-1)))$ in $L^+_p(\Gamma_0, |t-1|^{p-2-\alpha})$. As a result, the equivalence (1) $\Longleftrightarrow$ (3) and, in particular, the estimates (4.26) and (4.27), become corollaries of Theorem 4.8 and the estimates (4.22) and (4.23), respectively. The assertion on the sharpness of the estimates (4.26) and (4.27), as well as inequality (4.29), are consequences of Proposition 4.5 and inequality (4.25). Finally, the equivalence (1) $\Longleftrightarrow$ (2) is a corollary of Theorem 3.4. □

Remark 4.3 By Proposition 3.14,

$$\mathcal{L}^{\sigma}_{p,\varrho} = \mathcal{L}^{\sigma+}_{p,\varrho} \oplus \exp(-i\sigma x)\mathcal{L}^{\sigma+}_{p,\varrho}.$$

Hence, the estimates (4.26) and (4.27) allow us to describe the behavior at infinity of the whole class $\mathcal{L}^{\sigma}_{p,\varrho}$:

$$|f(z)| \le O(z)(1+\exp(\sigma \operatorname{Im} z))|z+i|^{-\frac{\alpha}{p}}, \quad \operatorname{Im} z \ge 0,$$

$$|f(z)| \le O(z)(1+\exp(-\sigma \operatorname{Im} z))|z-i|^{-\frac{\alpha}{p}}, \quad \operatorname{Im} z \le 0.$$

The space $\mathcal{L}_p^\sigma$ has been investigated by many authors. We have already observed earlier that $\mathcal{L}_p^\sigma \subset \mathcal{B}_\sigma$. It is also known (see Akhiezer [1], Chap. 4, § 84) that the functions of class $\mathcal{L}_p^\sigma$ on $\mathbb{R}$ tend to zero when $x \to \pm\infty$. Theorem 4.9 contains this result for $\alpha = 0$, showing in addition that the indicated decay can be arbitrarily slow. For $0 < \alpha < p-1$ the embedding $\mathcal{L}_{p,\varrho}^\sigma \subset \mathcal{B}_\sigma$ is preserved, and the functions from $\mathcal{L}_{p,\varrho}^\sigma$ decay on $\mathbb{R}$ no slower than $|x+i|^{-\alpha/p}$. Finally, for $-1 < \alpha < 0$ in $\mathcal{L}_{p,\varrho}^\sigma$ there are growing functions as well, and the space $\mathcal{L}_{p,\varrho}^\sigma$ is no longer a subclass of $\mathcal{B}_\sigma$.

Remark 4.4 By Proposition 3.1, for $\sigma > 0$,

$$L_p^+(\mathbb{R}, \varrho) = \mathcal{L}_{p,\varrho}^{\sigma+} \oplus \operatorname{im} T(\exp(i\sigma x)).$$

Returning to Theorem 4.7 for the case $\sigma > 0$ and using the fact that the system of functions $\{e_k\}$ of the form (1.30) is a biorthogonal basis in $\mathcal{L}_{p,\varrho}^{\sigma+}$, we see that the orthogonality conditions

$$\int_{\mathbb{R}} f(x) \left(\frac{1}{b^-} P^- \frac{1}{b^+} \bar{e}_k \right)(x)dx = 0, \quad k \in \mathbb{Z}, \tag{4.31}$$

are equivalent to conditions (4.8) and are necessary and sufficient for f to belong to $\operatorname{im} T(a)$.

4.3 Standard almost periodic discontinuities

Passing to the definition of discontinuities of almost periodic type, let us transfer the class of almost periodic functions, introduced in Section 4.1 for the real line $\mathbb{R}$, to the circle Γ_0 by means of the change-of-variable operator V:

$$(Vf)(t) = f\left(i\frac{1+t}{1-t} \right).$$

In connection with this, let us introduce the notations

$$\Pi_c(\Gamma_0) = V(\Pi_c(\mathbb{R})), \quad \Pi_w(\Gamma_0) = V(\Pi_w(\mathbb{R})), \quad \Pi_w^\pm(\Gamma_0) = V(\Pi_w^\pm(\mathbb{R})).$$

In particular, almost periodic polynomials in $\Pi_c(\Gamma_0)$ have the form

$$a(t) = \sum_{j=1}^{n} c_j \exp\left(\sigma_j \frac{t+1}{t-1} \right), \quad \sigma_j \in \mathbb{R}.$$

The almost periodic index is not affected by the action of the operator V and is here defined by the formula

$$\sigma(a) = \lim_{\theta \to 0} \frac{\theta}{4} \Big[\arg a(\exp(i\varphi)) \Big] \Big|_{\varphi=\theta}^{2\pi-\theta}, \quad 0 \le \varphi < 2\pi. \tag{4.32}$$

From Theorem 4.5 it follows that every function $a(t) \in \Pi_w(\Gamma_0)$ which satisfies the condition

$$\inf_{t\in\Gamma_0} |a(t)| > 0 \tag{4.33}$$

admits a generalized factorization of the form

$$a(t) = a^-(t)\exp\left(\sigma\frac{t+1}{t-1}\right)a^+(t), \tag{4.34}$$

where $\sigma = \sigma(a)$, $(a^+)^{\pm 1} \in \Pi_w^+(\Gamma_0)$, $(a^-)^{\pm 1} \in \Pi_w^-(\Gamma_0)$.

The standard almost periodic discontinuities will be defined here for closed oriented curves that are diffeomorphic to a circle. Namely, we say that a function $a(t) \in L_\infty(\Gamma)$ has a *standard almost periodic discontinuity* in the point $t_0 \in \Gamma$ if there exist a function $p_0(\tau) \in \Pi_c(\Gamma_0)$ and a diffeomorphism $\tau = w_0(t)$ of the curve Γ onto the circle Γ_0, such that w_0 preserves the orientation of Γ, $w_0(t_0) = 1$, the function $w_0(t)$ has a second derivative at t_0, and

$$\lim_{t\to t_0}(a(t) - p_0(w_0(t))) = 0. \tag{4.35}$$

Remark 4.5 The functions p_0 and w_0 figuring in the definition of a standard almost periodic discontinuity are not uniquely determined. Indeed, let $\Gamma = \Gamma_0$, $t_0 = 1$, $p_0(t) = \exp\left(\sigma_0\frac{t+1}{t-1}\right)$, $p_1(t) = \exp\left(\sigma_1\frac{t+1}{t-1}\right)$, where $\sigma_0 > 0$, $\sigma_1 > 0$. The general form of a fractional linear diffeomorphism of Γ_0 onto itself that preserves the point $t = 1$ is

$$w(t) = \frac{1-\overline{z}_0}{1-z_0}\cdot\frac{t-z_0}{1-\overline{z}_0 t}, \quad \text{where } z_0 \in D^+.$$

Here

$$w'(t) = \frac{1-|z_0|^2}{1-2\operatorname{Re} z_0 + |z_0|^2} > 0$$

and $w'(1) \neq 1$ for $z_0 \neq 0$. The reader is invited to verify that for $\sigma_1 = \sigma_0 w'(1)$ one has

$$\lim_{t\to 1}(p_1(w(t)) - p_0(t)) = 0,$$

and hence the pairs (p_1, w) and (p_0, y), where y is the identity mapping, define the same standard almost periodic discontinuity. For this reason, in the definition of the concept of a standard almost periodic discontinuity we do not use the term "(p_0, w_0)-characteristic" of the discontinuity (see Gohberg and Krupnik [3], p. 339). To characterize standard almost periodic discontinuities we will use the functional $\kappa_0(a)$ introduced below via equality (4.38) (see Proposition 4.9 and Theorem 4.12).

Our next goal is to construct a generalized factorization for functions that have a finite number of standard almost periodic discontinuities. But first we need a number of auxiliary results.

Let Γ_1 and Γ_2 be two smooth contours of class $\mathcal{T}$, and let $t = \alpha(u)$ be a diffeomorphism of Γ_1 onto Γ_2 (i.e., α is a continuously differentiable function mapping Γ_1 onto Γ_2 in a one-to-one manner such that $\alpha'(u) \neq 0$). Consider two Toeplitz operators, $T_{\Gamma_1}(b(u))$ and $T_{\Gamma_2}(b(\alpha^{-1}(t)))$, acting in the respective spaces $L_p(\Gamma_1, \varrho_1)$, $\varrho_1(u) = \prod_{k=1}^m |u - u_k|^{\alpha_k}$ and $L_p(\Gamma_2, \varrho_2)$, $\varrho_2(t) = \prod_{k=1}^m |t - \alpha(u_k)|^{\alpha_k}$. To decide whether the operators $T_{\Gamma_1}(b(u))$ and $T_{\Gamma_2}(b(\alpha^{-1}(t)))$ are simultaneously normally solvable or not one can use Theorem 4.11 given below. The proof of this theorem makes essential use of the following result.

Theorem 4.10 *Let $t = \alpha(u)$ be a diffeomorphism of the smooth contour $\Gamma_1 \in \mathcal{T}$ onto Γ_2. Then the operator defined by*

$$(K_{\Gamma_1,\Gamma_2} f)(u) = \frac{1}{2\pi i} \int_{\Gamma_1} \left(\frac{\alpha'(s)}{\alpha(s) - \alpha(u)} - \frac{1}{s - u} \right) f(s) ds$$

is compact in $L_p(\Gamma_1, \varrho_1)$.

From Theorem 4.10, the general theory of $\Phi_\pm$-operators, and the fact that one of the defect numbers of a Toeplitz operator is equal to zero (Proposition 1.4) one obtains the following result.

Theorem 4.11 *Let Γ_1, $\Gamma_2 \in \mathcal{T}$ be simple closed smooth contours and let $t = \alpha(u)$ be a diffeomorphism of Γ_1 onto Γ_2. Then if $T_{\Gamma_1}(b(u))$ is a Φ_+- (resp., Φ_--, Φ-) operator in the space $L_p(\Gamma_1, \varrho_1)$, then $T_{\Gamma_2}(b(\alpha^{-1}(t)))$ is a Φ_+- (resp., Φ_--, Φ-) operator in $L_p(\Gamma_2, \varrho_2)$ and $\operatorname{ind} T_{\Gamma_1}(b(u)) = \operatorname{ind} T_{\Gamma_2}(b(\alpha^{-1}(t)))$. If, in addition, $T_{\Gamma_1}(b(u))$ is left (resp., right, two-sided) invertible, then $T_{\Gamma_2}(b(\alpha^{-1}(t)))$ is also left (resp., right, two-sided) invertible.*

Proposition 4.6 *Suppose that a diffeomorphism $\tau = w_0(t)$ of the curve Γ onto the circle Γ_0 satisfies the conditions in the definition of a standard almost periodic discontinuity at the point $t_0 \in \Gamma$. Then the following representation holds on Γ:*

$$a(t) = \exp\left(\sigma \frac{w_0(t) + 1}{w_0(t) - 1} \right) = h_0(t) c_0(t), \quad \sigma \in \mathbb{R}, \tag{4.36}$$

where $c_0 \in GC(\Gamma)$, h_0 is given by (4.9) with $a_0 = 2\sigma / w_0'(t_0)$, and $h_0 \in GL_\infty(\Gamma)$.

Proof. Let us consider the function

$$b(t) = \sigma \frac{w_0(t) + 1}{w_0(t) - 1} - a_0 (t - t_0)^{-1}$$

and show that it is continuous in the point t_0. To this end we use the Taylor expansion of the function $w_0(t)$ about t_0:

$$w_0(t) = 1 + w_0'(t_0)(t - t_0) + \frac{w_0''(t_0)}{2} (t - t_0)^2 + \alpha(t)(t - t_0)^2,$$

where $\lim_{t\to\alpha_0}\alpha(t)=0$. Since $2\sigma=a_0w_0'(t_0)$, we have

$$b(t)=\sigma-\frac{2\sigma}{w_0'(t_0)}\cdot\frac{w_0''(t_0)+2\alpha(t)}{2w_0'(t_0)+(w_0''(t_0)+2\alpha(t))(t-t_0)}.$$

Further, since $w_0'(t_0)\neq 0$, the function $b(t)$ is continuous in the point t_0. Therefore, $b(t)\in C(\Gamma)$ and $c_0(t)=\exp b(t)=a(t)/h_0(t)$ is in $C(\Gamma)$ and does not vanish on Γ. Moreover, $|a(t)|=1$ for all $t\in\Gamma\setminus\{1\}$. We conclude that $h_0(t)=a(t)/c_0(t)\in GL_\infty(\Gamma)$. □

Remark 4.6 From Proposition 4.2 it follows that whenever on the curve Γ there exists a function $a(t)$ that has a standard almost periodic discontinuity in the point t_0, one of the functions $h_0(t)$ given by (4.9) or $[h_0(t)]^{-1}$ is a model function on Γ, and consequently Γ satisfies the conditions (4.10) and (4.11). Since here the mapping $\tau=w_0(t)$ preserves the orientation of Γ, $\arg a_0=\eta_\Gamma(t_0)$ when $\sigma>0$ and $\arg a_0=\eta_\Gamma(t_0)-\pi$ when $\sigma<0$.

Proposition 4.7 *Under the assumptions of Proposition 4.6, suppose that the function $p_0(\tau)\in\Pi_c(\Gamma_0)$ figuring in the definition of the standard almost periodic discontinuity is such that*

$$\inf_{\tau\in\Gamma_0}|p_0(\tau)|>0,\quad \sigma(p_0)=0. \tag{4.37}$$

Then the function $p_0(w_0(t))$ is an MNAF *in $L_p(\Gamma,\varrho)$.*

Proof. By Proposition 2.1, the operator $T(p_0)$ in $L_p^+(\Gamma_0,\varrho)$ is similar to the operator $T(p_0(\frac{x-i}{x+i}))$ in the space $L_p^+(\mathbb{R},\widetilde{\varrho})$, where $\widetilde{\varrho}(x)=|x-i|^{p-2}\varrho(\frac{x-i}{x+i})$,

$$\mathcal{K}^{-1}T(p_0)\mathcal{K}=T\left(p_0\left(\frac{x-i}{x+i}\right)\right),$$

and the isomorphisms $\mathcal{K}^{\pm1}$ have the form (4.30). Since condition (4.37) remains in force for the function $p_0(\frac{x-i}{x+i})\in\Pi_c(\mathbb{R})$, Theorem 4.7 tells us that the operator $T(p_0(\frac{x-i}{x+i}))$ is invertible in $L_p^+(\mathbb{R},\varrho)$ for all p, $1<p<\infty$, and all power weights $\varrho\in A_p(\mathbb{R})$. Therefore, the operator $T(p_0)$ is invertible in $L_p^+(\Gamma_0,\varrho)$ for all p, $1<p<\infty$, and all power weights $\varrho\in A_p(\Gamma_0)$. By Theorem 4.11, the operator $T(p_0(w_0(t)))$ is invertible in $L_p^+(\Gamma,\varrho)$ for all p, $1<p<\infty$, and all power weights $\varrho\in A_p(\Gamma)$. Hence, using Theorem 2.1 we conclude that $p_0(w_0(t))\in\mathrm{fact}(p,\varrho)$ and has a factorization with index zero for all p, $1<p<\infty$, and all power weights $\varrho\in A_p(\Gamma)$. By Theorem 4.6, the function $p_0(w_0(t))$ is an MNAF in $L_p(\Gamma,\varrho)$. □

Proposition 4.8 *Let $f(x)\in\Pi_c(\mathbb{R})$. Then for any two real numbers M_1 and M_2,*

$$\sup_{x\le M_1}|f(x)|=\sup_{x\ge M_2}|f(x)|=\sup_{x\in\mathbb{R}}|f(x)|$$

and

$$\inf_{x\le M_1}|f(x)|=\inf_{x\ge M_2}|f(x)|=\inf_{x\in\mathbb{R}}|f(x)|.$$

Proof. Let us show that $\sup_{x\in\mathbb{R}}|f(x)| = \sup_{x\le M_1}|f(x)|$. Fix $\varepsilon > 0$ and let $x_0 \in \mathbb{R}$ be any point such that $\sup_{x\in\mathbb{R}}|f(x)| - |f(x_0)| < \varepsilon$. By the definition of almost periodic functions, there exists a number $l = l(\varepsilon)$ such that $x_1 = x_0 - l < M_1$ and $|f(x_0) - f(x_1)| < \varepsilon$. Therefore,

$$\sup_{x\le M_1}|f(x)| \ge \sup_{x\in\mathbb{R}}|f(x)| - 2\varepsilon.$$

It remains to let ε go to zero. The other assertions of the proposition are established in much the same manner. □

Remark 4.7 It follows from Theorem 4.2 that, for any function $p \in \Pi_c(\mathbb{R})$ satisfying condition (4.2),

$$\sigma(p) = \lim_{\substack{l_1\to-\infty\\ l_2\to+\infty}} \frac{1}{l_1 - l_2}\Big[\arg p(x)\Big]\Big|_{x=l_1}^{l_2}.$$

But then for any function $p \in \Pi_c(\Gamma_0)$ satisfying condition (4.33) one has that

$$\sigma(p) = \lim_{\theta_1,\theta_2\to 0} \frac{\theta_1\theta_2}{2(\theta_1+\theta_2)}\Big[\arg p(\exp(i\varphi))\Big]\Big|_{\varphi=\theta_1}^{2\pi-\theta_2}.$$

Indeed, let $l_k = i(1+t_k)/(1-t_k)$, $k = 1, 2$, $t_1 = \exp(i\theta_1)$, $t_2 = \exp(i(2\pi - \theta_2))$. Then

$$\begin{aligned}\sigma(p) &= \lim_{\substack{l_1\to-\infty\\ l_2\to+\infty}} \frac{1}{l_2 - l_1}\left[\arg p\left(\frac{x-i}{x+i}\right)\right]\Big|_{x=l_1}^{l_2} \\ &= \lim_{t_1,t_2\to 1} \frac{i}{2}\cdot\frac{(t_2-1)(t_1-1)}{t_1-t_2}\Big[\arg p(t)\Big]\Big|_{t=t_1}^{t_2}.\end{aligned}$$

Since $t_k - 1 = 2\sin(\theta_k/2)(-\sin(\theta_k/2) + i\cos(\theta_k/2))$,

$$t_2 - t_1 = 2\sin((\theta_1+\theta_2)/2)(\sin((\theta_1-\theta_2)/2) - i\cos((\theta_1-\theta_2)/2)),$$

and continuing the chain of equalities written above, we obtain

$$\begin{aligned}\sigma(p) &= \lim_{\theta_1,\theta_2\to 0} \frac{\sin(\theta_1/2)\sin(\theta_2/2)}{\sin((\theta_1+\theta_2)/2)}\Big[\arg p(\exp(i\varphi))\Big]\Big|_{\varphi=\theta_1}^{2\pi-\theta_2} \\ &= \lim_{\theta_1,\theta_2\to 0} \frac{\theta_1\theta_2}{2(\theta_1+\theta_2)}\Big[\arg p(\exp(i\varphi))\Big]\Big|_{\varphi=\theta_1}^{2\pi-\theta_2}.\end{aligned}$$

Let $t_0 \in \Gamma$ and let the function $a \in GL_\infty(\Gamma)$ be continuous in a neighborhood of t_0, except, possibly, in the point t_0 itself. Let us introduce the real functional

$$\kappa_0(a) = \lim_{\delta\to 0} \frac{\delta}{4}\Big[\arg a(t)\Big]\Big|_{t=t''}^{t'}, \tag{4.38}$$

where $t', t'' \in \Gamma$, $t' \prec t_0 \prec t''$, $|t' - t_0| = |t'' - t_0| = \delta$. If the function a is continuous in t_0, then the limit (4.38) exists and is equal to zero. The same is true for a function a for which $|\arg a|$ does not grow too fast (for example, is bounded) in a neighborhood of the point t_0. The next result establishes a connection between the functional $\kappa_0(a)$ and the standard almost periodic discontinuities on Γ.

Proposition 4.9 *Suppose that the diffeomorphism $\tau = w_0(t)$ of the curve Γ onto the circle Γ_0 satisfies the conditions in the definition of a standard almost periodic discontinuity at the point $t_0 \in \Gamma$ and that $p(\tau) \in \Pi_c(\Gamma_0)$, $\inf_{\tau \in \Gamma_0} |p(\tau)| > 0$. Then $a(t) = p(w_0(t)) \in GL_\infty(\Gamma)$, $\kappa_0(a)$ exists, and*

$$\kappa_0(a) = \sigma(p)/|w_0'(t_0)|.$$

Proof. Let $\sigma(p) \neq 0$. By (4.34) and Proposition 4.6,

$$a(t) = p(w_0(t)) = \exp\{a_0(t - t_0)^{-1}\} p_0(w_0(t)) c_0(t),$$

where $a_0 = 2\sigma(p)/w_0'(t_0)$, $c_0(t) \in C(\Gamma)$, and the function $p_0(w_0(t))$ satisfies the conditions of Proposition 4.7. Clearly, $\kappa_0(c_0) = 0$. Let us show that

$$\kappa_0(\exp\{a_0(t - t_0)^{-1}\}) = \sigma(p)/|w_0'(t_0)|.$$

By Remark 4.6, $\arg a_0 = \eta_\Gamma(t_0)$ for $\sigma(p) > 0$ and $\arg a_0 = \eta_\Gamma(t_0) - \pi$ for $\sigma(p) < 0$. Put $\varphi = \eta_\Gamma(t_0)$ for brevity. Then $a_0 = \pm|a_0| \exp(i\varphi)$. Notice that $t' - t_0 = \delta \exp\{i\mu_1(\delta)\}$, $t'' - t_0 = \delta \exp\{i\mu_2(\delta)\}$, where $\mu_1(\delta) \to \varphi - \pi/2$, $\mu_2(\delta) \to \varphi + \pi/2$ when $\delta \to 0$. Therefore,

$$\begin{aligned}\kappa_0(\exp\{a_0(t - t_0)^{-1}\}) &= \lim_{\delta \to 0} \frac{\delta}{4} \left[\frac{\pm|a_0|}{|t - t_0|} \sin(\varphi - \arg(t - t_0)) \right]_{t=t''}^{t'} \\ &= \pm\frac{|a_0|}{4} \lim_{\delta \to 0} \Big[\sin(\varphi - \mu_1(\delta)) - \sin(\varphi - \mu_2(\delta)) \Big] = \pm\frac{|a_0|}{2} = \frac{\sigma(p)}{|w_0'(t_0)|}.\end{aligned}$$

It remains to observe that $\kappa_0(p_0(w_0(t))) = 0$, because $\sigma(p_0) = 0$ and, thanks to equality (4.3), the function $\arg p_0(w_0(t))$ is bounded on $\mathbb{R}$, and that

$$\kappa_0(a) = \kappa_0(\exp\{a_0(t - t_0)^{-1}\}) + \kappa_0(p_0(w_0(t))) + \kappa_0(c_0(t)). \qquad \square$$

Remark 4.8 From the proof of Proposition 4.9 it follows that in the case of a standard almost periodic discontinuity in the point t_0, the coefficient a_0 in the function $h_0(t) = \exp\{a_0(t - t_0)^{-1}\}$ is given by the formula

$$a_0 = \kappa_0(p(w_0)) \exp\{i\eta_\Gamma(t_0)\}.$$

We are now ready to construct generalized factorization with infinite index of functions with a finite number of standard almost periodic discontinuities.

Theorem 4.12 *Let the function $a \in GL_\infty(\Gamma)$ be continuous on the set $\Gamma \setminus \{t_j\}_{j=1}^n$ and have standard almost periodic discontinuities in the points t_j. Then*

$$a(t) = \left(\prod_{j=1}^{n} \exp\{a_j (t - t_j)^{-1}\} \right) b(t), \tag{4.39}$$

where $b \in \operatorname{fact}(p, \varrho)$ in each space $L_p(\Gamma, \varrho)$, $1 < p < \infty$, for all power weights $\varrho \in A_p(\Gamma)$, and

$$a_j = \kappa_j(a) \exp\{i\eta_\Gamma(t_j)\}, \tag{4.40}$$

where the functional $\kappa_j(a)$ is defined by formula (4.38) *at the point t_j.*

Proof. It follows from equality (4.35) and Proposition 4.8 that the almost periodic functions $p_j(\tau)$, $1 \le j \le n$, figuring in the definition of the standard almost periodic discontinuities of the function a obey condition (4.2). Therefore, the function

$$c(t) = \frac{a(t)}{\prod\limits_{j=1}^{n} p_j(w_j(t))}$$

is continuous and bounded away from zero on Γ. Further,

$$p_j(w_j(t)) = \exp\left(\sigma(p_j) \frac{w_j(t) + 1}{w_j(t) - 1} \right) p_{j,0}(w_j(t)),$$

where $p_{j,0}(\tau) \in \Pi_c(\Gamma_0)$ and $\sigma(p_{j,0}) = 0$. Using Proposition 4.6 and Remark 4.8, we obtain

$$\exp\left(\sigma(p_j) \frac{w_j(t) + 1}{w_j(t) - 1} \right) = \exp\{a_j (t - t_j)^{-1}\} c_j(t),$$

where $c_j \in GC(\Gamma)$ and a_j has the form (4.40). Therefore, the function admits a representation (4.39), where

$$b(t) = \left(\prod_{j=1}^{n} p_{j,0}(w_j(t)) c_j(t) \right) c(t).$$

By Proposition 4.7, the functions $p_{j,0}(w_j(t))$ are MNAFs. Since $c_j, c \in GC(\Gamma)$, it follows that $b \in \operatorname{fact}(p, \varrho)$ for all $1 < p < \infty$ and all power weights $\varrho \in A_p(\Gamma)$. □

Remark 4.9 It is readily seen that the index κ of the (p, ϱ)-factorization of the function b as well as the numbers a_j in the representation (4.39) are independent of the choice of the functions p_j and w_j that figure in the definition of standard almost periodic discontinuities.

Let us give the main qualitative result for Toeplitz operators whose symbol has a finite number of standard almost periodic discontinuities.

Theorem 4.13 *Suppose that the function $a \in GL_\infty(\Gamma)$ is continuous on the set $\Gamma \setminus \{t_j\}_{j=1}^n$ and has standard almost periodic discontinuities in the points t_j. Also let $\kappa_j(a) \neq 0$, $1 \le j \le n$. Then*

(1) *if $\kappa_j(a) > 0$, $1 \le j \le n$, the operator $T(a)$ is left invertible in $L_p^+(\Gamma, \varrho)$ and* $\dim \operatorname{coker} T(a) = \infty$;

(2) *if $\kappa_j(a) < 0$, $1 \le j \le n$, the operator $T(a)$ is right invertible in $L_p^+(\Gamma, \varrho)$ and* $\dim \ker T(a) = \infty$;

(3) *if $\kappa_j(a) > 0$, $1 \le j \le m$, and $\kappa_j(a) < 0$, $m+1 \le j \le n$, $m < n$, then the operator $T(a)$ is not normally solvable in any of the spaces $L_p^+(\Gamma, \varrho)$, and* $\dim \ker T(a) = \dim \operatorname{coker} T(a) = 0$.

Proof. Denote
$$h(t) = \prod_{j=1}^n \exp\{a_j(t-t_j)^{-1}\}. \tag{4.41}$$

If all $\kappa_j(a) > 0$, then $\arg a_j = \eta_\Gamma(t_j)$ and, by Propositions 4.2 and 3.4, $h \in \mathcal{U}$. Similarly, if all $\kappa_j(a) < 0$, then $\arg a_j = \eta_\Gamma(t_j) - \pi$, and consequently $h^{-1} \in \mathcal{U}$. Therefore, in the cases (1) and (2) the representation (4.39) is a generalized factorization of the function a. Here $\kappa = \operatorname{ind} b$ is, generally speaking, different from zero. By Property 3 of the class $\operatorname{fact}(\infty, p, \varrho)$, there exists a generalized factorization $a = \widetilde{h}\widetilde{b}$ or $a = \widetilde{h}^{-1}\widetilde{b}$ of the function a, where $\widetilde{h}$ is a model function and $\operatorname{ind} \widetilde{b} = 0$. Thus, we are in the conditions of Theorem 2.6, and so assertions (1) and (2) are valid.

It remains to show that (3) holds. Let $b = b^-(t-z_0)^\kappa b^+$ be a factorization of the function b, $z_0 \in \mathcal{D}_\Gamma^+$, and

$$h^+(t) = \prod_{j=1}^m \exp\{a_j(t-t_j)^{-1}\}, \quad h^-(t) = \prod_{j=m+1}^n \exp\{a_j(t-t_j)^{-1}\}.$$

Clearly, $h^+, (h^-)^{-1} \in L_\infty^+(\Gamma)$, $(h^+)^{-1}, h^- \in L_\infty^-(\Gamma)$. Let $f^+ \in \ker T(a)$. Then there exists a function $f^- \in L_p^-(\Gamma, \varrho)$ such that

$$b^-(t-z_0)^\kappa b^+ h^- h^+ f^+ = f^-. \tag{4.42}$$

First let us assume that $\kappa \ge 0$. Then from Section 2.4 it follows that $b^+ \in L_q^+(\Gamma, \varrho^{1-q})$ and $(b^-)^{-1} \in L_q^-(\Gamma, \varrho^{1-q}) \oplus \mathbb{C}$, where $q = p/(p-1)$. Therefore,

$$h^-\psi^+ = \psi^-, \tag{4.43}$$

where $\psi^+ = (t-z_0)^\kappa h^+ b^+ f^+ \in L_1^+(\Gamma)$ and $\psi^- = (b^-)^{-1} f^- \in L_1^-(\Gamma)$.

We claim that the functions $\psi^\pm(z)$ are analytic in all the points of the curve Γ, except for the set $M = \{t_{m+1}, \dots, t_n\}$. Indeed, let us surround the point $t_0 \in \Gamma \setminus M$

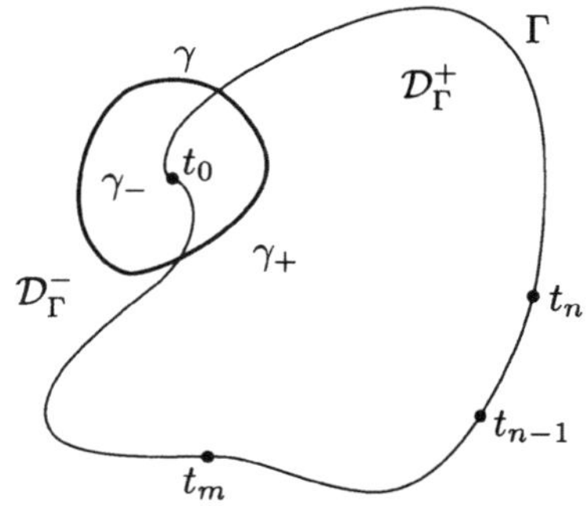

Figure 4.2

by a smooth contour γ such that $\overline{\mathcal{D}_\gamma^+} \cap M = \emptyset$ and such that the curve Γ divides the domain $\mathcal{D}_\gamma^+$ into two simply connected domains bounded by closed curves γ_+ and γ_- with $\mathcal{D}_{\gamma_+}^+ \subset \mathcal{D}_\Gamma^+$ and $\mathcal{D}_{\gamma_-}^+ \subset \mathcal{D}_\Gamma^-$ (see Figure 4.2).

Next, let us introduce the function

$$\psi(z) = \begin{cases} h^-(z)\psi^+(z), & \text{if } z \in \mathcal{D}_\Gamma^+, \\ \psi^-(z), & \text{if } z \in \mathcal{D}_\Gamma^-, \end{cases}$$

which is defined on $\mathbb{C} \setminus \Gamma$ and has interior and exterior nontangential limit values almost everywhere on Γ, which coincide thanks to equality (4.43).

Let us calculate the integral

$$\int_\gamma \psi(z)dz = \int_{\gamma_+} \psi(z)dz + \int_{\gamma_-} \psi(z)dz.$$

Since $\psi^+(z) \in L_1^+(\Gamma)$, one can verify, using the definition of the Smirnov space $E_1(\Gamma) = L_1^+(\Gamma)$ (see Section 2.3), that $\psi^+(z) \in L_1^+(\gamma_+)$. Therefore, $\psi(z) \in L_1^+(\gamma_+)$ ($h^-(z)$ is analytic in a neighborhood of the point t_0) and by the analogue of Proposition 1.1 for Γ the integral along γ_+ is equal to zero. Arguing in much the same way, one can show that the integral along γ_- is also equal to zero. Thus,

$$\int_\gamma \psi(z)dz = 0.$$

It is clear that in the last integral the contour γ can be replaced by any closed rectifiable curve contained in $\mathcal{D}_\gamma^+$. By Morera's theorem, $\psi(z)$ is analytic in $\mathcal{D}_\gamma^+$. Let us consider a neighborhood $\mathcal{O}(t_j)$ of any one of the points t_j, $1 \le j \le m$. Since

$$f^+(t)b^+(t)(t - z_0)^\kappa = (h^+)^{-1}\psi^+(t), \quad t \in \Gamma,$$

where $f^+b^+(t-z_0)^\kappa \in L_1^+(\Gamma)$, we see that $(h^+)^{-1}\psi^+ \in L_1^+(\Gamma)$. But $\psi^+(z)$ is analytic in $\mathcal{O}(t_k)$, and the function $(h^+(z))^{-1}$ grows exponentially when z approaches t_j nontangentially, $z \in \mathcal{D}_\Gamma^+$. Since the function $(t-t_j)^n \exp\{-a_j(t-t_j)^{-1}\}$ does not belong to $L_1^+(\Gamma)$ for any choice of the positive integer n, we conclude that $\psi^+ \equiv 0$, i.e., $\ker T(a) = \{0\}$.

The case $\kappa < 0$ reduces to the one already considered, with equality (4.43) replaced by the equality $(t-z_0)^\kappa h^- \psi^+ = \psi^-$.

Now let us show that $\dim \operatorname{coker} T(a) = 0$. Define the operator $A(a)$ by (2.9). Then from Proposition 2.2 it follows that

$$\dim \operatorname{coker} T(a) = \operatorname{codim} \overline{\operatorname{im} T(a)} = \operatorname{codim} \overline{\operatorname{im} A(a)} = \dim \ker A^*(a),$$

where the operator $A^*(a)$ is similar to the operator $C(a) = P_\Gamma^+ + P_\Gamma^- aI$. The proof of the equality $\ker C(a) = \{0\}$ in $L_p(\Gamma, \varrho)$ is carried out in the same way as the proof of the equality $\ker T(a) = \{0\}$ in $L_p^+(\Gamma, \varrho)$.

It remains to show that $\operatorname{im} T(a) \neq L_p^+(\Gamma, \varrho)$. The equation

$$[T(a)]f^+ = 1$$

is equivalent to the equality

$$a(t)f^+(t) = f^-(t), \qquad \text{where} \qquad f^- \in L_p^-(\Gamma, \varrho) \oplus \mathbb{C}.$$

Repeating the proof of the fact that the kernel of the operator $T(a)$ is trivial, we conclude that in the present case $f^+ = 0$, i.e., $1 \notin \operatorname{im} T(a)$. □

Remark 4.10 The functionals $\kappa_j(a)$ given by (4.38) differ from the almost periodic indices $\sigma_j(p_j)$ considered by Gohberg and Krupnik in [1], p. 341, by the positive factors $|w_j'(t_j)|^{-1}$, where w_j is the corresponding diffeomorphism of the curve Γ onto the circle Γ_0. In the construction of the qualitative theory of the operator $T(a)$ this difference is not essential. However, let us mention that in the description of $\ker T(a)$ in the case where the operator $T(a)$ is right invertible the number $\kappa_j(a)$ is responsible for the behavior of solutions near the point t_j and that in the language of entire functions it coincides with the type of that arbitrary entire function which gives the "contribution" to the space $\ker T(a)$.

4.4 Well-posed problems for the Toeplitz equation

From the preceding section it follows that a Toeplitz operator $T(a)$ whose symbol has standard almost periodic discontinuities has either an infinite-dimensional kernel or an infinite-dimensional cokernel. In connection with this, in the present section we will give two variants of well-posed problems for the equation $[T(a)]f = g$. We restrict ourselves to the case $\Gamma = \mathbb{R}$.

Let us consider the first variant of a well-posed problem for the Toeplitz equation $[T(a)]f = g$. In the case where the operator $T(a)$ is right invertible, the unique solution of the equation is singled out by supplementing the equation with the moment problem

$$\langle f, \varphi_k \rangle = \xi_k, \quad k \in \mathbb{Z},$$

where $\{\varphi_k\}$ is a specially chosen sequence of functionals in the dual space. In this case a crucial role is played by the choice of the basis in $\ker T(a)$ that gives the data $\{\xi_k\}$ and the system of functionals $\{\varphi_k\}$. Usually it is convenient that the system $\{\varphi_k\}$ be conjugate to a basis in $\ker T(a)$.

Let us illustrate the approach just described on the case of an operator $T(a)$ with a uniformly almost periodic symbol in the space $L_p^+(\mathbb{R}, \varrho)$. Recall that in this case $\varrho(x) \asymp |x|^\alpha$ near infinity, where $-1 < \alpha < p-1$. Let us introduce the space $l_{p,\delta}(\mathbb{Z})$ with the weight $\delta = \{(|k|+1)^\alpha\}_{k\in\mathbb{Z}}$. Suppose that equality (4.7) holds, where $\sigma < 0$, and let

$$\varphi_j = P^+ \bar{b} e_j, \quad j \in \mathbb{Z}, \tag{4.44}$$

where the functions e_j have the form (1.30). Since $b \in L_\infty(\mathbb{R})$, φ_j belongs to $L_q^+(\mathbb{R}, \varrho^{1-q})$.

Theorem 4.14 *Let $\sigma < 0$. Under the assumptions of Theorem 4.7, the problem*

$$\begin{cases} [T(a)]f = g, & g \in L_p^+(\mathbb{R}, \varrho), \\ \langle f, \varphi_j \rangle = \xi_j, & j \in \mathbb{Z}, \quad \{\xi_j\} \in l_{p,\delta}(\mathbb{Z}), \end{cases}$$

has a unique solution f in the space $L_p^+(\mathbb{R}, \varrho)$, which is given by

$$f = \frac{1}{b^+} P^+ \frac{1}{b^-} \left[\exp(i|\sigma|x) g + \sum_{k\in\mathbb{Z}} \xi_k e_k \right] \tag{4.45}$$

and depends continuously on the data. Here the functions e_k and φ_j have the form (1.30) *and* (4.44), *respectively.*

Proof. Applying Theorem 4.7 and then Theorem 4.9, we deduce that in the case at hand the general solution of the equation $[T(a)]f = g$ has the form

$$f = \frac{1}{b^+} P^+ \frac{1}{b^-} \left[\sum_{k\in\mathbb{Z}} c_k e_k + \exp(i|\sigma|x) g \right],$$

where $\{c_k\} \in l_{p,\delta}(\mathbb{Z})$. Since $b \in \text{fact}(p, \varrho)$,

$$\int_{\mathbb{R}} \left(\frac{1}{b^+} P^+ \frac{f}{b^-} \right)(x) \left(P^- b^+ b^- g \right)(x) dx = \int_{\mathbb{R}} \left(P^+ \frac{f}{b^-} \right)(x) (b^- g) dx = \int_{\mathbb{R}} f(x) g(x) dx$$

for all $f \in L_p^+(\mathbb{R}, \varrho)$ and all $g \in L_q^-(\mathbb{R}, \varrho^{1-q})$, $q = p/(p-1)$. Observing that $\overline{P^\pm} = P^\mp$ and recalling Remark 4.4, we obtain

$$\begin{aligned} \langle f, \varphi_j \rangle = \langle f, P^+ \bar{b} e_j \rangle &= \langle \sum_{k\in\mathbb{Z}} c_k e_k, e_j \rangle + \langle \exp(i|\sigma|x) g, e_j \rangle \\ &= c_j + \langle g, \exp(-i|\sigma|x) e_j \rangle = c_j, \end{aligned}$$

because $g \in L_p^+(\mathbb{R}, \varrho)$ and $\exp(-i|\sigma|x)e_j \in L_q^-(\mathbb{R}, \varrho^{1-q})$. Therefore, the requirement $\langle f, \varphi_j \rangle = \xi_j$, $j \in \mathbb{Z}$, singles out a unique solution of the equation under consideration, viz, the solution that has the form (4.45). Furthermore, applying Theorem 3.4 we get

$$\|f\| \le c \left(\|g\|_{L_p(\mathbb{R},\varrho)} + \|\{\xi_k\}\|_{l_{p,\delta}(\mathbb{Z})} \right),$$

which proves the continuous dependence of this unique solution on the data g and $\{\xi_k\}$. □

If the operator $T(a)$ is left invertible, then the equation $[T(a)]f = g$ has a solution only for right-hand sides g that lie in $\operatorname{im} T(a)$. The solvability of the equation for any $g \in L_p^+(\mathbb{R}, \varrho)$ can be ensured by adding to its right-hand side a specially chosen system of potentials from $L_p^+(\mathbb{R}, \varrho)$ with coefficients that are functionals of g. In the case considered here the following procedure proves sufficiently effective. Since the space $L_p^+(\mathbb{R}, \varrho)$ admits the direct sum decomposition

$$L_p^+(\mathbb{R}, \varrho) = \operatorname{im} T(a) \oplus \ker[T(a)]^{-1},$$

where $[T(a)]^{-1}$ is some left inverse of the operator $T(a)$, the problem reduces to finding a basis in $\ker[T(a)]^{-1}$ that has a convenient conjugate system. This basis is chosen as a system of potentials. The equation amended by means of this basis becomes solvable for all $g \in L_p^+(\mathbb{R}, \varrho)$, and its solution is given by the rule $f = [T(a)]^{-1}g$. Here we should add that it is advisable to construct from the very beginning a basis in $L_p^+(\mathbb{R}, \varrho) \ominus \operatorname{im} T(a)$ and a conjugate system of functional that is a basis in $\ker T^*(a)$. In this way we obtain, first, a description of the space $\operatorname{im} T(a)$ in terms of the orthogonality of the function g to the basis functionals from $\ker T^*(a)$, and, secondly, the necessary system of potentials together with a simple recipe for computing its coefficients. In this connection see Remark 4.4.

Theorem 4.15 *Let $\sigma > 0$. Under the assumptions of Theorem 4.7, for each given function $g \in L_p^+(\mathbb{R}, \varrho)$ there exists a unique sequence $\{\eta_k\} \in l_{p,\delta}(\mathbb{Z})$ such that the equation*

$$[T(a)]f = g + \sum_{j \in \mathbb{Z}} \eta_j P^+ b e_j \tag{4.46}$$

is solvable in $L_p^+(\mathbb{R}, \varrho)$. Its unique solution is given by the formula

$$\begin{aligned} f &= P^+ \frac{1}{\exp(i\sigma x) b^+} P^+ \frac{g}{b^-}, \\ \eta_j &= -\langle g, (\overline{b^-})^{-1} P^+ (\overline{b^+})^{-1} e_j \rangle, \quad j \in \mathbb{Z}, \end{aligned} \tag{4.47}$$

where the function e_j has the form (1.30).

Proof. From Theorem 4.7 it follows that

$$\begin{aligned} \ker[T(a)]^{-1} &= \ker \left[T(\exp(-i\sigma x)) \left(\frac{1}{b^+} P^+ \frac{1}{b^-} I \right) \right] \\ &= P^+ b I(\ker T(\exp(-i\sigma x))) = P^+ b I(\mathcal{L}_{p,\varrho}^{\sigma+}). \end{aligned}$$

Consequently,

$$f \in \ker[T(a)]^{-1} \Longleftrightarrow f = \sum_{k\in\mathbb{Z}} c_k P^+ b e_k,$$

where $\{c_k\} \in l_{p,\delta}(\mathbb{Z})$. Consider equation (4.46). By Theorem 4.7 and Remark 4.4, for its solvability it is necessary and sufficient that the orthogonality conditions (4.31) be satisfied:

$$\langle g + \sum_{k\in\mathbb{Z}} \eta_k P^+ b e_k, (\overline{b^-})^{-1} P^+ (\overline{b^+})^{-1} e_j \rangle = 0, \quad j \in \mathbb{Z}.$$

But

$$\langle P^+ b e_k, (\overline{b^-})^{-1} P^+ (\overline{b^+})^{-1} e_j \rangle = \delta_{kj}$$

and therefore the above orthogonality conditions are equivalent to equalities (4.47). Since

$$\begin{aligned}
\eta_j &= -\int_{\mathbb{R}} g(x) \left(\frac{1}{b^-} P^- \frac{\overline{e}_j}{b^+} \right)(x) dx \\
&= -\int_{\mathbb{R}} \left(\frac{1}{b^+} P^+ \frac{g}{b^-} \right)(x) \overline{e}_j(x) dx = -\langle P_\sigma \frac{1}{b^+} P^+ \frac{g}{b^-}, e_j \rangle,
\end{aligned}$$

where

$$P_\sigma(L_p^+(\mathbb{R},\varrho)) = \exp(i\sigma x) P^- \exp(-i\sigma x)(L_p^+(\mathbb{R},\varrho)) = \mathcal{L}_{p,\varrho}^{\sigma+},$$

Theorem 3.4 in conjunction with Remark 1.2 yields the conclusion that $\{\eta_k\} \in l_{p,\delta}(\mathbb{Z})$. □

Another approach, which goes back to the classical variant of well-posed Riemann problem, amounts to the fact that the reduction or enlargement of the class of solutions of the equation $[T(a)]f = g$ is achieved by requiring that the solution be equal to zero or infinity on a fixed infinite set of points of the domain $\mathcal{D}^+$. In the case where the operator $T(a)$ is right invertible, the set of zeros of the solution f must be sufficiently "thick" in order to ensure the uniqueness of the solution, and at the same time sufficiently "thin" in order to have good interpolation properties and to guarantee the solvability of the equation for arbitrary right-hand sides $g \in L_p^+(\mathbb{R},\varrho)$. On the contrary, in the case where the operator $T(a)$ is left invertible the set of poles of the solution f must be sufficiently "thin" in order to ensure uniqueness, and sufficiently "thick" to ensure solvability for any $g \in L_p^+(\mathbb{R},\varrho)$.

From the operational point of view, the crucial factor for us is the existence of appropriate "generators" for such point sets, i.e., of functions in $L_\infty^\pm(\Gamma)$ that have the indicated sets (and only them) as their zero sets. In the case of standard almost periodic discontinuities these aims are fulfilled in fully satisfactory manner by the class of Blaschke products.

Let $M_+ = \{z_k\}$ be the set of simple separated zeros of a sine-type function $s(z)$ of type $|\sigma|/2$, lying in the strip $0 < d_1 \le z_k \le d_2 < \infty$. Let us introduce the

Blaschke product

$$B^+(x) = \prod_{k\in\mathbb{R}} \frac{\overline{z}_k}{z_k} \cdot \frac{x - z_k}{x - \overline{z}_k}. \tag{4.48}$$

The following properties of the function B^+ make it a suitable candidate for the role of a "generator" of the set of zeros and poles for well-posed problems formulated for the equation $[T(a)]f = g$ in the case where $a \in \Pi_c(\mathbb{R})$.

Proposition 4.10 *Let $f \in L_p^+(\mathbb{R}, \varrho)$. Then $f(z) = 0$ for all $z \in M_+$ if and only if $f = B^+g$, where $g \in L_p^+(\mathbb{R}, \varrho)$ and B^+ has the form* (4.48).

Proof. This assertion is a simple corollary of Theorem 2.4 on the Riesz factorization. □

Theorem 4.16 *Under the assumptions of Theorem 4.7 for the case $\sigma < 0$, if the function $b = a\exp(-i\sigma x)$ admits a standard factorization, then the equation $[T(a)]f = g$ is solvable in $L_p^+(\mathbb{R}, \varrho)$ for any right-hand side $g \in L_p^+(\mathbb{R}, \varrho)$ and has a unique solution f that satisfies the condition $f(z) = 0$, $z \in M_+$, and depends continuously on g. This solution is given by*

$$f = \exp(-i\sigma x)\frac{1}{b^+}P^+\frac{g}{b^-} + \sum_{k\in\mathbb{Z}} f_k \frac{u_k}{b^+}, \tag{4.49}$$

where

$$f_k = -\left(b^+ \exp(-i\sigma x)P^+\frac{g}{b^-}\right)(z_k), \tag{4.50}$$

and u_k has the form (3.29).

Proof. From Theorems 4.7 and 3.4 it follows that in the case at hand the equation $[T(a)]f = g$ is solvable in $L_p^+(\mathbb{R}, \varrho)$ for any right-hand side $g \in L_p^+(\mathbb{R}, \varrho)$, and its general solution has the form (4.49), where $\{f_k\} \in l_{p,\delta}(\mathbb{Z})$, $\delta = \{|z_k|^\alpha, k \in \mathbb{Z}\}$. Moreover, (4.49) holds for all $z \in \Pi_+$. The requirement $f(z) = 0$, $z \in M_+$, forces the coefficients f_k to be of the form (4.50). Since the set M_+ satisfies the Carleson condition (3.23) and $b^+ \in GL_\infty^+(\mathbb{R})$, Theorem 3.3 gives that $\{f_k\} \in l_{p,\delta}(\mathbb{Z})$. Finally, the estimate

$$\|f\|_{p,\varrho} \le c_1\|g\|_{p,\varrho} + c_2\|\{f_k\}\|_{l_{p,\delta}} \le c_3\|g\|_{p,\varrho}$$

is a consequence of Theorem 3.4. □

Remark 4.11 From Theorem 4.16 it follows that in the case $\sigma < 0$ the operator $T(\exp(i\sigma x)B^+)$ is invertible in $L_p^+(\mathbb{R}, \varrho)$ for any p, $1 < p < \infty$, and any power weights $\varrho \in A_p(\mathbb{R})$.

Before we can turn to the case where the operator $T(a)$ is left invertible, we need an auxiliary statement. Let $\sigma > 0$ and $s^\pm(x) = \exp(\pm i\sigma x/2)s(x)$. Clearly, $s^\pm \in L_\infty^\pm(\mathbb{R})$. Together with the system of functions $\{u_k\}$ of the form (3.29), which constitutes a basis in $\mathcal{L}^{\sigma+}_{p,\varrho}$, let us consider the system of functions $\{v_n\}$ (see (3.40)),

$$v_n = v_{z_n}(x) = \frac{1}{2\pi i} \cdot \frac{\exp(i\sigma x - i\sigma \overline{z}_n) - 1}{x - \overline{z}_n}, \quad z_n \in M_+,$$

which are reproducing kernels for the subspace $\mathcal{M}(\exp(i\sigma x)) = \mathcal{L}^{\sigma+}_{p,\varrho}$.

Proposition 4.11 *The systems of functions $\{u_k\}$ and $\{v_k\}$ are biorthogonal. If $f \in L^+_p(\mathbb{R}, \varrho)$, then $\{\langle f, u_k\rangle\} \in l_{p,\delta}(\mathbb{Z})$, where $\delta = \{|z_k|^\alpha,\ k \in \mathbb{Z}\}$. For f to belong to $\operatorname{im} T(\exp(i\sigma x))$ it is necessary and sufficient that $\langle f, u_k\rangle = 0$ for all $k \in \mathbb{Z}$.*

Proof. Since v_n is a reproducing kernel for the space $\mathcal{L}^{\sigma+}_{p,\varrho}$, we have

$$\langle u_k, v_n\rangle = u_k(z_n) = \begin{cases} 1, & k = n, \\ 0, & k \neq n. \end{cases}$$

The inclusion $\{\langle f, u_k\rangle\} \in l_{p,\delta}(\mathbb{Z})$ was in fact established in the proof of Theorem 3.4. Let us prove the last assertion of the proposition. Suppose that $f \in \operatorname{im} T(\exp(i\sigma x))$. Then $f = \exp(i\sigma x) g$, where $g \in L^+_p(\mathbb{R}, \varrho)$, and

$$\langle f, u_k\rangle = \langle g, \exp(-i\sigma x) u_k\rangle = \int_{\mathbb{R}} \overline{g}(x) \frac{s^-(x)dx}{(s^+)'(z_k)(x - z_k)} = 0,$$

because $\overline{g} \in L^-_p(\mathbb{R}, \varrho)$ and $(x - z_k)^{-1} s^- \in L^-_q(\mathbb{R}, \varrho^{1-q})$, $1/p + 1/q = 1$. Conversely, suppose that $\langle f, u_k\rangle = 0$ for all $k \in \mathbb{Z}$. By Propositions 3.1 and 3.12, $L^+_p(\mathbb{R}, \varrho) = \ker T(\exp(-i\sigma x)) \oplus \operatorname{im} T(\exp(i\sigma x))$ and hence we can write $f = f_1 + \exp(i\sigma x) f_2$, where $f_1 \in \mathcal{L}^{\sigma+}_{p,\varrho}$, $f_2 \in L^+_p(\mathbb{R}, \varrho)$. Since the set $M_+ = \{z_k\}$ satisfies the Carleson condition (3.23) and the operator $T(\overline{B}^+ \exp(i\sigma x))$ is invertible in $L^+_p(\mathbb{R}, \varrho)$ (see Remark 4.11), Theorem 3.7 guarantees that the system of functions $\{v_n\}$ is a basis in $\mathcal{L}^{\sigma+}_{p,\varrho}$. Therefore,

$$f = \sum_{n \in \mathbb{Z}} c_n v_n + \exp(i\sigma x) f_2.$$

Applying the functionals u_k to both sides of this equality, we conclude that $c_n = 0$, $n \in \mathbb{Z}$, i.e., $f \in \operatorname{im} T(\exp(i\sigma x))$. □

Now let us introduce the space $\widetilde{L}^+_p(\mathbb{R}, \varrho)$ of all functions $f(z)$ that are meromorphic in Π_+ and satisfy the condition $B^+ f \in L^+_p(\mathbb{R}, \varrho)$, where B^+ has the form (4.48). This is a complete space with respect to the norm $\|f\| = \|B^+ f\|_{p,\varrho}$.

Theorem 4.17 *Let $\sigma < 0$. Under the assumptions of Theorem 4.7, if the function $b = a \exp(-i\sigma x)$ admits a standard factorization, then the equation $[T(a)]f = g$ is solvable in $\widetilde{L}^+_p(\mathbb{R}, \varrho)$ and has a unique solution for any right-hand side $g \in L^+_p(\mathbb{R}, \varrho)$, which depends continuously on g. This solution is given by the formula*

$$f = \frac{\exp(-i\sigma x)}{b^+} \left(P^+ \frac{g}{b^-} + \sum_{k \in \mathbb{Z}} f_k \frac{\overline{B^+(x)}}{x - \overline{z}_k} \right), \tag{4.51}$$

where

$$\{f_k\} = \{-\frac{1}{2\pi i}\langle B^+P^+\frac{g}{b^-}, u_k\rangle\} \in l_{p,\delta}(\mathbb{Z}), \tag{4.52}$$

and the function u_k *has the form* (3.29).

Proof. The equation $[T(a)]f = g$, where $f \in \widetilde{L}_p^+(\mathbb{R}, \varrho)$, is equivalent to the equation

$$P^+\overline{B^+}\exp(i\sigma x)b^-b^+\widetilde{f} = g, \text{ where } \widetilde{f} \in L_p^+(\mathbb{R}, \varrho).$$

Suppose that $\widetilde{f}$ is a solution of the latter. Since the operator $T(\overline{B^+}b)$ is right invertible in $L_p^+(\mathbb{R}, \varrho)$ and $\ker T(\overline{B^+}) = \mathcal{M}(B^+)$, Theorem 3.6 in conjunction with Proposition 3.18 implies that

$$\exp(i\sigma x)b^+\widetilde{f} = B^+P^+\frac{g}{b^-} + \sum_{k\in\mathbb{Z}}\frac{f_k}{x - \overline{z}_k}, \text{ where } \{f_k\} \in l_{p,\delta}(\mathbb{Z}).$$

By Proposition 4.11,

$$\langle B^+P^+\frac{g}{b^-} + \sum_{k\in\mathbb{Z}}\frac{f_k}{x-\overline{z}_k}, u_n\rangle = 0, \quad n \in \mathbb{Z}.$$

Further, since

$$\langle (x - \overline{z}_k)^{-1}, u_n\rangle = \begin{cases} 2\pi i, & k = n, \\ 0, & k \neq n, \end{cases}$$

the coefficients f_k are necessarily of the form (4.52), and

$$\widetilde{f} = \frac{\exp(-i\sigma x)}{b^+.}\left(B^+P^+\frac{g}{b^-} + \sum_{k\in\mathbb{Z}}\frac{f_k}{x-\overline{z}_k}\right).$$

This proves (4.51). It is readily verified that the converse is also true, i.e., for any $g \in L_p^-(\mathbb{R}, \varrho)$ the function f given by (4.51) belongs to $\widetilde{L}_p^+(\mathbb{R}, \varrho)$ and solves the equation $[T(a)]f = g$. Moreover, using Theorem 3.5, we have

$$\begin{aligned} \|f\| &= \left\|\frac{\exp(-i\sigma x)}{b^+}\left(B^+P^+\frac{g}{b^-} + \sum_{k\in\mathbb{Z}}\frac{f_k}{x-\overline{z}_k}\right)\right\|_{p,\varrho} \\ &\leq c_1\|g\|_{p,\varrho} + c_2\|\{f_k\}\|_{l_{p,\delta}} \leq c_3\|g\|_{p,\varrho}. \end{aligned}$$ □

Remark 4.12 By property (5) of the set of zeros of a sine-type function (see 3.7), the set of zeros and the set of poles of the solution f in Theorems 4.16 and 4.17 are stable under perturbations of a special type. For this kind of perturbations, the spaces described in Theorems 4.16 and 4.17 remain well-posedness classes for Toeplitz equations with symbols of the form (4.7).

4.5 General discontinuities of almost periodic type

When defining the standard almost periodic discontinuities, in addition to the requirement of continuous differentiability of the homeomorphism $\tau = w_0(t)$, imposed in order to guarantee that the normal solvability property of the Toeplitz operator is transferred from the circle Γ_0 to the contour Γ (Theorem 4.11), we asked that the function $w_0(t)$ have a second derivative in the point t_0. The latter requirement ensures that at the standard almost periodic discontinuity points representation (4.39) holds, in which the function $h(t) = \prod_{k=1}^{n} \exp\{a_k(t-t_k)^{-1}\}$ lies in $GL_\infty(\Gamma)$ and b lies in fact(p, ϱ). Examining the proof of Proposition 4.6, one readily observes that the condition $h \in GL_\infty(\Gamma)$ is satisfied whenever

$$\sup_{t\in\Gamma} \frac{|w_k'(t) - w_k'(t_k)|}{|t-t_k|} < \infty, \quad 1 \le k \le n.$$

Let us emphasize, however, that under these circumstances the function b does not necessarily admit a (p, ϱ)-factorization.

Before giving a relevant example, let us consider the algebra $PC(\Gamma)$ of all functions a that are continuous on Γ, except at a finite number of points $\{\tau_k\}_{k=1}^{n}$, where they have one-sided limits

$$a(\tau_k \pm 0) = \lim_{t\to\tau_k\pm 0} a(t).$$

Note that among the points τ_k there may also be nodes of the weight ϱ. Let us associate with such a function $a(t)$ the function of two variables

$$a_{p,\varrho}(t,\mu) = a(t+0)f(t,\mu) + a(t-0)(1-f(t,\mu)), \quad t\in\Gamma, \quad 0\le\mu<1,$$

where

$$f(t,\mu) = \begin{cases} \dfrac{\sin(\pi-\delta(t))\mu}{\sin(\delta(t))} \exp(i\delta(t)(\mu-1)), & \text{if } \delta(t)\in(0,\pi)\cup(\pi,2\pi), \\ \mu, & \text{if } \delta(t)=\pi, \end{cases}$$

and

$$\delta(t) = \begin{cases} \dfrac{2\pi}{p}, & \text{if } t\in\Gamma\setminus\{t_k\}_{k=1}^{m}, \\ \dfrac{2\pi(1+\alpha_k)}{p}, & \text{if } t\in\{t_k\}_{k=1}^{m}. \end{cases}$$

Here the points t_k and the numbers α_k are the nodes and the exponents of the weight ϱ, respectively. The range of the function $a_{p,\varrho}(t,\mu)$ is a closed curve in the complex plane (denoted here by $\Gamma_{p,\varrho}(a)$), which consist of the points $z = a(t)$, $t\in\Gamma$, and the arcs of circles that connect the points $a(\tau_k+0)$ and $a(\tau_k-0)$ in such a way that the angle under which the segment $[a(\tau_k+0), a(\tau_k-0)]$ is seen from the points of the corresponding arc equals $\delta(\tau_k)$. In the case of the space $L_2(\Gamma)$ the indicated arcs degenerate to segments.

Let us denote by γ_k complex numbers such that

$$\frac{a(\tau_k+0)}{a(\tau_k-0)} = \exp(i\gamma_k), \quad k=1,2,\dots,m.$$

Theorem 4.18 *Let $a \in PC(\Gamma)$. Then for the operator $T(a)$ to be normally solvable in the space $L_p^+(\Gamma,\varrho)$ it is necessary and sufficient that*

(1) $a(t\pm 0) \neq 0$, $t \in \Gamma$;

(2) *the numbers* $(\operatorname{Re}\gamma_k - \delta(\tau_k))/2\pi$ *be not integers.*

If conditions (1) *and* (2) *are satisfied, then* $a \in \operatorname{fact}(p,\varrho)$ *and its factorization index is equal to the number of turns that the curve* $\Gamma_{p,\varrho}(a)$ *makes around the origin.*

Note that conditions (1), (2) are equivalent to the requirement that $a_{p,\varrho}(t,\mu) \neq 0$, $t \in \Gamma$, $\mu \in [0,1)$.

Here now is the promised example. Suppose that the curve Γ is mapped onto Γ_0 by a diffeomorphism $\tau = w_0(t)$, $w_0(t) = 1$, which in a neighborhood of the point t_0 has the form

$$w_0(t) = \begin{cases} 1+(t-t_0)+ic_+(t-t_0)^2, & \text{if } t \in l_+, \\ 1+(t-t_0)+ic_-(t-t_0)^2, & \text{if } t \in l_-, \end{cases}$$

where $l_\pm$ are semi-neighborhoods of the point t_0 and the real numbers $c_\pm$ are selected so that

$$c_+ - c_- = -\frac{\pi(1+\alpha_0)}{p},$$

where α_0 is the exponent of the weight ϱ at the point t_0. Consider the function

$$a(t) = \exp\left(\frac{w_0(t)+1}{w_0(t)-1}\right).$$

Clearly, $a(t)$ is continuous on Γ, except in the point t_0, where it has an almost periodic discontinuity with $p_0(\tau) = \exp((\tau+1)/(\tau-1))$. Following the proof of Proposition 4.6, we obtain

$$a(t) = \exp(2(t-t_0)^{-1})c_0(t),$$

where

$$c_0(t) = \begin{cases} \exp\left(1 - \dfrac{2ic_+}{1+ic_+(t-t_0)}\right), & \text{if } t \in l_+, \\ \exp\left(1 - \dfrac{2ic_-}{1+ic_-(t-t_0)}\right), & \text{if } t \in l_-, \end{cases}$$

is a function continuous on $\Gamma \setminus \{t_0\}$ that has a discontinuity of the first kind in t_0. Since

$$c_0(t_0 \pm 0) = \exp(1 - 2ic_\pm), \quad \frac{c_0(t_0+0)}{c_0(t_0-0)} = \exp\left(i\frac{2\pi(1+\alpha_0)}{p}\right),$$

Theorem 4.18 shows that $T(c_0)$ is not a Φ-operator, and therefore $c_0 \notin \operatorname{fact}(p, \varrho)$. Thus, the conditions imposed on the behavior of the diffeomorphisms $w_k(t)$ in the points t_k are sharp with respect to smoothness in order for the function $a(t)$ to admit a generalized factorization (4.39). It is because of this circumstance that we termed the corresponding a.p. discontinuities *standard*.

At the same time, dropping the additional requirements on the behavior of the function $w_0(t)$ at the points t_k does not prevent us from constructing a theory of normal solvability of the operator $T(a)$. It is true, however, that in doing so we are, generally speaking, no longer able to predict what the structure of the inverse operators will be and neither will or to describe the subspaces $\ker T(a)$ and $\operatorname{im} T(a)$.

We say that the function $a(t) \in L_\infty(\Gamma)$ has *a discontinuity of almost periodic type with characteristic* (p_0, w_0) in the point $t_0 \in \Gamma$ if there exist a function $p_0(\tau) \in \Pi_c(\Gamma_0)$ and a diffeomorphism $\tau = w_0(t)$ of the curve Γ onto the curve Γ_0, preserving the orientation of Γ, such that $w_0(t_0) = 1$, and

$$\lim_{t \to t_0} (a(t) - p_0(w_0(t))) = 0.$$

The next theorem contains the main result of this section.

Theorem 4.19 *Assume that the function $a \in GL_\infty(\Gamma)$ is continuous on $\Gamma \setminus \{t_k\}_{k=1}^n$ and that is has discontinuities of almost periodic type with characteristics (p_k, w_k) at the points t_k, where $\sigma(p_k) \neq 0$, $1 \le k \le n$. Then the following assertions hold true.*

(1) *If $\sigma(p_k) > 0$, $1 \le k \le n$, then the operator $T(a)$ is left invertible in $L_p^+(\mathbb{R}, \varrho)$ and* $\dim \operatorname{coker} T(a) = \infty$.

(2) *If $\sigma(p_k) < 0$, $1 \le k \le n$, then the operator $T(a)$ is right in $L_p^+(\mathbb{R}, \varrho)$ invertible and* $\dim \ker T(a) = \infty$.

(3) *If $\sigma(p_k) > 0$, $1 \le k \le m$, and $\sigma(p_k) < 0$, $m+1 \le k \le n$, then the operator $T(a)$ is not normally solvable in any of the spaces $L_p^+(\mathbb{R}, \varrho)$, $1 < p < \infty$.*

Proof. In the course of the proof of Theorem 4.12 we have shown that

$$a(t) = \left(\prod_{k=1}^{n} p_k(w_k(t))\right) c(t),$$

where $c \in GC(\Gamma)$. For that purpose no additional smoothness properties of the functions $w_k(t)$ in the points t_k were needed. Let us introduce the functions

$$a_k(t) = p_k(w_k(t))\widetilde{c}_k(t),$$

where $\widetilde{c}_k \in GC(\Gamma)$ and, in some neighborhood of the point t_k,

$$\widetilde{c}_k(t) = \left(\prod_{j=1, j\neq k}^{n} p_j(w_j(t)) \right) c(t).$$

Consider the operators $T(a_k)$ and $T(\widetilde{a}_k)$, where

$$\widetilde{a}_k(\tau) = p_k(\tau)\widetilde{c}_k(w_k^{-1}(\tau)) \in GL_\infty(\Gamma_0).$$

The function $\widetilde{a}_k(\tau)$ has a unique standard almost periodic discontinuity in the point $t = 1$ and $\kappa_k(\widetilde{a}_k) = \sigma(p_k)$. By Theorem 4.13, if $\sigma(p_k) > 0$ the operator $T(\widetilde{a}_k)$ is left invertible and $\dim \operatorname{coker} T(\widetilde{a}_k) = \infty$, while if $\sigma(p_k) < 0$ the operator $T(\widetilde{a}_k)$ is right invertible and $\dim \ker T(\widetilde{a}_k) = \infty$. But then, in view of Theorem 4.11, the same is true about the operator $T(a_k)$. To complete the proof of the theorem it remains to use the following principle of separation of singularities. □

Theorem 4.20 *Let Γ be a simple closed smooth contour and let the function $a \in L_\infty(\Gamma)$ be continuous on the set $\Gamma \setminus M$, where $M = \{\tau_k\}_{k=1}^n$, and $\inf_{t\in\Gamma\setminus M} |a(t)| > 0$. Suppose that there exist functions a_k, $1 \le k \le n$, with the following properties:*

(1) *a_k is continuous on the set $\Gamma \setminus \{\tau_k\}$;*

(2) *$a_k(t) = a(t)$ in some neighborhood of the point t_k;*

(3) $\inf_{t\in\Gamma\setminus\{\tau_k\}} |a_k(t)| > 0$.

Then for the operator $T(a)$ to be right (resp., left) invertible in the space $L_p(\Gamma, \varrho)$ and to have an infinite-dimensional kernel (resp., cokernel) it is necessary and sufficient that all the operators $T(a_k)$, $1 \le k \le n$, be right (resp., left) invertible and that at least one of them have an infinite-dimensional kernel (resp., cokernel).

If the operators $T(a_k)$ with $1 \le k \le m < n$ are Φ_+-operators and at least one of them is not a Φ-operator, while for $m+1 \le k \le n$ the operators $T(a_k)$ are Φ_--operators and at least one of them is not a Φ-operator, then $T(a)$ is not normally solvable.

Semi-almost periodic discontinuities

A direct generalization of the almost-periodic functions considered in the preceding section is the class of functions that have a different almost periodic behavior when the arguments tends on $\mathbb{R}$ to plus infinity and minus infinity. The discontinuities defined by such functions are called semi-almost periodic. In this section we consider Toeplitz operators on the real line $\mathbb{R}$ whose symbols have discontinuities of the described type at infinity. The main step is here the construction of the corresponding model function, which is carried out with the help of the integral sine function $\mathrm{Si}(z)$, well known in the theory of special functions.

4.6 The class SAP

The set of all continuous functions on $\mathbb{R}$ which have (generally, different) limits as $x \to \pm\infty$ will be denoted by $C(\overline{\mathbb{R}})$. We will use SAP to denote the closed subalgebra of $L_\infty(\mathbb{R})$ generated by the union of $\Pi_c(\mathbb{R})$ (see Section 4.1) and $C(\overline{\mathbb{R}})$. Let us choose a pair of functions $u_\pm \in C(\overline{\mathbb{R}})$ with the following properties:

$$u_+(+\infty) = 1, \quad u_+(-\infty) = 0, \quad u_-(x) = 1 - u_+(x). \tag{4.53}$$

Theorem 4.21 *Any function $f \in SAP$ can be represented in the form*

$$f = f_0 + u_+ f_+ + u_- f_-, \tag{4.54}$$

where $f_\pm \in \Pi_c(\mathbb{R})$, $f_0 \in C(\overline{\mathbb{R}})$ and $f_0(\pm\infty) = 0$. The functions $f_\pm$ are uniquely determined and independent of the choice of the pair of functions $u_\pm$ of the form (4.53). *Moreover, if*

$$\inf_{x\in\mathbb{R}} |f(x)| > 0, \tag{4.55}$$

then $\inf\limits_{x\in\mathbb{R}} |f_\pm(x)| > 0$.

Proof. According to the definition of the algebra SAP, any function $f \in SAP$ can be represented as the uniform limit of a sequence of functions of the form $f_n = \sum\limits_{j=1}^{n} a_j b_j$, where $a_j \in \Pi_c(\mathbb{R})$ and $b_j \in C(\overline{\mathbb{R}})$. Let us write each function f_n in the form (4.54), that is,

$$f_n = f_{n,0} + u_+ f_{n+} + u_- f_{n-},$$

where

$$f_{n+}(x) = \sum_{j=1}^{n} a_j(x) b_j(+\infty), \quad f_{n-}(x) = \sum_{j=1}^{n} a_j(x) b_j(-\infty).$$

Since for each $\varepsilon > 0$ there are m and n large enough such that $\|f_n - f_m\|_\infty < \varepsilon$, we have

$$\sup_{x>M} |f_{n+}(x) - f_{m+}(x)| < 2\varepsilon.$$

Using Proposition 4.8, this yields $\|f_{n+} - f_{m+}\|_\infty < 2\varepsilon$, that is, the sequence $\{f_{n+}\}$ is fundamental in $\Pi_c(\mathbb{R})$. Similarly, the sequence $\{f_{n-}\}$ is fundamental in $\Pi_c(\mathbb{R})$. Denoting the limits of these sequences by f_+ and f_-, respectively, we arrive at representation (4.54). Clearly, the limits $f_\pm$ are independent of the choice of the functions $u_\pm$ satisfying (4.53).

The uniqueness of the functions $f_\pm$ in representation (4.54) follows from the fact that when $f \equiv 0$ one also has $f_+ = f_- \equiv 0$. Indeed, if $f \equiv 0$, then, in much the same manner as above, for any $\varepsilon > 0$ there is an $M > 0$ for which

$$\sup_{x>M} |f_+(x)| = \|f_+\|_\infty < 2\varepsilon.$$

Now let us prove the last claim of the theorem. Indeed, for any $\varepsilon > 0$ there is an $M > 0$ such that $\sup_{x>M} |f(x) - f_+(x)| < \varepsilon$. But then, denoting $\Delta = \inf_{x\in\mathbb{R}} |f(x)| > 0$, we have $\inf_{x>M} |f_+(x)| > \Delta - \varepsilon$. Using again Proposition 4.8, we conclude that, for ε small enough, $\inf_{x\in\mathbb{R}} |f_+(x)| > \Delta - \varepsilon > 0$. In a similar manner one verifies that $\inf_{x\in\mathbb{R}} |f_-(x)| > 0$. □

To each function $f \in SAP$ that satisfies condition (4.55) we associate the pair of numbers

$$\mu_+(f) = \sigma(f_+), \quad \mu_-(f) = \sigma(f_-), \tag{4.56}$$

where $\sigma(f)$ is the almost periodic index of a function f of the form (4.4). If $\mu_+(f) = \mu_-(f) = 0$, then $\log f_\pm \in \Pi_c(\mathbb{R})$ (Theorem 4.4). In this case we associate to f yet another pair of numbers, namely,

$$\lambda_\pm(f) = \lambda(f_\pm) \stackrel{\text{def}}{=} \exp\{\mathbf{M}[\log f_\pm]\}, \tag{4.57}$$

where

$$\mathbf{M}(p) = \lim_{T\to\infty} \frac{1}{2T} \int_{-T}^{T} p(x)dx,$$

is the mean value of a uniformly almost periodic function, which exists for any $p \in \Pi_c(\mathbb{R})$. Clearly, $\lambda(f \cdot g) = \lambda(f) \cdot \lambda(g)$ and $\lambda(f) = f$ whenever f is a constant.

An example of a function in SAP is

$$f(x) = \frac{1}{2}[1 + \operatorname{sign} x] \exp(i\mu_+ x) + \frac{1}{2}[1 - \operatorname{sign} x] \exp(i\mu_- x), \tag{4.58}$$

where $\mu_\pm \in \mathbb{R}$. Here $f_+(x) = \exp(i\mu_+ x)$, $f_-(x) = \exp(i\mu_- x)$.

In the next section we will construct a model function whose behavior at infinity imitates the behavior of a function of the form (4.58).

4.7 Model function

Let us consider the integral sine function[1]

$$\mathrm{Si}(z) = \frac{1}{\pi}\int_0^z \frac{\sin\tau d\tau}{\tau}, \qquad z \in \mathbb{C}.$$

$\mathrm{Si}(z)$ is an entire function of exponential type, $\mathrm{Si}(-z) = -\mathrm{Si}(z)$, $z \in \mathbb{C}$, and its asymptotics at infinity has the form

$$\mathrm{Si}(z) = \begin{cases} \frac{1}{2} - \frac{e^{iz}}{2\pi z}(1 + O(\frac{1}{|z|})) - \frac{e^{-iz}}{2\pi z}(1 + O(\frac{1}{|z|})), & |\arg z| < \pi, \\ -\frac{1}{2} - \frac{e^{iz}}{2\pi z}(1 + O(\frac{1}{|z|})) - \frac{e^{-iz}}{2\pi z}(1 + O(\frac{1}{|z|})), & 0 < \arg z < 2\pi. \end{cases} \tag{4.59}$$

From (4.59) one derives the following properties of $\mathrm{Si}(z)$.

Proposition 4.12 *Let $x \in \mathbb{R}$. Then*

(1) $\lim_{x\to\pm\infty} \mathrm{Si}(x) = \pm\frac{1}{2}$ *and for any p, $1 < p < \infty$, and any weight ϱ of the form* (1.1)

$$\left(\mathrm{Si}(x) - \frac{1}{2}\operatorname{sign} x\right) \in L_p(\mathbb{R}, \varrho);$$

(2) $\exp(\pm ix)\,\mathrm{Si}(x) \in H^\infty(\Pi_\pm)$.

Now let us introduce the entire functions

$$E_\pm(\mu z) = \left(\mathrm{Si}(\mu z) \pm \frac{1}{2}\right)\exp(i\mu z), \qquad \mu \in \mathbb{R}. \tag{4.60}$$

From Proposition 4.12 it follows that

$$E_\pm(\mu z) \in H^\infty(\Pi_+) \text{ if } \mu > 0, \quad E_\pm(\mu z) \in H^\infty(\Pi_-) \text{ if } \mu < 0, \tag{4.61}$$

and for $\mu > 0$,

$$\begin{cases} \lim_{x\to+\infty}(E_+(\mu x) - \exp(i\mu x)) = 0, & \lim_{x\to-\infty} E_+(\mu x) = 0, \\ \lim_{x\to+\infty} E_-(\mu x) = 0, & \lim_{x\to-\infty}(E_-(\mu x) - \exp(i\mu x)) = 0. \end{cases} \tag{4.62}$$

Therefore, the functions $E_\pm(\mu x)$ belong to SAP (for example, $E_+(\mu x) = f_0(x) + u_+(x)\exp(i\mu x)$). However, they cannot be used directly as model functions, because, due to the properties of $\mathrm{Si}(z)$, the function $E_\pm(\mu x)$ have infinitely many zeros which, for $\mu > 0$, tend to $-\infty$ in the case of $E_+(\mu x)$ and to $+\infty$ in the case of $E_-(\mu x)$. Nevertheless, one can choose positive numbers $v_\pm$ such that

$$h_{\mu_\pm}(x) = E_\pm(\mu_\pm x) + v_\pm \tag{4.63}$$

will be model functions for any $\mu_\pm > 0$.

[1] The customary definition does not include the factor $1/\pi$, which is used here for technical reasons.

Let us study in more detail the behavior of the functions $h_{\mu\pm}(z)$. Put $\Pi_\delta = \{z \in \mathbb{C} : \operatorname{Im} z > \delta\}$, $\delta \in \mathbb{R}$, and let $\{z_{\pm k}\}$ be the set of zeros of the function $h_{\mu\pm}(z)$.

Theorem 4.22 *Given any $\varepsilon > 0$, one can find numbers $v_\pm \in (0, \varepsilon)$ such that the functions $h_{\mu\pm}$ given by (4.63) are model functions on $\mathbb{R}$. Moreover,*

(1) *for any $\delta \in \mathbb{R}$, $h_{\mu\pm}(z) \in H^\infty(\Pi_\delta)$,*

(2) *for any $\varphi \in (0, \pi)$, $\lim_{r\to\infty} h_{\mu\pm}(r\exp(i\varphi)) = v_\pm$.*

Also, for any $d \in \mathbb{R}$,

$$\lim_{x\to\pm\infty}(h_{\mu\pm}(x+id) - \exp(-\mu_\pm d)\exp(i\mu_\pm x)) = v_\pm,$$

$$\lim_{x\to\mp\infty} h_{\mu\pm}(x+id) = v_\pm,$$

and for any $\varphi \in (-\pi, 0)$,

$$\lim_{r\to\infty}(|h_{\mu\pm}(r\exp(i\varphi))|r\exp(2r\mu_\pm\sin\varphi)) = \frac{1}{2\pi}.$$

(3) *The zeros of the functions $h_{\mu\pm}(z)$, $z_{\pm k}$, $k \in \mathbb{N}$, are simple and have the following asymptotics for $k \to +\infty$:*

$$z_{\pm k} = \pm\frac{2\pi k}{\mu_\pm} \pm \frac{\pi}{\mu_\pm} - i\frac{\log v_\pm}{\mu_\pm} + O(k^{-1}). \tag{4.64}$$

(4) *If one denotes $U_\pm(\varepsilon) = \bigcup_{k=1}^\infty\{z \in \mathbb{C} : |z - z_{\pm k}| < \varepsilon\}$, then*

$$\inf_{z\in\mathbb{C}\setminus U_\pm(\varepsilon)} |h_{\mu\pm}(z)| > 0.$$

Proof. We will consider only the function h_{μ_+}, the case of the function h_{μ_-} being similar. The inclusion $h_{\mu_+} \in H^\infty(\Pi_+)$ follows from (4.61). Let us show that the number v_+, $0 < v_+ < \varepsilon$, can be chosen in such a way that $h_{\mu_+}^{-1} \in L_\infty(\mathbb{R})$. Indeed, from (4.59) it follows that for $M > 0$ large enough and any v_+, $0 < v_+ < 1$,

$$\inf_{|x|>M} |h_{\mu_+}(x)| > 0.$$

But the set

$$U_M = \{z \in \mathbb{C} : z = E_+(\mu_+ x),\ x \in [-M, M]\}$$

is closed and contains only finitely many real points, since

$$E_+(\mu_+ x) = \exp(i\mu_+ x)\left(\operatorname{Si}(\mu_+ x) + \frac{1}{2}\right)$$

and the function $\mathrm{Si}(\mu_+ x)$ is real-valued. That set of real points has the explicit form

$$\{0, \pm(\mathrm{Si}(\pi k) + \frac{1}{2}), \qquad \text{where} \quad |k| \leq \frac{\mu_+ M}{\pi}, k \in \mathbb{Z}\}.$$

It follows that in any interval $(0, \varepsilon)$ one can find a number v_+ which is separated from the set U_M. But then, for such a v_+, $\inf_{|x|\leq M} |h_{\mu_+}(x)| > 0$, i.e., $h_{\mu_+}^{-1} \in L_\infty(\mathbb{R})$.

Assertions (1) and (2) are straightforward consequences of (4.59).

The form of the asymptotics of the zeros $z_{\pm k}$ of the function $h_{\mu_\pm}(z)$ follows from (4.63) and (4.59). Its principal term $\zeta_{\pm k}$ is a solution of the equation $\exp(i\mu_\pm z) + v_\pm = 0$ and $|\zeta_{\pm k}| \asymp |k|$, $k \in \mathbb{Z} \setminus \{0\}$. For large values of $|z|$ the equation $h_{\mu_\pm}(z) = 0$ has the form $\exp(i\mu_\pm z) + v_\pm + O(|z|^{-1}) = 0$. Hence, by applying Rouché's theorem to it in sufficiently small neighborhoods of the points $\zeta_{\pm k}$ and for sufficiently large values of $|k|$ one obtains formulas (4.64). Furthermore, the numbers $v_\pm$ can be chosen so that all the zeros $z_{\pm k}$ are simple. Indeed, otherwise one could find an interval $I_\pm \subset \mathbb{R}$ such that for any $v_\pm \in I_\pm$ the functions $h_{\mu_\pm}(z)$ are model functions and have multiple (of course, different) zeros. But then the function $h'_{\mu_\pm}(z)$, which does not depend on $v_\pm$, would have a continuum of zeros, which is not possible.

Assertion (4) follows from assertion (3) and equalities (4.59). □

Remark 4.13 Without modifying the proof of Theorem 4.22 in substantial manner one can show that, for any given $\varepsilon > 0$, $v_\pm$ may be chosen to be complex numbers with the property that $\arg v_\pm$ take any prescribed value in $[0, 2\pi)$ and $|v_\pm| \in (0, \varepsilon)$.

Theorem 4.22 enables us to construct a model function in the form of a Blaschke product.

Proposition 4.13 *The functions $h_{\mu_\pm}$ of the form* (4.63) *admit a Riesz factorization of the form*

$$h_{\mu_\pm} = F_{\mu_\pm} B_{\mu_\pm},$$

where

$$B_{\mu_\pm}(x) = \prod_{k=1}^{\infty} \frac{|1 + z_{\pm k}^2|}{1 + z_{\pm k}^2} \cdot \frac{x - z_{\pm k}}{x - \overline{z}_{\pm k}}$$

is the Blaschke product constructed from the zeros of $h_{\mu_\pm}(z)$ and

$$F_{\mu_\pm}(x) = h_{\mu_\pm} \overline{B_{\mu_\pm}(x)}$$

are outer functions, $F_{\mu_\pm}, F_{\mu_\pm}^{-1} \in H^\infty(\Pi_+)$.

Proof. The function h_{μ_+} belongs to $H^\infty(\Pi_+)$, and $h_{\mu_+}^{-1}$ belongs to $L_\infty(\mathbb{R})$. By Theorem 2.2, the outer part $F_{\mu_\pm}$ of the Riesz factorization of the function $h_{\mu_\pm}$ satisfies the condition $F_{\mu_+}^{\pm 1} \in H^\infty(\Pi_+)$. Next, $h_{\mu_+}(z)$ is an entire function. Therefore, assuming that the inner part of the factorization contains a singular factor $E(z)$, we infer that the support of the singular measure that generates this factor contains at most one point, and so $E(z) = \exp(i\sigma z)$, $\sigma \geq 0$. But then for $\sigma > 0$,

$$\lim_{y\to+\infty} |h_{\mu_+}(iy)| = \lim_{y\to+\infty} |F_{\mu_+}(iy)||B_{\mu_+}(iy)| \exp(-\sigma y) = 0,$$

which contradict assertion (2) of Theorem 4.22. We conclude that $\sigma = 0$ and $E(z) \equiv 1$. The case of the function h_{μ_-} is dealt with in a similar manner. □

4.8 Generalized factorization of SAP functions

The model functions $h_{\mu_\pm}$ constructed in the preceding section enable us to study generalized factorization in the class SAP. First let us prove the following auxiliary result.

Proposition 4.14 *Suppose $f \in SAP$, condition* (4.55) *is satisfied, $\mu_+(f) = \mu_-(f) = 0$ and $\lambda_+(f) = \lambda_-(f)$, where the functionals $\mu_\pm$ and $\lambda_\pm$ are given by* (4.56) *and* (4.57), *respectively. Then the function f is an* MNAF.

Proof. The function f admits the representation (4.54),

$$f = f_0 + u_+ f_+ + u_- f_-.$$

With no loss of generality we may assume that $\lambda_+(f) = \lambda_-(f) = 1$. By the assumptions of the proposition, f_+ can be approximated in the metric of $L_\infty(\mathbb{R})$ by a sequence of almost periodic polynomials of the form

$$f_{n+}(x) = \sum_{j=1}^{n} c_j \exp(\alpha_j x), \quad \alpha_j \in \mathbb{R},$$

for which $\sigma(f_{n+}) = 0$, $\lambda(f_{n+}) = 1$. By Theorem 4.21, $\inf_{x\in\mathbb{R}} |f_{n+}(x)| > 0$. Hence, by Theorem 4.4, $g_{n+}(x) = \log(f_{n+}(x)) \in \Pi_{\mathrm{w}}(\mathbb{R})$ and the function g_{n+} can be written in the form

$$g_{n+}(x) = \sum_{j=1}^{\infty} d_j \exp(i\beta_j x), \qquad \text{where } \beta_j \in \mathbb{R}, \quad \sum_{j=1}^{\infty} |d_j| < \infty.$$

Since

$$\mathbf{M}(\exp(i\beta_j t)) = \begin{cases} 0, & \text{if } \beta_j \neq 0, \\ 1, & \text{if } \beta_j = 0, \end{cases}$$

and $\mathbf{M}(g_{n+}) = \mathbf{M}(\log f_{n+}) = 0$, one has $\beta_j \neq 0$, $j \in \mathbb{N}$. Consequently,

$$g_{n+}(x) = g_n^+(x) - g_n^-(x) = \sum_{\beta_j>0} d_j \exp(i\beta_j x) + \sum_{\beta_j<0} d_j \exp(i\beta_j x).$$

Let us introduce the functions

$$a_n^+(x) = \sum_{\beta_j>0} d_j E_+(\beta_j x), \quad a_n^-(x) = \sum_{\beta_j<0} d_j E_-(\beta_j x),$$

where $E_\pm$ are given by (4.60). By (4.61), $a_n^\pm \in H^\infty(\Pi_\pm)$, and (4.62) implies that

$$\lim_{x\to+\infty}(g_{n+}(x) - a_n^+(x) - a_n^-(x)) = 0, \quad \lim_{x\to-\infty} a_n^\pm(x) = 0.$$

Finally, let us introduce the function

$$F_{n+}(x) = \exp\{a_n^+(x)\}\exp\{a_n^-(x)\}.$$

Notice that this equality is actually a standard factorization of the function F_{n+} and also that

$$\lim_{x\to+\infty}(f_{n+}(x) - F_{n+}(x)) = 0, \quad \lim F_{n+}(x) = 1.$$

Similarly, approximating the almost periodic function f_- by a sequence of almost periodic polynomials $\{f_{n-}\}$ and repeating the procedure for the function F_{n+}, we construct the function

$$F_{n-}(x) = \exp\{b_n^+(x)\}\exp\{b_n^-(x)\},$$

which satisfies the conditions

$$\lim_{x\to-\infty}(f_{n-}(x) - F_{n-}(x)) = 0, \quad \lim_{x\to-\infty} F_{n-}(x) = 1.$$

Now consider the functions

$$f_n = f_0 + u_+ f_+ + u_- f_-.$$

Clearly,

$$\lim_{n\to\infty}\|f - f_n\| = 0,$$

where

$$f_n(x) = c_n(x)F_{n+}(x)F_{n-}(x),$$

and the function $c_n(x)$ is continuous on the real line $\mathbb{R}$ ($c_n(x) \in C(\overline{\mathbb{R}})$, $c_n(\pm\infty) = 1$). By Theorem 2.1 and assertion (3) of Theorem 4.6, the function f_n is a MNAF. But then, by assertion (2) of Theorem 4.6, f is also an MNAF. □

Theorem 4.23 *Let $a \in SAP$ and let condition* (4.55) *be satisfied. If $\mu_\pm = \mu_\pm(a) = 0$, then*

$$a = hb, \tag{4.65}$$

where $h \in C(\overline{\mathbb{R}})$, $h(\pm\infty) = \lambda_\pm(a)$, and b is an MNAF.

If, however, $\mu_+^2 + \mu_-^2 \neq 0$, then there exist numbers $v_\pm$ such that

$$a = h_{|\mu_+|}^{\varepsilon_+} h_{|\mu_-|}^{\varepsilon_-} b, \tag{4.66}$$

where $b \in \mathrm{fact}(p, \varrho)$; moreover, if both $\mu_\pm \neq 0$, then $\varepsilon_\pm = \mathrm{sign}\, \mu_\pm$ and the functions $h_{|\mu_\pm|}$ are given (4.63), while if $\mu_+ = 0$ (resp., $\mu_- = 0$), then $h_{|\mu_+|}^{\varepsilon_+} \equiv v_+$ (resp., $h_{|\mu_-|}^{\varepsilon_-} \equiv v_-$).

Proof. Assume first that $\mu_\pm = 0$. Pick an arbitrary nondegenerate function $h \in C(\overline{\mathbb{R}})$ satisfying $h(\pm\infty) = \lambda_\pm(f)$, and consider the function $b = ah^{-1}$. Since $a \in SAP$, $b \in SAP$ as well. Using representation (4.54) for the function a,

$$a = a_0 + u_+ a_+ + u_- a_-,$$

we see that

$$\begin{aligned} b &= b_0 + u_+ b_+ + u_- b_-, \quad b_\pm = h^{-1}(\pm\infty) a_\pm, \\ b_0 &= h^{-1} a_0 + u_+ a_+ (h^{-1} - h^{-1}(+\infty)) + u_- a_- (h^{-1} - h^{-1}(-\infty)). \end{aligned}$$

Since $\mu_\pm(b) = 0$ and $\lambda_\pm(b) = 1$, Proposition 4.14 shows that b is a MNAF.

Now let $\mu_+^2 + \mu_-^2 \neq 0$. Consider the function

$$g = g_0 + v_-^{-\varepsilon_-} u_+ g_+ + v_+^{-\varepsilon_+} u_- g_-,$$

where $g_\pm(x) = \exp(-i\mu_\pm x) a_\pm(x)$, $g_0 \in C(\overline{\mathbb{R}})$ and $g_0(\pm\infty) = 0$, with g_0 selected so that $\inf_{x\in\mathbb{R}} |g(x)| > 0$. The function

$$b = ah_{|\mu_+|}^{-\varepsilon_+} h_{|\mu_-|}^{-\varepsilon_-}$$

can be written as $b = cg$, where

$$c = bg^{-1} = ah_{|\mu_+|}^{-\varepsilon_+} h_{|\mu_-|}^{-\varepsilon_-} g^{-1}.$$

The function c lies in $C(\mathbb{R})$. Let us examine its behavior at infinity. Using assertion (2) of Theorem 4.22 in the case $\mu_+ \neq 0$ we conclude that, when $x \to +\infty$, $c(x)$ is equal to

$$\begin{aligned} &\frac{\exp(i\mu_+ x) g_+(x) + o_1(x)}{(\exp(i|\mu_+|x) + v_+ + o_2(x))^{\varepsilon_+} (v_- + o_3(x))^{\varepsilon_-} (v_-^{-\varepsilon_-} g_+(x) + o_4(x)))} \\ &= \exp(i\mu_+ x)(\exp(i|\mu_+|x) + v_+)^{-\varepsilon_+} + o_5(x), \end{aligned}$$

where $o_i(x)$, $i = 1, \ldots, 5$, are functions which tend to zero as $x \to +\infty$. If in the choice of $|v_+|$ for the function $h_{|\mu_+|}$ we take ε sufficiently small (Theorem 4.22), we can ensure that $|c(x) - 1| < \varepsilon$ for sufficiently large positive values of x. Similarly, considering the behavior of the function $c(x)$ as $x \to -\infty$ in the case when $\mu_- \neq 0$, we can choose $|v_-|$ so that $|c(x)-1| < \varepsilon$ for all negative values of x with sufficiently large modulus. We conclude that the numbers $|v_\pm|$ can be chosen so that for any $\varepsilon > 0$ one can find an $M_\varepsilon > 0$ for which

$$\sup_{|x|>M_\varepsilon} |c(x) - 1| < \varepsilon. \tag{4.67}$$

But then the function $c(x)$ is locally sectorial and, by Theorem 2.1, $c \in \text{fact}(p, \varrho)$.

Since $\sigma(g_\pm) = 0$, with every choiche of the number $\arg v_+$ we can associate an appropriate number $\arg v_-$ such that

$$v_-^{-\varepsilon_-} \lambda_+(g) = v_+^{-\varepsilon_+} \lambda_-(g).$$

That this can be done without violating the conditions $h_{|\mu_-|}^{-1} \in L_\infty(\mathbb{R})$ and (4.67) follows from Proposition 4.13. It is possible that in this process we need to make somewhat smaller only one of the numbers $|v_\pm|$. The result of this is that the function g satisfies the conditions of Proposition 4.14, and hence is an MNAF. Therefore, if $\mu_+\mu_- \neq 0$, then $b = cg \in \text{fact}(p, \varrho)$.

Assume, for the sake of definiteness, that $\mu_+ = 0$, $\mu_- \neq 0$. Defining the function c as above, let us choose a function $c_+ \in GC(\overline{\mathbb{R}})$ such that $c_+(+\infty) = v_+^{-\varepsilon_+}$, $c_+(-\infty) = 1$ and for sufficiently large $M_\varepsilon > 0$,

$$\sup_{|x|>M_\varepsilon} |c(x) - c_+(x)| < \varepsilon.$$

Modifying, if necessary, $\arg v_+$, we can ensure that the function c_+ satisfies the conditions of Theorem 4.18. By that theorem, $c_+ \in \text{fact}(p, \varrho)$, and then Proposition 2.3 guarantees that $c \in \text{fact}(p, \varrho)$. Thus, representation (4.66) also holds in the case where one of the numbers $\mu_\pm$ is equal to zero. □

The conclusions of the preceding theorem enable us to construct a theory of normal solvability for Toeplitz operators with symbols in SAP. Recall that, according to property (3) of the class $\text{fact}(\infty, p, \varrho)$ (see Section 2.7), we may assume that in (4.65), (4.66) the (p, ϱ)-factorization of the function $b(x)$ has the form $b(x) = b_+(x)b_-(x)$.

Theorem 4.24 *Let $a \in SAP$. Then for the operator $T(a)$ to be normally solvable in the space $L_p^+(\mathbb{R}, \varrho)$, $1 < p < \infty$, it is necessary that $\inf_{x\in\mathbb{R}} |a(x)| > 0$. If the latter condition is satisfied, then the factorization (4.65) or (4.66) holds.*

(1) *If $\mu_\pm(a) = 0$, then $T(a)$ is a Φ-operator in $L_p^+(\mathbb{R}, \varrho)$ if and only if the number*

$$\frac{1}{2\pi} \arg \frac{\lambda(a_+)}{\lambda(a_-)} - \frac{1+\alpha}{p}$$

is not an integer. If this condition is not satisfied, then the operator $T(a)$ is not normally solvable in $L_p^+(\mathbb{R}, \varrho)$ for the given p. Here α is the exponent of the weight ϱ at infinity.

(2) *If* $\mu_\pm \geq 0$ *and* $\mu_+^2 + \mu_-^2 \neq 0$, *then* $T(a)$ *is left invertible in* $L_p^+(\mathbb{R}, \varrho)$ *and* $\dim \operatorname{coker} T(a) = \infty$. *The operator*

$$[T(a)]^{-1} = P^+(b^+)^{-1}h_{\mu_+}^{-1}h_{\mu_-}^{-1}P^+(b^-)^{-1}I$$

is a left inverse for $T(a)$, *and for* f *to lie in* $\operatorname{im} T(a)$ *it is necessary and sufficient that*

$$\int_{\mathbb{R}} f(x)\frac{1}{b^-(x)}\left(P^-\frac{1}{b^+h_{\mu_+}h_{\mu_-}(\tau+i)^n}\right)(x)dx = 0, \qquad n \in \mathbb{N}.$$

(3) *If* $\mu_\pm \leq 0$ *and* $\mu_+^2 + \mu_-^2 \neq 0$, *then* $T(a)$ *is right invertible in* $L_p^+(\mathbb{R}, \varrho)$ *and* $\dim \ker T(a) = \infty$. *The operator*

$$[T(a)]^{-1} = (b^+)^{-1}h_{|\mu_+|}h_{|\mu_-|}P^+(b^-)^{-1}I$$

is a right inverse of $T(a)$, *and for* f *to lie in* $\ker T(a)$ *it is necessary and sufficient that*

$$f = a^{-1}b^-P^-((b^-)^{-1})ag,$$

where $g \in \ker T(h_{|\mu_+|}^{-1}h_{|\mu_-|}^{-1})$.

(4) *If* $\mu_+\mu_- < 0$, *then* $T(a)$ *is not normally solvable in any of the spaces* $L_p^+(\mathbb{R}, \varrho)$, $1 < p < \infty$, *and* $\dim \ker T(a) = \dim \operatorname{coker} T(a) = 0$.

Proof. Assertion (1) is a consequence of the representation (4.65) of the symbol and Theorem 4.18. Assertions (2) and (3) are consequences of the representation (4.66) of the symbol and Theorem 2.6.

Now let us prove (4). To this end, we introduce the function $a_0(x)$ by

$$a_0(x) = \begin{cases} \lambda_+ \exp(i\mu_+x), & x \geq 0, \\ c(x)\exp(i\mu_-x), & x \leq 0, \end{cases}$$

where $\lambda_+ = \lambda(\exp(-i\mu_+x)a_+)$ and the function $c(x)$ is continuous and bounded away from zero for $x \leq 0$ and has limit values $c(0) = \lambda_+$, $c(-\infty) = \lim_{x\to-\infty} c(x) = \lambda_- = \lambda(\exp(-i\mu_-x)a_-)$. Clearly, $a_0 \in SAP$ and $\mu_\pm(a_0) = \mu_\pm(a)$. Furthermore,

$$a = a_0b, \tag{4.68}$$

where $b = aa_0^{-1}$ belongs to SAP and satisfies the assumptions of Proposition 4.14, as a result of which it admits the factorization

$$b(x) = b^-(x)\left(\frac{x-i}{x+i}\right)^\kappa b^+(x)$$

in any space $L_p(\mathbb{R}, \varrho)$, $1 < p < \infty$.

Let $\varphi \in \ker T(a)$, i.e., $[T(a)]\varphi = P^+ a_0 b \varphi = 0$. By the properties of the (p, ϱ)-factorization, $[b^-(x)(x-i)]^{-1} \in L_q(\mathbb{R}, \varrho^{1-q})$ and if $g \in L_p(\mathbb{R}, \varrho)$, then, by Hölder's inequality,

$$\int_{\mathbb{R}} \frac{|g(x)|dx}{|b^-(x)||x-i|} \leq \|g\|_{p,\varrho} \|[b^-(x)(x-i)]^{-1}\|_{q,\varrho^{1-q}}.$$

In conjunction with condition (2) of the definition of the class $\mathrm{fact}(p, \varrho)$, this implies that the operator $P^+ \frac{1}{b^-} I$ maps $L_p^+(\mathbb{R}, \varrho)$ into $L_1^+(\mathbb{R}, |x-i|^{-1})$. Therefore, if we apply this operator to both sides of the equality $[T(a)]\varphi = 0$ we obtain the following equality in the space $L_1^+(\mathbb{R}, |x-i|^{-1})$:

$$\lambda_+ P^+ \left(\left(\frac{x-i}{x+i} \right)^\kappa \exp(i\mu_+ x)\psi \right) = P^+ \psi_-, \tag{4.69}$$

where

$$\psi(x) = b^+(x)\varphi(x) \in L_1^+(\mathbb{R}, |x-i|^{-1}),$$

and

$$\psi_- = \frac{1}{2}[1 - \operatorname{sign} x] \left(\frac{x-i}{x+i} \right)^\kappa [c(x)\exp(i\mu_- x) - \lambda_+ \exp(i\mu_+ x)]\psi(x).$$

Let $\kappa \leq 0$. Subtracting the principal part of the Laurent series of the function $\left(\frac{z-i}{z+i}\right)^\kappa \exp(i\mu_+ z)\psi(z)$ at the point $z = i$ and using the fact that $P^+(L_1^+(\mathbb{R}, |x-i|^{-1})) = L_1^+(\mathbb{R}, |x-i|^{-1})$, we deduce from (4.69) that, almost everywhere on $\mathbb{R}$,

$$\lambda_+ \exp(i\mu_+ x)\psi(x) = \left(\frac{x+i}{x-i} \right)^\kappa (P^+ \psi_-)(x) + \sum_{j=1}^{|\kappa|} \frac{c_j}{(x+i)^j}, \tag{4.70}$$

where $c_j \in \mathbb{C}$. By the Sokhotskiĭ-Plemelj formulas, the function $P^+ \psi_-$ admits an analytic continuation to $\mathbb{C} \setminus (-\infty, 0]$ and is representable therein as a Cauchy integral:

$$(P^+ \psi_-)(z) = \frac{1}{2\pi i} \int_{-\infty}^0 \frac{\psi_-(\tau)d\tau}{\tau - z}.$$

Taking here $z = R\exp(i\theta)$ and letting $R \to \infty$ for $|\theta| < \pi$, we obtain

$$|(P^+ \psi_-)(z)| \leq \frac{1}{2\pi} \sup_{-\infty<\tau<0} \left| \frac{\tau - i}{\tau - z} \right| \cdot \|\psi_-\|_{L_1(\mathbb{R}, |x-i|^{-1})} \leq \text{const}.$$

Now from (4.70) it follows that the function ψ admits an analytic continuation to the domain $\mathbb{C} \setminus M$, where $M = (-\infty, 0] \cup \{i\}$, and for all θ, $|\theta| < \pi$, and large values of R,

$$|\exp(i\mu_+ R\exp(i\theta))\psi(R\exp(i\theta))| \leq \text{const}. \tag{4.71}$$

At the same time, the function $\psi(z)$ is representable in Π_+ as a Cauchy integral over $\mathbb{R}$ with a density of class $L_1(\mathbb{R}|x-i|^{-1})$, and consequently, for $\theta \in (0,\pi)$,

$$|\psi(R\exp(i\theta))| \le \text{const}. \tag{4.72}$$

To show that $\psi \equiv 0$, let us introduce the function

$$F(z) = \psi(z)\exp(((i\mu_+/2) + \varepsilon)z),$$

where $\varepsilon > 0$. Fix a number θ_0, $\theta_0 \in (0,\pi/2)$. By (4.71),

$$\begin{aligned} |F(R\exp(i\theta))| &\le \text{const}\,|\exp((-i\mu_+/2+\varepsilon)R\exp(i\theta))| \\ &\le \text{const}\exp(R(\varepsilon\cos\theta + (\mu_+/2)\sin\theta)). \end{aligned}$$

It follows that for $\varepsilon = (\mu_+/2)\tan\theta_0$ the function $F(z)$ is bounded on the ray $\arg z = -\theta_0$. At the same time, thanks to (4.72),

$$|F(R\exp(i\theta_0))| \le |\psi(R\exp(i\theta_0))||\exp((i\mu_+/2+\varepsilon)R\exp(i\theta_0))| \le \text{const},$$

i.e., $F(z)$ is bounded on the sides of the sector $K_0 = \{z \in \mathbb{C} : |\arg z| \le \theta_0\}$. Since

$$|F(z)| \le \text{const}\cdot\exp(\varepsilon|z|)$$

for all $z \in K_0$, and the opening of the angle K_0 is $2\theta_0 < \pi$, the Phragmén-Lindelöf principle (see, e.g., Shabat [1], p. 287) guarantees that the function $F(z)$ is bounded in the whole sector K_0. Therefore, if $\kappa > 0$, then

$$|\psi(x)| \le \text{const}\cdot\exp(-\varepsilon x).$$

It is known (see Garnett [1], p. 72) that if a function $\psi(x)$ which does not vanish identically belongs to $L_1^+(\mathbb{R},|x-i|^{-1})$, then

$$\int_{\mathbb{R}} \frac{\log|\psi(x)|}{1+x^2}dx > -\infty.$$

We therefore conclude that $\psi = \varphi = 0$ a.e. The case $\kappa > 0$ can be reduced to the one just considered upon observing that

$$\left(\frac{x-i}{x+i}\right)^{\kappa}\psi(x) \in L_1^+(\mathbb{R},|x-i|^{-1}).$$

Thus, when $\mu_+ > 0$ and $\mu_- < 0$, $\ker T(a) = \{0\}$. The same conclusion is reached in a similar manner when $\mu_+ < 0$ and $\mu_- > 0$. Since $T^*(a) = T(\overline{a})$ where $\overline{a} \in SAP$ and $\mu_+(\overline{a})\mu_-(\overline{a}) < 0$, one simultaneously has $\operatorname{coker} T(a) = \{0\}$.

It remains to show that $\operatorname{im} T(a) \ne L_p^+(\mathbb{R},\varrho)$. To do this let us consider the equation

$$[T(a)]\varphi = P^+[(x+i)^{-1}b^-] \tag{4.73}$$

in $L_p^+(\mathbb{R},\varrho)$ and show that it has no solution. Applying the operator $P^+\frac{1}{b^-}I$ to both sides, we arrive at the following equation in $L_1^+(\mathbb{R},|x-i|^{-1})$:

$$P^+\left(\frac{x-i}{x+i}\right)^\kappa a_0(x)\psi(x)=\frac{1}{x+i}, \tag{4.74}$$

where, as above, $\psi=b^+\varphi\in L_1^+(\mathbb{R},|x-i|^{-1})$. Arguing as we did above for the equation (4.69), we conclude that in the present case, too, a function ψ satisfying (4.74) must vanish a.e. Therefore, equation (4.73) is not solvable and $\operatorname{im}T(a)\neq L_p^+(\mathbb{R},\varrho)$. □

Remark 4.14 In view of Proposition 4.13, in the investigation of an operator $T(a)$ with a symbol of the form (4.66) the functions $h_{|\mu\pm|}$ given (4.63) can be replaced by Blaschke products, and therefore the space $\ker T(a)$ can be described by means of the model subspace $\mathcal{M}(B_a)$, with

$$B_a(x)=\prod_{|k|=1}^{\infty}\frac{|1+z_k^2|}{1+z_k^2}\cdot\frac{x-z_k}{x-\overline{z}_k}, \tag{4.75}$$

where $z_k=z_{+k}$ for $k\geq 1$ and $z_k=z_{-k}$ for $k\leq -1$, and $z_{\pm k}$ are the zeros of the functions $h_{|\mu\pm|}$ in Π_+ (see (4.64)). Here the role of the function b^+ will be played by the product $F_{|\mu_+|}^{-1}F_{|\mu_-|}^{-1}b^+$, where $F_{|\mu\pm|}$ are the outer functions produced by the Riesz factorization (2.25) of the functions $h_{|\mu\pm|}$.

4.9 Model subspaces

In this section we give a detailed description of the model subspaces corresponding to the case of semi-almost periodic discontinuities considered above. First let us employ Theorem 3.2 to describe the subspace $\ker T(h_{|\mu_+|}^{-1}h_{|\mu_-|}^{-1})$.

Theorem 4.25 *Let $\mu_\pm>0$ and let the functions $h_{\mu\pm}$ be defined by* (4.63). *Then for a function φ to lie in $\ker T(h_{\mu_+}^{-1}h_{\mu_-}^{-1})$ in the space $L_p^+(\mathbb{R},\varrho)$ it is necessary and sufficient that φ lie in $L_p^+(\mathbb{R},\varrho)$, admit an analytic continuation to the entire complex plane $\mathbb{C}$, and satisfy the following estimates at infinity:*

$$|\varphi(z)|\leq o(|z|)|z|^{-\frac{\alpha}{p}},\quad \operatorname{Im}z\geq 0,$$

$$|\varphi(z)|\leq o(|z|)|z|^{-\frac{\alpha}{p}}|h_{\mu_+}(z)||h_{\mu_-}(z)|,\quad \operatorname{Im}z\leq 0.$$

Here $\lim_{|z|\to\infty}o(|z|)=0$ and α is the exponent of the weight $\varrho(x)$ at infinity.

Proof. Let $v(t)=i(1+t)/(1-t)$, $v^{-1}(x)=(x-i)/(x+i)$, where $t\in\Gamma_0$ and $x\in\mathbb{R}$. Put

$$(\mathcal{K}f)(t)=\frac{f(v(t))}{t-1},\quad (\mathcal{K}^{-1}g)(x)=\frac{2ig(v^{-1}(x))}{x+i}.$$

By Proposition 3.11,

$$\ker T(h_{\mu_+}^{-1}h_{\mu_-}^{-1}) = \mathcal{K}^{-1}(\mathcal{M}(h)),$$

where $h(t) = h_{\mu_+}(v(t))h_{\mu_-}(v(t))$ is a model function on Γ_0, which admits analytic continuation to $\mathbb{C} \setminus \{1\}$. By assertion (1) of Theorem 4.22, there exists a pair of circles $\gamma_\pm$ such that $\gamma_\pm \subset D^\pm$, $\gamma_\pm$ are (h, Γ_0)-admissible contours, and the estimates

$$d(z, \gamma_\pm) \geq \text{const}\,|z-1|^2, \quad z \in \Gamma_0,$$

hold. The circles $\gamma_\pm$ are images of lines $\operatorname{Im} z = c_\pm$ under the mapping $\zeta = (z - i)/(z+i)$ for real numbers $c_+(> 0)$ and $c_-(< 0)$ of sufficiently small modulus. By Theorem 3.2, for ψ to lie in $\mathcal{M}(h)$ in the space $L_p^+(\Gamma_0, \varrho_0) = \mathcal{K}(L_p^+(\mathbb{R}, \varrho))$, where $\varrho_0(t) = |t-1|^{p-2}\varrho(i(1+t)/(1-t))$, it is necessary and sufficient that ψ lie in $L_p^+(\Gamma_0, \varrho_0)$, admit an analytic continuation to $\mathbb{C} \setminus \{1\}$, and obey the following estimates in the neighborhood of the point $\zeta = 1$:

$$|\psi(\zeta)| \leq o(\zeta)|\zeta - 1|^{\frac{\alpha-p}{p}}, \quad \zeta \in \overline{\mathcal{D}^+},$$

$$|\psi(\zeta)| \leq o(\zeta)|\zeta - 1|^{\frac{\alpha-p}{p}}|h(\zeta)|, \quad \zeta \in \overline{\mathcal{D}^-},$$

where $\lim_{\zeta\to1} o(\zeta) = 0$. To complete the proof it remains to apply the operator $\mathcal{K}^{-1}$ to the function ψ. □

Now we can use Remark 4.14 to describe the model subspace $\mathcal{M}(B_a)=\ker T(B_a^{-1})$, where B_a is a Blaschke product of the form (4.75). Recall the notation $V_\varepsilon = \bigcup_{|j|\geq1} c_\varepsilon(\overline{z}_j)$, where $c_\varepsilon(\overline{z}_j) = \{z : |z - \overline{z}_j| \leq \varepsilon\}$.

Theorem 4.26 *Let the function B_a have the form* (4.75). *Then for a function φ to lie in $\mathcal{M}(B_a)$ in the space $L_p^+(\mathbb{R}, \varrho)$ it is necessary and sufficient that φ lie in $L_p^+(\mathbb{R}, \varrho)$, admit an analytic continuation to the domain $\mathbb{C} \setminus \{\overline{z}_j\}$, have only poles of the first order in the points $z = \overline{z}_j$, and obey the estimates*

$$|\varphi(z)| \leq o(|z|)|\varepsilon|^{-1}|z|^{-\frac{\alpha}{p}}, \quad z \in \mathbb{C} \setminus V_\varepsilon,$$

$$|\varphi(z)| \leq o(|z|)|z - \overline{z}_j|^{-1}|z|^{-\frac{\alpha}{p}}, \quad z \in c_\varepsilon(\overline{z}_j), \quad |j| \geq 1,$$

where $\lim_{|z|\to\infty} o(|z|) = 0$ and α is the same as in Theorem 4.25.

Proof. By Theorem 4.22, the zeros of the function $B_a(z)$ are separated and satisfy the condition $0 < h_1 \leq \operatorname{Im} z_j \leq h_2 < \infty$, $|j| \geq 1$. By Proposition 3.20, there exists a pair of lines $\operatorname{Im} z = c_\pm$, where $c_+ > 0$, $c_- < 0$, for which

$$\sup_{c_- \leq \operatorname{Im} z \leq c_+} |B_a(z)|^{\pm1} < \infty.$$

Arguing as in the proof of Theorem 4.25, we conclude that $\varphi(z)$ admits an analytic continuation to the domain $\mathbb{C} \setminus \{\overline{z}_j\}$ and satisfies the estimates

$$|\varphi(z)| \leq o(|z|)|z|^{-\frac{\alpha}{p}}, \quad \operatorname{Im} z \geq 0,$$

$$|\varphi(z)| \leq o(|z|)|z|^{-\frac{\alpha}{p}}|B_a(z)|, \quad \operatorname{Im} z \leq 0.$$

To complete the proof it remains to apply Proposition 3.20. □

Remark 4.15 According to the same Proposition 3.20, the zeros of a Blaschke product satisfy the Carleson condition (3.23). Hence, by Theorem 3.6, the systems of functions $\{w_k\}$ given by (3.33) and $\{e_k\}$ given by (3.34) constitute bases in the subspace $\mathcal{M}(B_a)$ of $L_p(\mathbb{R}, \varrho)$.

Whirl points of power type

In this section we consider situations in which the argument of the symbol, when we approach a discontinuity point, becomes unbounded and tends to $\pm\infty$ like a power. From this point of view the results given here generalize the results of the preceding two sections, where the power in question had exponent 1. Let us mention that the formulation of such a problem, which is traditional in the theory of the Riemann boundary value problem with infinite index, goes back to N. V. Govorov.

Throughout this section Γ will be a smooth curve of class $\mathcal{T}$. Let $t_0 \in \Gamma$ and let ω be a neighborhood of t_0, $\omega \subset \Gamma$. The point t_0 divides ω into two semi-neighborhoods ω_- and ω_+, which follow one after another in the positive orientation of Γ ($\omega_- \prec \omega_+$). Let the function $a(t)$ be defined on Γ and be bounded away from zero on the set $\omega \setminus \{t_0\}$. If for some continuous branch of the function $\arg a(t)$ at least one of the limits

$$A_\pm = \lim_{t \to t_0 \pm 0} \arg a(t)$$

is equal to $+\infty$ or $-\infty$, then we say that t_0 is a *whirl point* of the function $a(t)$. If $\arg a(t)$ is unbounded in each (resp., only one) of the semi-neighborhoods ω_- and ω_+, then t_0 is called a *two-sided* (resp., *one-sided*) whirl point of $a(t)$. Here, as usual, $t \to t_0 \pm 0$ means that $t \to t_0$ with $t \in \omega_\pm$.

4.10 Two-sided whirls

Consider a Toeplitz operator $T(a)$ whose symbol a lies in $GL_\infty(\Gamma)$, is continuous on $\Gamma \setminus \{t_0\}$, and has a second-kind discontinuity in the point t_0, $t_0 \in \Gamma$, specified by the equalities

$$\lim_{t \to t_0 \pm 0} (\arg a(t) \mp \sigma_\pm |t - t_0|^{-\lambda_\pm}) = c_\pm, \tag{4.76}$$

where $\lambda_\pm$, $\sigma_\pm$, $c_\pm \in \mathbb{R}$, $\sigma_\pm \neq 0$, $0 < \lambda_\pm < 1$. From (4.76) it follows that t_0 is a two-sided whirl point for $a(t)$, and the exponents of the whirl to the left and to the right of t_0 are, generally speaking, different. Let us point out that condition (4.76) defines a discontinuity only for $\arg a(t)$, while the function $|a(t)|$ is bounded away from zero and hence admits a standard factorization with index zero (Proposition 2.3).

In the case $\sigma_+\sigma_- > 0$, an analytic function in $\mathcal{D}_\Gamma^+$ which satisfies condition (4.76) can be constructed as follows. Let φ denote the angle made by the inward normal to the curve Γ at the point t_0 and the positively oriented real axis. Then the angle $\theta_\pm$ that the semi-tangent $l_\pm$ to Γ at t_0 makes with the positively oriented real axis is equal to $\theta_\pm = \varphi \mp \pi/2$. Consider the function

$$h(z) = \exp\{-i\varepsilon c_+\}h_+(z) + \exp\{-i\varepsilon c_-\}h_-(z), \tag{4.77}$$

where

$$h_\pm(z) = \exp\{\mp i a_\pm (z-t_0)^{-\lambda_\pm}\}, \tag{4.78}$$

$$a_\pm = |\sigma_\pm| \exp\{i\lambda_\pm \theta_\pm\}, \quad \varepsilon = \operatorname{sign} \sigma_+ = \operatorname{sign} \sigma_-. \tag{4.79}$$

Pick single-valued branches of the functions $h_\pm(z)$ by making a cut from the point t_0 to infinity inside the domain $\mathcal{D}_\Gamma^-$ such that near t_0 the cut coincides with the outward normal to Γ at t_0. We define $(z-t_0)^{-\lambda_\pm}$ as $r^{-\lambda_\pm}\exp\{-i\lambda_\pm\theta\}$ for $z - t_0 = r\exp\{i\theta\}$, where $r > 0$ and $|\varphi - \theta| < \pi$.

Let us examine the behavior of the functions $h_\pm(z)$ near the point $z = t_0$.

Proposition 4.15 (1) *In the vicinity of the point $z = t_0$, the behavior of the functions $h_\pm(z)$ given by* (4.78) *is characterized by the equality*

$$\log|h_\pm(z)| = \pm|\sigma_\pm| r^{-\lambda_\pm} \sin(\lambda_\pm(\theta_\pm - \theta)), \quad |\varphi - \theta| < \pi. \tag{4.80}$$

In particular, the functions $h_\pm(z)$ are bounded in the half-plane $\theta_+ \le \theta \le \theta_-$.

(2) *The function $h(z)$ given by* (4.77) *has a countable set of zeros $\{z_n\}$, whose asymptotics for $n \to \infty$ is as follows:*

(a) *if $\lambda_+ = \lambda_- = \lambda$, then*

$$z_n = t_0 + [(2\pi)^{-1}(a_+ + a_-)]^{\lambda^{-1}} n^{-\lambda^{-1}} + O\left(n^{-1-\lambda^{-1}}\right), \tag{4.81}$$

where the coefficients $a_\pm$ have the form (4.79)*;*

(b) *if $\lambda_+ \ne \lambda_-$, then*

$$z_n = t_0 + e^{i\theta_\pm}\left[\alpha_\pm n^{-\lambda_\pm^{-1}} + \beta_\pm e^{i\lambda_\mp \pi} n^{(\lambda_\mp - \lambda_\pm - 1)\lambda_\pm^{-1}}\right] + O(n^{-\delta_\pm}), \tag{4.82}$$

where $\alpha_\pm = (|\sigma_\pm|(2\pi)^{-1})^{\lambda_\pm^{-1}}$, $\delta_\pm = \min(1+\lambda_\pm^{-1}, 2 - \lambda_\pm^{-1}(2\lambda_\mp - 1))$, $\beta_\pm = \lambda_\pm^{-1}|\sigma_\mp||\sigma_\pm|^{(1-\lambda_\mp)\lambda_\pm^{-1}}(2\pi)^{(\lambda_\mp - \lambda_\pm - 1)\lambda_\pm^{-1}}$, with the upper (resp., lower) signs corresponding to the case $\lambda_+ > \lambda_-$ (resp., $\lambda_+ < \lambda_-$).

Proof. Formulas (4.80) are derived by elementary calculations using the explicit form (4.78)–(4.79) of the functions $h_\pm(z)$.

The equation $h(z) = 0$ is equivalent to the set of equations

$$a_+(z-t_0)^{-\lambda_+} + a_-(z-t_0)^{-\lambda_-} = \varepsilon(c_- - c_+) + (2n+1)\pi, \quad n \in \mathbb{Z}. \tag{4.83}$$

When $\lambda_+ = \lambda_- = \lambda$, the solution of (4.83) is given by

$$z_n = t_0 + \left(\frac{a_+ + a_-}{(2n+1)\pi + \varepsilon(c_- - c_+)} \right)^{\lambda^{-1}}. \tag{4.84}$$

The numbers n must be chosen so that $z_n \in \mathcal{D}_\Gamma^+$. Since

$$\lambda\theta_+ < \arg(a_+ + a_-) < \lambda\theta_-, \tag{4.85}$$

$z_n \in \mathcal{D}_\Gamma^+$ for n large enough, and from (4.84), (4.85) it follows that the set of negative values of n for which z_n can lie in $\mathcal{D}_\Gamma^+$ is finite. In the case $\lambda_+ > \lambda_-$ the change of variables $\zeta = (z - t_0)^{-\lambda_+}$ in (4.83) leads to the equation

$$\zeta + \frac{a_-}{a_+}\zeta^\mu = \frac{\varepsilon(c_- - c_+) + (2n+1)\pi}{a_+}, \qquad \mu = \frac{\lambda_-}{\lambda_+} < 1, \tag{4.86}$$

with the condition

$$-\lambda_+(\pi + \varphi) < \arg\zeta < \lambda_+(\pi - \varphi).$$

For any $n \in \mathbb{Z}$ of sufficiently large modulus equation (4.86) has a unique solution, whose asymptotic behavior for large $|n|$ is found in the standard way:

$$\zeta_n = (z_n - t_0)^{-\lambda_+} = \frac{2\pi n}{a_+} - \frac{a_-}{a_+}\left(\frac{2\pi n}{a_+}\right)^\mu + O(n^{2\mu-1}) + O(1). \tag{4.87}$$

Negative values of n of large modulus are discarded on the same grounds as in the case $\lambda_+ = \lambda_-$, and from (4.87) we deduce that for $n \to \infty$,

$$z_n = t_0 + [a_+ + (2\pi n)^{-1}]^{\lambda_+^{-1}} + \frac{a_-}{a_+\lambda_+}\left(\frac{2\pi n}{a_+}\right)^{(\lambda_- - \lambda_+ - 1)\lambda_+^{-1}} + O(n^{-\delta_+}).$$

This yields the asymptotics (4.82) and the fact that $z_n \in \mathcal{D}_\Gamma^+$. The case $\lambda_- > \lambda_+$ is dealt with in a similar manner. □

Remark 4.16 If $\lambda_+ = \lambda_- = \lambda$, then the zeros z_n of the function $h(z)$ lie on the ray $z = t_0 + r\exp(i\theta)$, where $\theta = \lambda^{-1}\arg(a_+ + a_-) \in (\theta_+, \theta_-)$; for $|\sigma_+| = |\sigma_-|$ this ray coincides with the inward normal to Γ at the point t_0. If, however, $\lambda_+ \neq \lambda_-$, then the zeros z_n lie on a certain curve γ, which is tangent either to l_+ (when $\lambda_+ > \lambda_-$) or to l_- (when $\lambda_- > \lambda_+$). Moreover, for $z \in \gamma$ the following two-sided estimate holds in a small neighborhood of the point t_0:

$$d(z, \gamma) \asymp |z - t_0|^{1 + |\lambda_+ - \lambda_-|}.$$

Let us formulate necessary and sufficient conditions for a function h of the form (4.77), 4.78), (4.79) to be a model function on the curve Γ.

Proposition 4.16 *For a function h of the form* (4.77)–(4.79) *to be a model function on the curve* Γ *it is necessary and sufficient that none of the zeros of* $h(z)$ *lie on* Γ *and there exist a neighborhood* ω *of the point* t_0, $\omega \subset \Gamma$ *such that*

$$d(t, l_\pm) \leq \text{const}\,|t - t_0|^{1+\lambda_\pm}, \quad t \in \omega_\pm. \tag{4.88}$$

Proof. First let us show that condition (4.88) is necessary and sufficient for the function $h_\pm(t)$ to be bounded and bounded away from zero in some semi-neighborhood $\omega_\pm$ of t_0. Let $r = |t-t_0|$, $\psi(t) = \theta_+(t) - \theta(t)$, and $\theta(t) = \arg(t-t_0)$. Then it follows from (4.80) that

$$|h_+(t)| = \exp\{|\sigma_+|r^{-\lambda_+}\sin(\lambda_+\psi(t))\}.$$

Consequently, the function $|\log|h_+(t)||$ is bounded if and only if the function $r^{-\lambda_+}|\sin(\lambda_+\psi(t))|$ is bounded. But $d(t, l_+) = r|\sin\psi(t)|$ whenever $t \in \omega_+$. Hence, the boundedness of the function $|\log|h_+(t)||$ for $t \in \omega_+$ is equivalent to condition (4.88). The case of the function $h_-(t)$ is dealt with similarly.

In addition, from (4.88) it follows that $\lim_{t\to t_0\mp 0} h_\pm(t) = 0$. Let us emphasize that for the latter relation to hold, apart from smoothness near the point $t = t_0$, no condition on Γ is required.

Thus, let us assume that (4.88) is satisfied. Then the function $h_+(t)$ (resp., $h_-(t)$) is bounded and bounded away from zero on the arc ω_+ (resp., ω_-) and it tends to zero as $t \to t_0$ on the arc ω_- (resp., ω_+). Therefore, in some two-sided neighborhood ω of t_0, a function $h(t)$ of the form (4.77) is bounded and bounded away from zero. On $\Gamma\setminus\omega$, $h(t)$ is continuous, and, by the assumptions of the proposition, it has no zeros there. Consequently, $h \in GL_\infty(\Gamma)$.

Further, from Proposition 4.15 and the fact that any path terminating at t_0 and being located in $\overline{\mathcal{D}_\Gamma^+}$ outside of the semi-neighborhood $\theta_+ \leq \arg(z - t_0) \leq \theta_-$ (if such a path exists) satisfies conditions (4.88), it follows that $h \in L_\infty^+(\Gamma)$, i.e., h is a model function on Γ.

Conversely, let $h \in L_\infty^+(\Gamma) \cap GL_\infty(\Gamma)$. The $h(t)$ must have no zeros on Γ, and since $\lim\limits_{t\to t_0+0} h_-(t) = 0$, the function $h_+(t)$ does not vanish on ω_+. In other words, condition (4.88) is satisfied for $t \in \omega_+$. In much the same way one can prove that this condition is necessarily satisfied for $t \in \omega_-$. □

The next proposition describes the class $\mathcal{T}(h, \Gamma)$ of admissible deformations of the curve Γ for the case at hand.

Proposition 4.17 *Let* Γ *be a smooth curve of class* $\mathcal{T}$ *and let the function h given by* (4.77)–(4.79) *be a model function on* Γ. *For* γ *to belong to the class* $\mathcal{T}(h, \Gamma)$ *it is necessary and sufficient that h be a model function on* γ *and that the sets zeros of* $h(z)$ *lying in* $\mathcal{D}_\Gamma^+$ *and* $\mathcal{D}_\gamma^+$, *respectively, coincide.*

Furthermore, there exists a pair of curves $\gamma_\pm \in \mathcal{T}(h, \Gamma)$, *contained in* $\overline{\mathcal{D}_\Gamma^\pm}$, *respectively, for which the following inequalities hold in some neighborhood* ω *of the point* t_0, $\omega \subset \Gamma$:

$$d(t, \gamma_\pm) \geq \text{const}\,|t - t_0|^{1+\lambda}, \quad t \in \omega, \quad \lambda = \max(\lambda_+, \lambda_-). \tag{4.89}$$

Proof. The first assertion of the proposition is a straightforward consequence of Proposition 4.16 and the definition of an (h,Γ)-admissible curve (see Section 3.2).

Next, consider the family of all curves $\{\gamma_0\}$ that pass through the point t_0, have a common tangent l with the curve Γ at t_0, and satisfy the condition $d(z,l) = c|z-t_0|^{1+\lambda}$, $\lambda = \max(\lambda_+, \lambda_-)$, for $z \in \gamma_0$ in some neighborhood of t_0. It follows from Proposition 4.16 and Remark 4.16 that this family is a sufficiently rich subclass of $\mathcal{T}(h,\Gamma)$. Hence, to prove the second assertion of the proposition it suffices to choose in this family some curves $\gamma_{c_\pm} \in \mathcal{T}(h,\Gamma)$ that lie in the respective domain $\overline{\mathcal{D}_\Gamma^\pm}$ for sufficiently large values of the numbers $c_\pm$. □

We are now ready to describe the model subspace $\mathcal{M}(h)$.

Proposition 4.18 *Let Γ be a smooth curve of class $\mathcal{T}$, let the function h given by* (4.77)–(4.79) *be a model function on Γ, and let $\{z_n\}_{n=1}^{\infty}$ be the zeros of $h(z)$ in the domain $\mathcal{D}_\Gamma^+$.*

Then for a function f^+ to belong to $\mathcal{M}(h)$ in the space $L_p^+(\Gamma,\varrho)$, $1<p<\infty$, where the weight ϱ has the form (2.2), *it is necessary and sufficient that the following conditions be satisfied:*

(a) $f^+(t) \in L_p^+(\Gamma,\varrho)$;

(b) *the function $f^-(t) = h^{-1}f^+(t)$ admits an analytic continuation to the domain $\mathbb{C}\setminus(\{t_0\}\cup\{z_n\}_{n=1}^{\infty})$ and vanishes at infinity;*

(c) *each point $z=z_n$, $n\in\mathbb{N}$, is either a regular point or a pole of the function $f_-(z)$, while $z=t_0$ is a regular, essentially singular or nonisolated singular point of the function $f_-(t)$, in a neighborhood of which the following estimate holds:*

$$|f^\pm(z)| \le o(z)|z-t_0|^{-\frac{1+\lambda+\alpha}{p}}, \quad z \in \overline{\mathcal{D}_\Gamma^\pm}, \tag{4.90}$$

where α, the exponent of the weight $\varrho(t)$ at the point t_0, may be equal to zero, $\lim_{z\to t_0} o(z) = 0$ and $\lambda = \max(\lambda_+,\lambda_-)$.

Proof. This is a direct corollary of Theorem 3.2 and (4.89). □

Let us consider the case $\Gamma = \Gamma_0$. Since in this case condition (4.88) is satisfied for every point $t_0 \in \Gamma_0$ and every $\lambda_\pm$, $0<\lambda_\pm<1$, the function h is a model function on Γ_0 (if some zero $z_n \in \Gamma_0$ is present, then instead of $h(t)$ one considers the function $h(t)/(t-z_n)^s$, where s is the order of the zero).

In the present case the function h admits a Riesz factorization. Hence, in the study of discontinuities of the form (4.76) one can use Blaschke products as model functions.

Proposition 4.19 *Suppose that the function h given by* (4.77), (4.78), (4.79) *does not vanish on Γ_0. Then*

$$h = FB,$$

with $F \in GH^\infty(\mathbb{D}^+)$ *and*

$$B(t) = \prod_{n=1}^{\infty} \frac{\overline{z}_n}{|z_n|} \cdot \frac{z_n - t}{1 - \overline{z}_n t}, \tag{4.91}$$

where z_n, $n \in \mathbb{N}$, *are the zeros of* $h(z)$ *in* $\mathbb{D}^+$.

Proof. By Proposition 4.16, h is a model function on Γ_0, i.e., $h \in H^\infty(\mathbb{D}^+) \cap GL_\infty(\Gamma_0)$. Therefore, its outer part F belongs to $GH^\infty(\mathbb{D}^+)$ (Theorem 2.2). Since $h(t)$ has a unique singular point $t = t_0$ on Γ_0, the singular inner function E corresponding to h must be of the form

$$E(t) = \exp\left(\mu \frac{t + t_0}{t - t_0}\right), \quad \mu \geq 0$$

(see Section 2.6). From Proposition 4.15 it follows that there exists a ray $z = t_0 + r\exp(i\theta)$, $\theta \in (\theta_+, \theta_-)$, such that for r sufficiently small,

$$|h(t)| > \exp\{-c_1 r^{-\lambda}\}, \qquad \text{where } c_1 > 0, \quad 0 < \lambda < 1.$$

At the same time, observing that $\theta_0 = \arg t_0 = \theta_+ - \pi/2$, we have

$$|E(z)| \leq c_2 |\exp\{2\mu t_0 (z - t_0)^{-1}\}| = c_2 \exp\{2\mu \cos(\theta - \theta_0) r^{-1}\},$$

that is

$$|h(z)| = c_3 \exp\{-c_4 r^{-1}\}, \quad \text{where } c_3 > 0, \quad c_4 \geq 0.$$

It follows that $\mu = 0$ and that the inner part of the function h reduces to the Blaschke product (4.91), constructed from the zeros of $h(z)$ lying in the domain $\mathbb{D}^+$. □

The next result describes the model subspace $\mathcal{M}(B)$ in "complex" terms.

Proposition 4.20 *Let the function B have the form* (4.91). *Then for a function f^+ to belong to the subspace $\mathcal{M}(B)$ in $L_p^+(\Gamma_0, \varrho)$, $1 < p < \infty$, where the weight ϱ has the form* (2.2), *it is necessary and sufficient that $f^+(t)$ belong to $L_p^+(\Gamma_0, \varrho)$ and admit an analytic continuation to the domain $\mathbb{C} \setminus (\{t_0\} \cup \{\overline{z}_n\}_{n=1}^{\infty})$. If this is the case, then $f^+(t)$ must vanish at infinity, each point $\overline{z}_n$, $n \in \mathbb{N}$, is either a regular point or a pole, and t_0 is a regular, essentially singular, or nonisolated singular point, in a neighborhood of which the estimates*

$$\begin{cases} |f^+(z)| \leq o(z)|z - t_0|^{-\frac{1+\lambda+\alpha}{p}}, & z \in \overline{\mathbb{D}^+}, \\ |f^+(z)| \leq o(z)|z - t_0|^{-\frac{1+\lambda+\alpha}{p}} |B(z)|, & z \in \overline{\mathbb{D}^-}, \end{cases} \tag{4.92}$$

hold with α, λ and $o(z)$ as in Proposition 4.18.

Proof. This is a corollary of Theorem 3.2. The estimates (4.92) follow from the estimates (4.90) upon observing that $f^+ \in \mathcal{M}(B)$ if and only if $Ff^+ \in \mathcal{M}(h)$, where F is the outer part of the function h. □

Remark 4.17 The description of the subspace $\mathcal{M}(B)$ on Γ_0 is somewhat simpler than that of the subspace $\mathcal{M}(h)$ on an arbitrary smooth contour Γ, because the only singularities of the function $B(z)$ in the domain $\mathbb{D}^-$ are poles, while the function $h(z)$ has a cut in $\mathbb{D}_\Gamma^-$ which singles out a single-valued branch. Let us remark that in the general case one can choose a meromorphic function as model function. By Propositions 4.16 and 4.17, in $\mathcal{T}(h,\Gamma)$ there exists a circle Γ_+ with the property that the sets of all the zeros of $h(z)$ in $\mathcal{D}_\Gamma^+$ and in $\mathcal{D}_{\Gamma_+}^+$ coincide up to a finite subset. Using these zeros, we may construct a meromorphic function in the domain $\mathbb{C}\setminus\{t_0\}$ that serves as the analogue of the Blaschke product for the circle Γ_+. Since, by Theorem 3.1, $\ker T_\Gamma(h^{-1}) = \ker T_{\Gamma_+}(h^{-1})$ in the case of an arbitrary contour Γ on which h is a model function, the space $\ker T_\Gamma(h^{-1})$ can be described in the terms of Proposition 4.20.

Now let us state the main result of this section.

Theorem 4.27 *Let Γ be a smooth contour of class $\mathcal{T}$ which satisfies conditions (4.88) and let the function h given by (4.77), (4.78), (4.79) have no zeros on Γ.*

If the function $a(t) \in GL_\infty(\Gamma)$ is continuous on the set $\Gamma\setminus\{t_0\}$ and has a discontinuity of type (4.76) at t_0, where $\sigma_+\sigma_- > 0$, then a admits the representation

$$a = h^\varepsilon b, \tag{4.93}$$

where $\varepsilon = \operatorname{sign}\sigma_+ = \operatorname{sign}\sigma_-$ and $b \in \operatorname{fact}(p,\varrho)$.

Proof. From equality (4.78) it follows that for $t\in\omega_+$ the argument of the function $h_+(t)$ is given by the formula

$$\arg h_+(t) = -|\sigma_+| r^{-\lambda_+}\cos(\lambda_+\psi(t)),$$

where, as above, $r = |t-t_0|$, $\psi(t) = \theta_+ - \arg(t-t_0)$. Observing that $d(t,l_+) = r|\sin\psi(t)|$, $t\in\omega_+$, and using the estimate (4.88), we see that, for $t\in\omega_+$,

$$\begin{aligned} |\arg h_+(t) + |\sigma_+|r^{-\lambda_+}| &\le c_1 r^{-\lambda_+}\sin^2\frac{\lambda_+\psi(t)}{2} \\ &\le c_2 r^{-2-\lambda_+}[d(t,l_+)]^2 \le c_3 r^{\lambda_+}, \end{aligned}$$

where $c_{1,2,3}$ are absolute constants. Therefore,

$$\lim_{t\to t_0+0}(\arg h_+(t) + |\sigma_+|r^{-\lambda_+}) = 0.$$

Taking into account that

$$\lim_{t\to t_0+0}[h_-(t)h_+^{-1}(t)] = 0,$$

we conclude that

$$\begin{aligned} &\lim_{t\to t_0+0}(\arg a(t) - \arg h^{-\varepsilon}(t)) \\ &= \lim_{t\to t_0+0}[(\arg a(t) - \arg h_+^{-\varepsilon}(t) + \varepsilon\arg(e^{-i\varepsilon c_+} + e^{-i\varepsilon c_-}h_-(t)h_+^{-1}(t))] \\ &= \lim_{t\to t_0+0}[c_+ + \sigma_+ r^{-\lambda_+} - \varepsilon^2 c_+] = 0. \end{aligned}$$

Similarly, using the fact that

$$\lim_{t\to t_0-0}(\arg h_-(t) - |\sigma_-|r^{-\lambda_-}) = 0, \qquad \lim_{t\to t_0-0} h_+(t)h_-^{-1}(t) = 0,$$

one can show that

$$\lim_{t\to t_0-0}(\arg a(t) - \arg h^{-\varepsilon}(t)) = 0.$$

Thus, the function $c(t) = \arg a(t) - \arg h^{-\varepsilon}(t)$ belongs to $C(\Gamma)$ and, consequently, $a(t)$ can be written in the form (4.93), where

$$b(t) = |a(t)h^{\varepsilon}(t)| \exp\{ic(t)\}.$$

Since the function $|a(t)h^{\varepsilon}(t)|$ is bounded away from zero on Γ, Theorem 2.1 in conjunction with Proposition 2.1 shows that $b \in \operatorname{fact}(p, \varrho)$. □

Let $\zeta_0 \in \mathcal{D}_\Gamma^+$ and let the (p, ϱ)-factorization of the function b figuring in (4.93) have the form

$$b = (t - \zeta_0)^{\kappa} b^+ b^-, \quad \text{where } \kappa = \operatorname{ind} b.$$

Then (see property (3) in Section 2.7)

$$a = h_1^{-\varepsilon} b_1, \tag{4.94}$$

where the model function h_1 on Γ has the form

$$h_1 = \begin{cases} h(t)(t-\zeta_0)^{|\kappa|}, & \text{if } \varepsilon\kappa \le 0, \\ h(t)\prod\limits_{j=1}^{|\kappa|}(t - z_j)^{-1}, & \text{if } \varepsilon\kappa > 0, \end{cases}$$

and $b_1 \in \operatorname{fact}(p, \varrho)$, with $\operatorname{ind} b_1 = 0$. Moreover, $b_1 = b_1^+ b_1^-$, where $b_1^+ = b^+$,

$$b_1^- = \begin{cases} b^- & \text{if } \varepsilon\kappa \le 0, \\ b^- \prod\limits_{j=1}^{|\kappa|}\left(\frac{t-z_j}{t-\zeta_0}\right)^{\varepsilon} & \text{if } \varepsilon\kappa > 0, \end{cases}$$

and z_j, $j = 1, 2, \ldots, |\kappa|$, are the first $|\kappa|$ zeros of the function $h(z)$ in the domain $\mathcal{D}_\Gamma^+$.

Theorem 4.28 *Let Γ be a smooth contour of class $\mathcal{T}$ which satisfies conditions* (4.88).

If the function $a(t) \in GL_\infty(\Gamma)$ is continuous on $\Gamma \setminus \{t_0\}$ and has a discontinuity of the form (4.76) *in $t = t_0$, where $\sigma_+\sigma_- > 0$, then a has a representation* (4.94).

Moreover, if $\operatorname{sign}\sigma_+ = \operatorname{sign}\sigma_- = 1$, *then the operator $T(a)$ is right invertible in $L_p^+(\Gamma, \varrho)$ and* $\dim\ker T(a) = \infty$. *One of the right inverses of $T(a)$ has the form*

$$[T(a)]^{-1} = P_\Gamma^+ h_1 (b_1^+)^{-1} P_\Gamma^+ (b_1^-)^{-1} I,$$

and for f *to lie in* $\ker T(a)$ *it is necessary and sufficient that*

$$f = h_1(b_1^+)^{-1} P_\Gamma^- b_1^+ h_1^{-1} g^+,$$

where $g^+ \in \ker T(h_1^{-1})$.

If $\operatorname{sign}\sigma_+ = \operatorname{sign}\sigma_- = -1$, *then the operator* $T(a)$ *is left invertible in* $L_p^+(\Gamma, \varrho)$ *and* $\dim \operatorname{coker} T(a) = \infty$. *One of the left inverses of* $T(a)$ *has the form*

$$[T(a)]^{-1} = P_\Gamma^+ h_1^{-1}(b_1^+)^{-1} P_\Gamma^+ (b_1^-)^{-1} I,$$

and for f *to lie in* $\operatorname{im} T(a)$ *it is necessary and sufficient that*

$$\int_\Gamma [(b_1^-)^{-1}(P_\Gamma^-((\tau - z_0)^{-n} h_1^{-1}(\tau)(b_1^+(\tau))^{-1}))](t) f(t) dt = 0,$$

where $n \in \mathbb{N}$, $z_0 \in \mathcal{D}_\Gamma^-$.

Proof. All the assertions of Theorem 4.28 concerning the operator $T(a)$ are corollaries of Theorem 2.6 and representation (4.94) of the symbol a. □

Remark 4.18 In the theory of the Riemann boundary value problem with infinite index the whirl points for which

$$A_\pm = \lim_{t \to t_0 \pm 0} \arg a(t) = \pm\infty$$

are referred to as points of index plus infinity, whereas the points for which $A_\pm = \mp\infty$ are called points of index minus infinity. It follows from Theorem 4.28 that if t_0 is a point of index plus (resp., minus) infinity for the symbol a, then the operator $T(a)$ is right (resp., left) invertible.

4.11 One-sided whirls

The aim of the present section is to give some extensions of the conditions in representation (4.76), with the aim of covering the case of one-sided whirls. Specifically, let, as above, $0 < \lambda_+ < 1$, but $0 \le \lambda_- \le 1$, that is, we bring into consideration the cases

$$0 < \lambda_+ < 1, \quad \lambda_- = 1, \tag{4.95}$$

$$0 < \lambda_+ < 1, \quad \lambda_- = 0, \tag{4.96}$$

which differ essentially from the cases already considered. Under condition (4.95) and (4.96), the function $h_-(t)$ is different from zero on both sides relative to the point t_0 ($t \in \omega_\pm$), while before this was true only for $t \in \omega_-$ ($\lim_{t \to t_0+0} h_-(t) = 0$).

To circumvent the difficulties introduced by these cases, we will modify the model function (4.77) slightly and put

$$h(z) = \delta \exp\{-i\varepsilon c_+\} h_+(z) + \exp\{-i\varepsilon c_-\} h_-(z). \tag{4.97}$$

Here the functions $h_\pm(z)$ have the form (4.78), (4.79), the number δ is 1 if $0 < \lambda_- < 1$, while in the cases (4.95), (4.96) we choose $\delta > 0$ large enough to ensure that $h^{-1} \in L_\infty(\Gamma)$. That such a choice of δ in (4.97) is possible is shown below in Proposition 4.21.

The investigation of the operator $T(a)$ in the cases (4.95) and (4.96) is carried out following the scheme of the preceding section.

Proposition 4.21 *Let the function $h(t)$ be given by* (4.77), (4.78), (4.79). *Suppose that there exists a neighborhood $\omega \subset \Gamma$ of the point $t_0 \in \Gamma$ such that*

$$d(t,l) \leq \text{const}\,|t - t_0|^2, \quad t \in \omega. \tag{4.98}$$

Then there exists a sufficiently large number δ such that the function $h(t)$ given by (4.97) *is a model function on Γ.*

Proof. In the case (4.95), if condition (4.98) is satisfied, the function $h_-(t)$ is bounded and bounded away from zero in both semi-neighborhoods $\omega_\pm$ of t_0. The same is valid for the function h_- in the case (4.96), because it degenerates into a constant. It follows that in the situations considered here the behavior of h_+ and h_- near the point t_0 can be described as follows. On ω_+ both h_+ and h_- are bounded and bounded away from zero, and so is the function h_- on ω_-; moreover, $\lim_{t \to t_0 - 0} h_+(t) = 0$. Therefore, by choosing δ large enough one can ensure that the function $h(t)$ given by (4.97) is bounded away from zero in some neighborhood of t_0. At the same time, modifying δ if necessary one can ensure that the analytic function $h(t)$ is nondegenerate on the remaining parts of the contour. □

Remark 4.19 One can show that under the assumptions of Proposition 4.21 (which is an analogue of Proposition 4.16) the function $h(z)$ has an infinite number of zeros in $\mathcal{D}_\Gamma^+$. Moreover, it is not difficult to construct the asymptotics of these zeros in the cases (4.95), (4.96), which are similar to formulas (4.82) in Proposition 4.15, but we will not dwell further on this topic.

Propositions 4.17, 4.18, 4.19, 4.20, as well as their proofs, remain valid in the cases considered here. The same is true for the statement of Theorem 4.27, but the proof requires the (p,ϱ)-factorizability of the function

$$b(t) = |a(t)||h(t)|^\varepsilon \exp\{ic(t)\},$$

because here the function $c(t)$ is no longer continuous at the point t_0. Let us give this proof, using Theorem 2.1 on the (p,ϱ)-factorizability of locally sectorial functions.

The function $c(t)$ is continuous on the set $\Gamma \setminus \{t_0\}$. Hence, it suffices to exhibit a neighborhood ω of t_0, $\omega \subset \Gamma$, such that for $t \in \omega$ the range of the function $\exp\{ic(t)\}$ lies in a sufficiently narrow sector with the vertex at the origin. As in the proof of Theorem 4.27, one shows that

$$\lim_{t \to t_0 - 0} (\arg a(t) - \arg h^{-\varepsilon}(t)) = 0.$$

If now $t \in \omega_+$, then

$$\begin{aligned} c(t) &= \arg a(t) - \arg h^{-\varepsilon}(t) \\ &= \arg a(t) - \arg[\delta \exp\{-i\varepsilon c_+\} h_+(t)]^{-\varepsilon} - \arg\left[1 + \frac{\exp\{-i\varepsilon c_-\} h_-(t)}{\delta \exp\{-i\varepsilon c_+\} h_+(t)}\right]^{-\varepsilon}. \end{aligned}$$

Since the function $h_-(t)$ is bounded on ω_+, while the function $h_+(t)$ is bounded away from zero there, by choosing δ large enough one can make the last term arbitrarily small in modulus. Recalling (see the proof of Theorem 4.27) that

$$\lim_{t \to t_0+0} (\arg a(t) + \varepsilon \arg[\delta \exp\{-i\varepsilon c_+\} h_+(t)]) = 0,$$

we conclude that for t close to t_0 the function $c(t)$ is sufficiently small in modulus. By Theorem 2.1, $\exp\{ic(t)\} \in \mathrm{fact}(p, \varrho)$. Therefore, $b \in \mathrm{fact}(p, \varrho)$.

References and comments

4.1–4.3 As already mentioned in the introduction, discontinuities of almost periodic type appeared for the first time in the work of Gohberg and Feldman [1]–[3], in the study of integro-difference Wiener-Hopf equations. Independently and almost at the same time close results were obtained by Coburn and Douglas [1]. The foundation of the approach proposed in the aforementioned works is the factorization Theorem 4.5. Subsequently, almost periodic discontinuities in the theory of singular integral equations have attracted the attention of a large number of researchers (see Berkovich and Konyshkova [1], Berkovich [1], Gaponenko and Dybin [1], Konyshkova [1], Skomaha [1], Kulagina [1], [2]).

Singular integral operators in the space $L_{p,\varrho}$ whose coefficients have discontinuities of almost periodic type were studied in the papers of Gohberg and Sementsul [1] and Sementsul [1]. These authors introduced the concept of a general almost periodic singularity of a measurable function and constructed a theory of one-sided invertibility of singular integral operators with continuous coefficients that have a finite number of almost periodic discontinuities (without describing the structure of the inverse operators and the subspaces $\ker A$ and $\operatorname{im} A$).

Dybin [4] and Grudsky and Dybin [1], [2] have isolated the class of standard almost periodic discontinuities and have elaborated a constructive theory of invertibility of singular integral operators in weighted Lebesgue spaces on arbitrary closed smooth contours for this case. The case of general A_p-weights and infinitely many discontinuities was considered by Dybin in [13].

The concept of a multiplier that does not affect factorizability (MNAF; see Theorem 4.6) was introduced by Spitkovsky, who constructed the corresponding theory (Spitkovsky [1], [4]; see also Litvinchuk and Spitkovsky [1]). Theorem 4.9 was proved by S. Grudsky.

Theorem 4.10 was obtained by Grudsky [1], who used ideas and methods of the well-known work of Calderón [2]. The corollary of this result, Theorem 4.11, allows us to carry over the results on normal solvability of operators from the unit circle to an arbitrary contour not only in the case of almost periodic discontinuities, but also for arbitrary symbols from L_∞.

The book of Böttcher, Karlovich, and Spitkovsky [3] contains a nice introduction to the basic properties of almost periodic functions and, in particular, a proof of Theorem 4.2, which is known as Bohr's theorem.

4.4 This section follows the papers of Dybin [6], [13] and Dybin and Dodokhova [1], [2].

4.5 Theorem 4.18 is due to Gohberg and Krupnik [1]–[3]. We note that already in 1960 Widom [4] discovered the appearance of circular arcs in the spectra of singular integral operators on $L_p(\mathbb{R})$ with piecewise continuous coefficients. The history and theory of Toeplitz operators with piecewise continuous symbols on the spaces $L_p^+(\Gamma, \varrho)$ with a power weight ϱ is well documented in the monographs Böttcher and Silbermann [2], Clancey and Gohberg [1], Duduchava [1], Gohberg and Krupnik [3], Litvinchuk and Spitkovsky [1], Roch and Silbermann [1], Simonenko and Chin′ Ngok Min′ [1]. The case of general Carleson curves Γ and general Muckenhoupt weights is considered in Böttcher and Karlovich [1], Böttcher, Karlovich and Rabinovich [1], Böttcher and Seybold [1] (discrete Wiener-Hopf operators), Böttcher and Spitkovsky [2] (Wiener-Hopf integral operators), and Spitkovsky [8]. Piecewise quasicontinuous symbols combine discontinuities of the second kind with certain "slow" oscillations. Toeplitz operators with such symbols were first studied by Sarason [4] (in $L_p^+(\Gamma_0)$) and subsequently by Böttcher and Silbermann [2] (in $L_p^+(\Gamma_0)$) and Böttcher and Spitkovsky [1] (in $L_p^+(\Gamma_0, \varrho)$). In this region we also mention the works of Böttcher, Karlovich and Rabinovich [2]–[3] and Rabinovich [1]–[3].

4.6–4.9 The problem of developing a theory that comprises both Toeplitz operators with piecewise continuous symbols and Toeplitz operators with almost periodic symbols was suggested to Sarason by Gohberg and led to the invention of the class SAP. Sarason [3] then worked out the theory of normal solvability in the space L_2 of Toeplitz operators whose symbols have discontinuities of semi-almost periodic type on the real line. In the works of Saginashvili [1]–[3] and of Duduchava and Saginashvili [1], [2] these results were generalized to singular integral operators and Wiener-Hopf operators acting in L_p, $1 < p < \infty$. A constructive theory of such operators was developed by Grudsky [3]. Let us mention also the works of Xia [1], [2]. In Böttcher, Karlovich and Spitkovsky [1], the operators are studied in spaces with general (A_p)-weights.

Let us also emphasize that the Toeplitz operators with SAP symbols reveal many properties of general Toeplitz operators that remain in the dark when attention is restricted to symbols in $C + H^\infty$, PSC, or AP. For example, if a_n is a sequence of functions in $C + H^\infty$, PC, or AP which converges uniformly (that is, in the L^∞ norm) to some function a, then the spectra of $T(a_n)$ converge to the spectrum of $T(a)$ in the Hausdorff metric. However, if a_n is taken from SAP, then the spectra of $T(a_n)$ may converge to something that is significantly smaller than the spectrum of $T(a)$. More about the mysteries of Toeplitz operators with SAP symbols can be found in the papers of Böttcher, Grudsky, Spitkovsky [1]–[5] and in Chapter 4 of the book of Böttcher, Karlovich, and Spitkovsky [3].

Matrix problems with semi-almost periodic symbols were studied in some particular cases, which, however, are important for a number of applications, by Novokshenov [1]. A similar approach to these problems and applications to problems in diffraction theory were given by Babeshko in [1].

The theory of matrix operators with almost periodic or semi-almost periodic symbols is much more involved than its scalar counterpart. The pioneering work in this direction was done by Karlovich [1], [4], Spitkovsky [6], [9], and Karlovich and Spitkovsky [1]–[6]. An important role in these investigations is played by the so-called almost periodic factorization, which is studied in detail in Böttcher, Karlovich, and Spitkovsky [3]. The papers of Bastos, Karlovich, Spitkovsky, and Tishin [1] and Bastos, Karlovich, dos Santos, and Tishin [1], [2] are remarkable recent contributions to the problem of almost periodic factorization. The Fredholm theory of Toeplitz operators with matrix-valued SAP symbols on $L^p_+(\mathbb{R}, \varrho)$ with $1 < p < \infty$ and power weights ϱ was only recently completed by Böttcher, Karlovich, and Spitkovsky [2]. The book of Böttcher, Karlovich, and Spitkovsky [3] is a comprehensive treatise of what is presently known on almost periodic factorization and on matrix Toeplitz operators, and it also discusses many problems in this field that are still open.

4.10–4.11 These sections follow the paper of Grudsky [4]. Singular integral operators whose symbols have whirl points of power type were also investigated in the papers of Semenko [1] and Monakhov and Semenko [1]–[3]. The Riemann boundary value problem with infinite index was studied by Govorov [2]–[8] and his followers (Tolochko [1]–[3], Sandrygaĭlo [1], [2], Plaksa [1], [2], Rogozin [1], Danilov [1], Ostrovskiĭ [1]–[3]). Problems close to the ones considered here were also investigated by Alekna [1], [2], Chibrikova [1], [2], Yurov [1], [2]. Let us also mention a series of studies devoted to various aspects of the theory of Toeplitz operators with infinite index: Böttcher [1], Kats [3], Kats and Zapuskalova [1], Volevich and Gindikin [1], [2], Vladimirov [1]–[3], Hayashi [1], Nakazi [1]–[3].

The method employed in Sections 4.1–4.11 is also applicable to whirls of logarithmic or power-logarithmic types. The article Böttcher and Grudsky [1] may serve as an introduction to the fascinating phenomena one encounters in the wealth of Toeplitz operators generated by symbols with whirls of different kinds.

Chapter 5

Generalized Factorization of u-periodic Functions and Matrix Functions

In this chapter we confine ourselves to the setting of the space $L_p(\Gamma_0)$, $1 < p < \infty$. First, we will give the needed definitions and facts from the theory of matrix Toeplitz operators. Then we will introduce the concept of the generalized factorization of a matrix valued function (matrix function for short) and formulate the corresponding invertibility theory for the operator $T(a)$ with matrix symbol a. Conceptually, the main section of this chapter is Section 5.3, devoted to the theory of generalized factorization of u-periodic matrix functions. After that we will consider new classes of Toeplitz operators with infinite index, first in the scalar and then in the matrix case.

5.1 Block Toeplitz operators

Let $L_p^{(n)}(\Gamma_0)$ (resp., $L_p^{(n\times n)}(\Gamma_0)$), $1 \le p \le \infty$, denote the space of vector functions $\vec{f} = \{f_j\}_{j=1}^n$ (resp., matrix functions $F = (f_{kj})_{k,j=1}^n$) of order n whose elements f_j (resp., f_{kj}) belong to $L_p(\Gamma_0)$, equipped with the norm

$$\|\vec{f}\|_{p,n} = \left(\sum_{j=1}^n \|f_j\|_p^2\right)^{1/2} \qquad (\text{resp., } \|F\|_{p,n} = \left(\sum_{k,j=1}^n \|f_{kj}\|_p^2\right)^{1/2}).$$

We denote by S and $P^{\pm}$ the Cauchy singular integral operator and the analytic projectors in the spaces $L_p^{(n)}(\Gamma_0)$, $L_p^{(n\times n)}(\Gamma_0)$, acting elementwise. Further, let us introduce the spaces $L_p^{(n)\pm}(\Gamma_0) = P^{\pm}(L_p^{(n)}(\Gamma_0))$, $L_p^{(n\times n)\pm}(\Gamma_0) = P^{\pm}(L_p^{(n\times n)}(\Gamma_0))$,

$1 \le p < \infty$, and the Hardy spaces $H_\infty^{(n)}(\mathbb{D}^\pm)$ and $H_\infty^{(n\times n)}(\mathbb{D}^\pm)$ of vector and matrix functions whose elements are analytic and bounded in the respective domain $\mathbb{D}^\pm$.

Let $A \in L_\infty^{(n\times n)}(\Gamma_0)$. Then, as in the case $n = 1$, the Toeplitz operator $T(A)$ acts in the space $L_p^{(n)+}(\Gamma_0)$ by the rule

$$T(A) = P^+AI. \tag{5.1}$$

The next result is the matrix analogue of Proposition 1.6.

Proposition 5.1 *For an operator $T(A)$ of the form (5.1) with symbol $A \in L_\infty^{(n\times n)}(\Gamma_0)$ to be a Φ_+- or Φ_--operator in the space $L_p^{(n)+}(\Gamma_0)$, $1 < p < \infty$, it is necessary that*

$$\operatorname{ess\,inf}_{t\in\Gamma_0} |\det A(t)| > 0. \tag{5.2}$$

The concept of p-factorization plays the same important role in the theory of block Toeplitz operators as in the scalar case. We say that a matrix function $A(t) \in L_\infty^{(n\times n)}(\Gamma_0)$ admits a *p-factorization with respect to the contour* Γ_0 if $A(t)$ is representable in the form

$$A(t) = A^-(t)D(t)A^+(t), \tag{5.3}$$

where $D(t) = \operatorname{diag}[t^{\kappa_1}, t^{\kappa_2}, \dots, t^{\kappa_n}]$, κ_j are integers, $\kappa_1 \ge \kappa_2 \ge \dots \ge \kappa_n$, and the matrix functions $A^\pm(t)$ satisfy the following conditions:

(1) $A^+ \in L_q^{(n\times n)+}(\Gamma_0)$, $[A^+]^{-1} \in L_p^{(n\times n)+}(\Gamma_0)$, $A^- \in L_p^{(n\times n)-}(\Gamma_0) \oplus \mathbb{C}^{(n\times n)}$, $[A^-]^{-1} \in L_q^{(n\times n)-}(\Gamma_0) \oplus \mathbb{C}^{(n\times n)}$, where $q = p/(p-1)$, $1 < p < \infty$; here $\mathbb{C}^{(n\times n)}$ denotes the subspace of constant matrix functions of order n;

(2) the operator $A^-S[A^-]^{-1}$ is bounded in $L_p^{(n)}(\Gamma_0)$.

The numbers κ_j are called the *partial factorization indices*. The set of all matrix functions $A \in L_\infty^{(n\times n)}(\Gamma_0)$ that admit a p-factorization with respect to Γ_0 will be denoted by $\operatorname{fact}_n p$.

Theorem 5.1 *Let $A \in L_\infty^{(n\times n)}(\Gamma_0)$. Then the operator $T(A)$ given by (5.1) is a Φ-operator in $L_p^{(n)+}(\Gamma_0)$, $1 < p < \infty$, if and only if $A \in \operatorname{fact}_n p$. If this is the case, then*

$$\dim\ker T(A) = \sum_{\kappa_j<0} |\kappa_j|, \qquad \dim\operatorname{coker} T(A) = \sum_{\kappa_j>0} \kappa_j, \qquad \operatorname{ind} T(A) = -\sum_{j=1}^{n} \kappa_j.$$

Let us give a number of sufficient conditions for membership in the class $\operatorname{fact}_n p$.

A matrix function $A \in L_\infty^{(n\times n)}(\Gamma_0)$ is said to be *p-sectorial* (or to belong to the class $G_p^{(n\times n)}(\Gamma_0)$) if

$$A(t) = V^-(t)D(t)A_0(t)V^+(t), \tag{5.4}$$

where $V^\pm \in GH_\infty^{(n\times n)}(\mathbb{D}^\pm)$, D has the same form as in (5.3), and $A_0(t) \in L_\infty^{(n\times n)}(\Gamma_0)$ has unitary values for almost all $t \in \Gamma_0$ and its spectrum lies in the sector

$$|\arg t| < \pi[\max(p,q)]^{-1}, \quad q = p/(p-1).$$

Proposition 5.2 (1) $G_2^{(n\times n)}(\Gamma_0) = \mathrm{fact}_n\, 2$.

(2) *$G_p^{(n\times n)}(\Gamma_0) \subset \mathrm{fact}_n\, r$ for $1 < p < \infty$ and $\min(p,q) \le r \le \max(p,q)$.*

(3) *If $A \in G_p^{(n\times n)}(\Gamma_0)$ and $\kappa_j = 0$, $j = 1,2,\dots,n$ (in particular, if $A = A_0$ in (5.4)), then the factorization (5.3) has the form $A = A^-A^+$ and does not depend on r, $\min(p,q) \le r \le \max(p,q)$.*

Remark 5.1 The class $G_p^{(n\times n)}(\Gamma_0)$ is sufficiently wide. In the case $n = 1$ it coincides with the set of functions that admit an r-factorization for all r in the segment $[\min(p,q), \max(p,q)]$ (Krupnik [3], §12). In particular,

$$GC(\Gamma_0) \subset G_p^{(1\times 1)}(\Gamma_0).$$

Incidentally, from Krupnik's work referred to above it follows that the class $G_p(\Gamma_0)$ defined in Section 2.5 actually coincides with $G_p^{(1\times 1)}(\Gamma_0)$.

We call representation (5.3) a *standard factorization* of the matrix function A if

$$A^\pm \in GH_\infty^{(n\times n)}(\mathbb{D}^\pm). \tag{5.5}$$

As in the scalar case (Theorem 2.2), nondegenerate matrix functions whose elements belong to some decomposing algebra $\mathfrak{A}(\Gamma_0)$ of continuous functions admit a standard factorization. Let us denote

$$\mathfrak{A}^{(n\times n)\pm}(\Gamma_0) = P^\pm(\mathfrak{A}^{(n\times n)}(\Gamma_0)).$$

Theorem 5.2 *Let $\mathfrak{A}(\Gamma_0)$ be a decomposing algebra and let $A \in G\mathfrak{A}^{(n\times n)}(\Gamma_0)$. Then for $T(A)$ to be a Φ-operator in $L_p^{(n\times n)+}(\Gamma_0)$ it is necessary and sufficient that*

$$\det A(t) \ne 0, \quad t \in \Gamma_0.$$

If the last condition is satisfied, then A admits a standard factorization, in which

$$A^+ \in G\mathfrak{A}^{(n\times n)+}(\Gamma_0), \quad A^- \in G(\mathfrak{A}^{(n\times n)-}(\Gamma_0) \oplus \mathbb{C}^{(n\times n)}),$$

and the number $\kappa = \sum_{j=1}^{n} \kappa_j$ *coincides with the topological index of the function* $\det A(t)$*:*

$$\kappa = \frac{1}{2\pi} \arg \det A(t)\big|_{\Gamma_0}.$$

In contrast to the scalar case, the factorization (5.3) cannot be explicitly constructed even for continuous matrix functions. The general theory enables us only to calculate the number κ, while algorithms for calculating the partial indices κ_j and the factors $A^{\pm}$ are known only for some special classes of matrix functions, for example, for rational matrix functions.

5.2 Generalized factorization of matrix functions

By definition, a matrix function $A \in L_\infty^{(n\times n)}(\Gamma_0)$ is said to admit a $(\{u_j\}_{j=1}^n, p)$-*factorization* if it can be written in the form

$$A = A^- \Lambda A^+, \tag{5.6}$$

where

$$\Lambda = \operatorname{diag}[u_1^{\varepsilon_1}, u_2^{\varepsilon_2}, \dots, u_n^{\varepsilon_n}],$$

u_j are inner functions,

$$\varepsilon_j = \begin{cases} 1, & \text{if } 1 \le j \le n_1, \\ 0, & \text{if } n_1 < j \le n_2, \\ -1, & \text{if } n_2 < j \le n, \end{cases} \qquad 0 \le n_1 \le n_2 \le n,$$

and the factors $A^{\pm}$ satisfy conditions (1) and (2) of definition (5.3).

Let us introduce the matrix functions

$$\Lambda_+ = \operatorname{diag}[u_1, \dots, u_{n_1}, 1, \dots, 1],$$

$$\Lambda_- = \operatorname{diag}[1, \dots, 1, u_{n_2+1}^{-1}, \dots, u_n^{-1}]$$

and remark that $\Lambda = \Lambda_- \Lambda_+$; here $\Lambda_+ = E$ when $n_1 = 0$ and $\Lambda_- = E$ when $n_2 = n$.

If u is an inner function, we will denote by $M_j(u)$, $j = 1, \dots, n$, the subspace of $L_p^{(n)+}(\Gamma_0)$ consisting of the vector functions $\vec{f} = \{f_k\}_{k=1}^n$ with the property that $f_k = 0$ if $k \neq j$ and $f_j \in M(u)$ in $L_p^+(\Gamma_0)$ (see Section 3.1). $M_j(u)$ will also be referred to as a *model subspace*.

Theorem 5.3 *Let the matrix function* $A \in L_\infty^{(n\times n)}(\Gamma_0)$ *admit a* $(\{u_j\}_{j=1}^n, p)$*-factorization in the space* $L_p^{(n)}(\Gamma_0)$*. Then the operator* $T(A)$ *given by* (5.1) *has a generalized inverse in* $L_p^{(n)+}(\Gamma_0)$*. In particular, if* $n_1 = 0$ *then* $T(A)$ *is right invertible, if* $n_2 = n$ *then* $T(A)$ *is left invertible, and if* $n_1 = 0$ *and* $n_2 = n$ *then* $T(A)$ *is* *invertible in* $L_p^{(n)+}(\Gamma_0)$. *In any of these cases the corresponding inverse operator can be written in the form*

$$[T(A)]^{-1} = P^+(A^+)^{-1}\Lambda^{-1}P^+(A^-)^{-1}I. \tag{5.7}$$

Furthermore, if there exists an index j, $1 \leq j \leq n$, *such that* $\varepsilon_j = -1$ (*resp.,* $\varepsilon_j = 1$) *and* u_j *is not a finite Blaschke product, then* $\dim\ker T(A) = \infty$ (*resp.,* $\dim\operatorname{coker} T(A) = \infty$). *If, however, for any* j, $1 \leq j \leq n$, *either* $\varepsilon_j = 0$ *or* u_j *is a finite Blaschke product, then* $T(A)$ *is a* Φ*-operator and*

$$\dim\ker T(A) = \frac{1}{2\pi}\left(\arg\prod_{j=n_2+1}^{n} u_j\right)\Bigg|_{\Gamma_0},$$

$$\dim\operatorname{coker} T(A) = \frac{1}{2\pi}\left(\arg\prod_{j=1}^{n_1} u_j\right)\Bigg|_{\Gamma_0},$$

$$\operatorname{ind} T(A) = -\frac{1}{2\pi}\left(\arg\prod_{j=1}^{n} u_j^{\varepsilon_j}\right)\Bigg|_{\Gamma_0}.$$

Proof. Let $n_1 = 0$. Then $\varepsilon_j \leq 0$ for all j, $1 \leq j \leq n$, and $\Lambda \in H_\infty^{(n\times n)}(\mathbb{D}^-)$, $\Lambda^{-1} \in H_\infty^{(n\times n)}(\mathbb{D}^+)$. Therefore, on $L_p^{(n)+}(\Gamma_0)$,

$$\begin{aligned} T(A)[T(A)]^{-1} &= P^+A^-\Lambda A^+P^+(A^+)^{-1}\Lambda^{-1}P^+(A^-)^{-1}I \\ &= P^+A^-P^+(A^-)^{-1}I = I, \end{aligned}$$

i.e., the operator $[T(A)]^{-1}$ given by (5.7) is a right inverse for $T(A)$. The case $n_2 = n$ is similar. Now let $0 < n_1 \leq n_2 < n$. Consider the linear manifolds

$$\mathfrak{L}_1 = \left\{\vec{f} \in L_p^{(n)+}(\Gamma_0) : \vec{f} = (A^+)^{-1}\vec{g},\ \vec{g} = (g_1, \dots, g_{n_1}, 0, \dots, 0),\ g_j \in H_\infty(\mathbb{D}^+)\right\},$$

and

$$\mathfrak{L}_2 = \left\{\vec{f} \in L_p^{(n)+}(\Gamma_0) : \vec{f} = (A^+)^{-1}\vec{h}, \vec{h} = (0,\dots,0,h_{n_1+1},\dots,h_n), h_j \in H_\infty(\mathbb{D}^+)\right\}.$$

We claim that, on $\mathfrak{L}_1$,

$$T(A)[T(A)]^{-1}T(A) = T(A). \tag{5.8}$$

Indeed,

$$\begin{aligned} &(T(A)[T(A)]^{-1}T(A))(A^+)^{-1}\vec{g} \\ &= T(A)P^+(A^+)^{-1}\Lambda^{-1}P^+(A^-)^{-1}P^+A^-\Lambda_+\vec{g} \\ &= T(A)P^+(A^+)^{-1}\Lambda^{-1}\Lambda_+\vec{g} = T(A)(A^+)^{-1}\vec{g}. \end{aligned}$$

Similarly, one can show that equality (5.8) holds on $\mathfrak{L}_2$.

Next, observe that $\operatorname{clos}(\mathfrak{L}_1 \oplus \mathfrak{L}_2) = L_p^{(n)+}(\Gamma_0)$. First let us prove this for $n = 1$, i.e., show that $\operatorname{clos}[(A^+)^{-1}(H_\infty(\mathbb{D}^+))] = L_p^+(\Gamma_0)$. Suppose this is not the case. Then one can find a nonzero element $f \in L_q^+(\Gamma_0)$, $q = p/(p-1)$, such that $\langle (A^+)^{-1}\psi, f\rangle = 0$ for all $\psi \in H_\infty(\mathbb{D}^+)$. Hence, $\langle \psi, \overline{(A^+)^{-1} f}\rangle = 0$ for all $\psi \in H_\infty(\mathbb{D}^+)$, and since $g = \overline{(A^+)^{-1} f} \in L_1(\Gamma_0)$, we conclude that $g \in L_1^-(\Gamma_0)$. But then $f = \overline{A^+ g} \in L_q^-(\Gamma_0)$, that is, $f = 0$ almost everywhere on Γ_0. The case $n > 1$ is dealt with in analogous manner. In view of property (2) in definition (5.3), the operator $[T(A)]^{-1}$ is bounded in $L_p^{(n)}(\Gamma_0)$, and since $\operatorname{clos}(\mathfrak{L}_1 \oplus \mathfrak{L}_2) = L_p^{(n)+}(\Gamma_0)$, relation (5.8) holds on $L_p^{(n)+}(\Gamma_0)$. Therefore, the operator of $[T(A)]^{-1}$ given by (5.7) is a generalized inverse for $T(A)$.

Now let us show that the subspace $\ker T(A)$ is infinite-dimensional whenever for some j, $1 \le j \le n$, $\varepsilon_j = -1$ and the function u_j is not a finite Blaschke product. Indeed, in this case, for any z, $|z| < 1$, $(u_j(t) - u_j(z))/(t - z) \in (M(u_j) \cap H_\infty(\mathbb{D}^+))$, i.e., the subspace $M(u_j) \cap H_\infty(\mathbb{D}^+)$ is infinite-dimensional. But then so is the subspace $(A^+)^{-1}(M_j(u_j) \cap H_\infty^{(n)}(\mathbb{D}^+))$, which, as one can readily verify, is embedded in $\ker T(A)$. Thus, $\dim\ker T(A) = \infty$. If now for some j, $1 \le j \le n$, $\varepsilon_j = 1$ and the function u_j is not a finite Blaschke product, then one can show that $\dim\operatorname{coker} T(A) = \infty$ by passing to the adjoint operator $T^*(A)$.

The last assertion of our theorem follows from Theorem 5.1. □

Theorem 5.3 does not provide a description of the solvability properties of the matrix Toeplitz equation

$$T(A)\vec{f} = \vec{g} \tag{5.9}$$

in the space $L_p^{(n)+}(\Gamma_0)$. In the general case such a description is quite complicated (see Spitkovsky [7], [8]). Here we will restrict ourselves to a particular case that will suffice in our future considerations.

Theorem 5.4 *Under the assumptions of Theorem* 5.3, *let* $A^\pm \in GH_\infty^{(n\times n)}(\mathbb{D}^\pm)$. *Then the subspace* $\ker T(A)$ *has the form*

$$\ker T(A) = (A^+)^{-1}\left(\bigoplus_{j=n_2+1}^{n} M_j(u_j)\right).$$

Furthermore, for $\vec{g}$ *to lie in* $\operatorname{im} T(A)$ *it is necessary and sufficient that*

$$(P^+(A^-)^{-1}\vec{g})_j \in \operatorname{im} T(u_j) \text{ in } L_p^+(\Gamma_0), \quad j = 1, 2, \ldots, n_1,$$

where $(\vec{g})_j = g_j$ *is the* j*-th component of the vector* $\vec{g} = \{g_j\}_{j=1}^n$. *If* $\vec{g} \in \operatorname{im} T(A)$, *then in all cases the solution* $\vec{f}$ *of the equation* (5.9) *can be written in the form*

$$\vec{f} = [T(A)]^{-1}\vec{g} + \vec{f_0},$$

where $\vec{f_0} \in \ker T(A)$ *and the operator* $[T(A)]^{-1}$ *has the form* (5.7).

Proof. Let $\vec{f} \in L_p^{(n)+}(\Gamma_0)$. Then $T(A)\vec{f} = 0$ if and only if $P^+\Lambda_-\Lambda_+(A^+\vec{f}) = 0$ i.e., if and only if

$$\begin{cases} (A^+\vec{f})_j = 0, & j = 1, \dots, n_2, \\ (A^+\vec{f})_j \in M(u_j) \text{ in } L_p^+(\Gamma_0), & j = n_2+1, \dots, n. \end{cases}$$

This means that $\ker T(A) = (A^+)^{-1}(\bigoplus_{j=n_2+1}^{n} M_j(u_j))$, as claimed. If $\vec{f}, \vec{g} \in L_p^{(n)+}(\Gamma_0)$, then $T(A)\vec{f} = \vec{g}$ if and only if

$$P^+\Lambda_-\Lambda_+(A^+\vec{f}) = P^+(A^-)^{-1}\vec{g},$$

which is equivalent to $(P^+(A^-)^{-1}\vec{g})_j \in \operatorname{im} T(u_j)$ in the space $L_p^+(\Gamma_0)$ for all $j = 1, \dots, n_1$. □

Remark 5.2 From Theorem 5.4 it follows that in the situation considered here the solution of the matrix Toeplitz equation reduces to the solution of a set of scalar equations of the kind studied in the preceding chapters.

5.3 u-periodic matrix functions

Let $F(t) \in L_p^{(n\times n)}(\Gamma_0)$, $1 \le p \le \infty$, and let $u(t)$ be an inner function, $u \in H_\infty(\mathbb{D}^+)$. Consider the mapping γ_u acting according to the rule

$$(\gamma_u F)(t) = F(u(t)), \quad t \in \Gamma_0.$$

The operator γ_u is a well-known object of study, called a *composition operator*. The following result of the theory of composition operators summarizes those properties of compositions operators that will be needed below.

Theorem 5.5

(1) *The mapping γ_u is a bounded linear operator in the space $L_p^{(n\times n)}(\Gamma_0)$, $1 \le p < \infty$. The subspaces $L_p^{(n\times n)+}(\Gamma_0)$ and $L_p^{(n\times n)-}(\Gamma_0) \oplus \mathbb{C}^{(n\times n)}$ are invariant under γ_u.*

(2) *The map γ_u is an endomorphism of the algebra $L_\infty^{(n\times n)}(\Gamma_0)$. The subalgebras $H_\infty^{(n\times n)}(\mathbb{D}^\pm)$ are invariant under γ_u.*

(3) *If $u \ne$ const, then for any matrix function $F \in L_p^{(n\times n)}(\Gamma_0)$, $1 \le p \le \infty$, the following estimates hold:*

$$\left(\frac{1-|u(0)|}{1+|u(0)|}\right)^{1/p} \|F\|_{p,n} \le \|\gamma_u F\|_{p,n} \le \left(\frac{1+|u(0)|}{1-|u(0)|}\right)^{1/p} \|F\|_{p,n}. \tag{5.10}$$

Proof. First let us establish the estimates (5.10), from which the boundedness of the operator γ_u in $L_p^{(n\times n)}(\Gamma_0)$, $1\le p\le\infty$, follows as a corollary. Let $n=1$, $p=2$. Representing the function $F(t)$ by its Fourier series,

$$F(t)=\sum_{j=-\infty}^{\infty}F_jt^j,\quad \|F\|_2^2=\sum_{j=-\infty}^{\infty}|F_j|^2<\infty,$$

we obtain

$$\|\gamma_uF\|_2^2=\frac{1}{2\pi i}\int_{\Gamma_0}F(u(t))\overline{F(u(t))}\,\frac{dt}{t}=\frac{1}{2\pi i}\sum_{k,j=-\infty}^{\infty}F_k\overline{F}_j\int_{\Gamma_0}(u(t))^{k-j}\frac{dt}{t}.$$

Since $u^{-1}(\infty)=\overline{u(0)}$ (see Section 2.6),

$$\frac{1}{2\pi i}\int_{\Gamma_0}\frac{u^j(t)}{t}dt=\begin{cases}u^j(0), & j=1,2,\dots,\\ \overline{u}^{|j|}(0), & j=-1,-2,\dots,\end{cases}$$

and consequently

$$\|\gamma_uF\|_2^2=2\sum_{k>j}\mathrm{Re}(F_k\overline{F}_ju^{k-j}(0))+\sum_{k=-\infty}^{\infty}|F_k|^2,$$

i.e., $\|\gamma_uF\|_2$ is completely determined by the value $u(0)$. To calculate this norm we take the simplest inner function

$$B(t)=\frac{u(0)-t}{(1-\overline{u(0)}t)}.$$

Clearly, $B(0)=u(0)$ and the function $B(t)$ maps Γ_0 onto itself in one-to-one manner. Therefore,

$$\|\gamma_uF\|_2^2=\frac{1}{2\pi}\int_{\Gamma_0}|F(B(t))|^2|dt|=\frac{(1-|u(0)|)^2}{2\pi}\int_{\Gamma_0}\frac{|F(\tau)|^2|d\tau|}{|1-\overline{u(0)}\tau|^2}.$$

Observing that

$$\frac{1}{1-|u(0)|^2}=\sup_{\tau\in\Gamma_0}\frac{1}{|1-\overline{u(0)}\tau|},\qquad \frac{1}{1+|u(0)|^2}=\inf_{\tau\in\Gamma_0}\frac{1}{|1-\overline{u(0)}\tau|},$$

we obtain the estimates (5.10) for the case at hand. Their validity for $n=1$, $1\le p<\infty$, follows from the equality $\|F\|_p^p=\|F^{p/2}\|_2^2$. For $p=\infty$ we use the fact that $\lim_{p\to\infty}\|F\|_p=\|F\|_\infty$ whenever $F\in L_\infty(\Gamma_0)$ and, consequently, (5.10) takes on the form $\|\gamma_uF\|_\infty=\|F\|_\infty$. For $n>1$ inequalities (5.10) follow from the definition of the norm in $L_p^{(n\times n)}(\Gamma_0)$.

Since $u^k\in L_p^+(\Gamma_0)$ and $u^{-k}\in L_p^-(\Gamma_0)\oplus\mathbb{C}$ for any positive integer k, the map γ_u takes all polynomials in positive (resp., negative) powers of t into $L_p^+(\Gamma_0)$ (resp., $L_p^-(\Gamma_0)\oplus\mathbb{C}$), $1\le p\le\infty$. From this observation and the boundedness of γ_u in $L_p^{(n\times n)}(\Gamma_0)$ one obtains all the assertions of the theorem concerning invariant subspaces of γ_u. □

Remark 5.3 From the left estimate in (5.10) it follows that the kernel of the operator γ_u in $L_p^{(n\times n)}(\Gamma_0)$, $1 \le p \le \infty$, is trivial. Furthermore, (5.10) shows that for any inner function u the operator γ_u is an isometry of the algebra $L_\infty^{(n\times n)}(\Gamma_0)$ and that the condition $u(0) = 0$ is necessary and sufficient for the map γ_u to be an isometry in $L_p^{(n\times n)}(\Gamma_0)$, $1 \le p \le \infty$.

The next result, which is a corollary of Theorem 5.5, forges the latter into the main tool for constructing a generalized factorization of matrix functions of the form $F(u(t))$.

Theorem 5.6 *Suppose that the matrix function $A(t) \in L_\infty^{(n\times n)}(\Gamma_0)$ admits a standard factorization* (5.3), (5.5). *Then for any inner function $u(t) \in H_\infty(\mathbb{D}^+)$ the matrix function $A(u(t))$ admits a $(\{u_j\}, p)$-factorization of the form*

$$A(u(t)) = A^-(u(t))D(u(t))A^+(u(t)),$$

where $A^\pm(u(t)) \in GH_\infty^{(n\times n)}(\mathbb{D}^\pm)$, $D(u(t)) = \operatorname{diag}[u^{\kappa_1}(t), \dots, u^{\kappa_n}(t)]$, $u_j = u^{|\kappa_j|}$.

Remark 5.4 Let us mention that Theorems 5.3 and 5.4 yield a theory of normal solvability of Toeplitz operators $T(\gamma_u A)$ in $L_p^{(n\times n)+}(\Gamma_0)$, $1 < p < \infty$, for matrix functions $A(t)$ that satisfy the assumptions of Theorem 5.6. Those assumptions are satisfied by many known classes of matrix functions, in particular, by the decomposing subalgebras of continuous functions.

Theorems 5.2 and 5.6 have the following consequence.

Proposition 5.3 *Let $A(t) \in \mathfrak{A}^{(n\times n)}(\Gamma_0)$, where $\mathfrak{A}(\Gamma_0)$ is a decomposing subalgebra of the algebra of continuous functions, and let $\det A(t) \ne 0$, $t \in \Gamma_0$. Then the matrix function $A(u(t))$ admits a $(\{u_j\}, p)$-factorization for every $1 < p < \infty$ with $u_j(t) = u(t)^{|\kappa_j|}$ for $j = 1, \dots, n$.*

Let us mention that any nondegenerate matrix function $A(t)$ with elements from one of the algebras $H_\mu(\Gamma_0)$, $W(\Gamma_0)$, or $R(\Gamma_0)$ satisfies the conditions of Proposition 5.3. Moreover, in the case of the algebra $R(\Gamma_0)$ the factorization of the matrix function $A(u(t))$ can be obtained in a constructive manner.

Let us give one more corollary of Theorem 5.6, which concerns MNAFs (see Section 4.1).

Proposition 5.4 *Let $d(t)$ be an* MNAF *and let the factorization index of $d(t)$ be zero. Then the composition $d(u(t))$ is also an* MNAF *and its factorization index is equal to zero.*

Proof. This is a straightforward consequence of Theorem 4.6 (the equivalence of assertions (1) and (2)). □

Let us consider the class of matrix functions $G_p^{(n\times n)}(\Gamma_0)$. If $A \in G_p^{(n\times n)}(\Gamma_0)$, then under the action of the homomorphism γ_u representation (5.4) does not, generally speaking, go over into factorization (5.6). Nevertheless, the representation of the symbol $A(u(t))$ obtained in this manner allows us to construct a theory of normal solvability of the operator $T(A(u(t)))$ in the case when all nonzero partial indices of the matrix function A have the same sign.

Theorem 5.7 *Let $A \in G_p^{(n\times n)}(\Gamma_0)$ and let all the nonzero partial indices of the matrix function $A(t)$ have the same sign. Further, let $r \in [\min(p,q), \max(p,q)]$ and let $u(t)$ be an arbitrary inner function. Then the Toeplitz operator $T(A(u(t)))$ in $L_r^{(n)+}(\Gamma_0)$ is*

(1) *right invertible, if $\kappa_j \le 0$ for all $1 \le j \le n$;*

(2) *left invertible, if $\kappa_j \ge 0$ for all $1 \le j \le n$;*

(3) *invertible, if $\kappa_j = 0$ for all $1 \le j \le n$.*

In all three cases the corresponding inverse operator can be written in the form

$$[T(A(u))]^{-1} = [A_0^+(u)V^+(u)]^{-1}P^+[V^-(u)D(u)A_0^-(u)]^{-1}I, \tag{5.11}$$

where $A_0 = A_0^- A_0^+$ is the r-factorization of the matrix function A_0 figuring in (5.4).

Proof. The matrix function $A_0(u(t))$ is unitary valued on Γ_0 and its spectrum lies in the sector $|\arg t| < \pi[\max(p,q)]^{-1}$. By assertion (3) of Proposition 5.2, $A_0(u(t))$ admits an r-factorization with zero partial indices: $A_0(u(t)) = \widetilde{A}^-(t)\widetilde{A}^+(t)$, where the matrices $\widetilde{A}^\pm(t)$ satisfy conditions (1) and (2) of definition (5.3). On the other hand, $A_0(u(t)) = A_0^-(u(t))A_0^+(u(t))$, where the matrix functions $A_0^\pm(u(t))$ satisfy condition (1) of definition (5.3) for all values of r in the segment considered here. Since such a representation of the matrix $A_0(u(t))$ is unique up to a constant matrix factor (see Litvinchuk and Spitkovsky [1]), the operator $A_0^-(u(t))S[A_0^-(u(t))]^{-1}I$ is bounded in $L_r^{(n)}(\Gamma_0)$. This shows that the operator defined by the right-hand side of equality (5.11) is bounded in $L_r^{(n)+}(\Gamma_0)$. Verifying that this operator is the corresponding right, left, or two-sided inverse of the operator $T(A(u))$ is elementary. □

As a corollary of Theorem 5.7, one can strengthen Theorem 5.6 for the case $n = 1$. Notice that in the case $n = 1$ the factorization (5.6) is a particular case of the (h,p)-factorization (2.30), namely, the case $h(t) = u_1(t)$.

Theorem 5.8 *Let $a \in G_p(\Gamma_0)$. Then for every inner function $u(t) \in H_\infty(\mathbb{D}^+)$ the composite function $a(u)$ admits a (u,p)-factorization*

$$a(u) = a^-(u)u^\kappa a^+(u),$$

where κ is the factorization index and $a^-(u)a^+(u)$ is the r-factorization of the function $a(u)u^{-\kappa}$, which is the same for all r in the segment $[\min(p,q), \max(p,q)]$, $q = p/(p-1)$.

5.4 Infinite index of logarithmic type

Let us apply the factorization theory for u-periodic functions elaborated in Sections 5.2 and 5.3 to the case of discontinuities of "infinite index" type. In this section we restrict our attention to the symbol $a(t)$ of a scalar Toeplitz operator with one whirl point of logarithmic type. For the function $a(t)$, we will effectively construct a generalized factorization of the form (2.30) with a Blaschke product playing the role of the model factor h.

To demonstrate how the factorization theory of u-periodic functions can be used to strength-en Theorem 2.10 on the generalized factorization of a function a with a logarithmic whirl point, we show that the theorem remains valid for $p \neq 2$.

Theorem 5.9 *Let $a(t) \in GL_\infty(\Gamma_0) \cap C(\Gamma_0 \setminus \{1\})$ and suppose that at the point $t = 1$ the function $a(t)$ has a discontinuity of the second kind such that*

$$\lim_{\theta \to \pm 0} (\arg a(\exp\{i\theta\}) \pm \pi \log |\theta|^{-1}) = 0. \tag{5.12}$$

Then all the conclusions of Theorem 2.10 hold true in the case of the space $L_p(\Gamma_0)$, where $1 < p < \infty$.

Proof. Returning to the proof of Theorem 2.10 and using the notations introduced therein, we recall that

$$\lim_{\theta \to +0} [\arg B(\exp\{i\theta\}) + A(\theta)] = 0 \tag{5.13}$$

and that, as $n \to \infty$, the sequence

$$\alpha_n(\varepsilon) = [A(\exp\{-n - \varepsilon\}) - \pi(n + \varepsilon)] \tag{5.14}$$

converges uniformly on $[0, 1]$ to a continuous function $\alpha(\varepsilon)$ which satisfies the condition $\alpha(0) = \alpha(1)$. First let us show that $\alpha(\varepsilon) = -\alpha(1 - \varepsilon)$. Indeed,

$$\begin{aligned}
\alpha(\varepsilon) + \alpha(1 - \varepsilon) &= \lim_{n \to \infty} (\alpha_n(\varepsilon) + \alpha_n(1 - \varepsilon)) \\
&= \lim_{n \to \infty} \Bigg[2 \sum_{k=1}^{\infty} [\arctan \exp\{n + \varepsilon - k + 1/2\} \\
&\qquad + \arctan \exp\{n - \varepsilon - k + 3/2\}] - \pi(2n + 1) \Bigg] \\
&= \lim_{n \to \infty} \left[\left(2 \sum_{k=1}^{2n+1} + 2 \sum_{k=2n+2}^{\infty} \right) (\ldots) - \pi(2n + 1) \right] \\
&= 2 \lim_{n \to \infty} \sum_{j=0}^{\infty} (\arctan \exp\{-n - j - 3/2 + \varepsilon\} \\
&\qquad + \arctan \exp\{-n - j - 1/2 - \varepsilon\}) = 0.
\end{aligned}$$

To calculate the sum $\sum_{k=1}^{2n+1}(\ldots)$ we used the identity

$$\arctan x + \arctan x^{-1} = \pi/2.$$

We extend the function $\alpha(\varepsilon)$ by periodicity from the segment $[0,1]$ to $\mathbb{R}$ by setting $\alpha(x) = \alpha(x-[x])$, $x \in \mathbb{R}$. Since $\alpha(0) = \alpha(1)$, the function $\alpha(x)$ is continuous on $\mathbb{R}$. Furthermore, $\alpha(x)$ is odd, because

$$\alpha(\varepsilon) = -\alpha(1-\varepsilon), \quad \varepsilon \in [0,1].$$

Denoting $|x| = n+\varepsilon$, $\theta = \theta(x) = \operatorname{sign} x \cdot \exp\{-|x|\}$, we deduce from (5.13) and (5.14) that, on $\mathbb{R}_+ = [0,\infty)$,

$$\arg B(\exp\{i\theta(x)\}) = -\pi x - \alpha(x) + c_0(x), \tag{5.15}$$

where the function $c_0(x)$ is continuous and vanishes at ∞. Since the functions $\arg B(\exp\{i\theta(x)\})$ and $\pi x + \alpha(x)$ are defined on $\mathbb{R}\setminus\{0\}$ and are odd, equality (5.15) can be extended to $\mathbb{R}_- = (-\infty, 0]$. The function $c_0(x)$ is also odd on $\mathbb{R}\setminus\{0\}$, and hence is continuous on $\mathbb{R}_-$ and vanishes at $-\infty$.

Let us consider the function $u(x) = -\pi x - \alpha(x) \in C(\mathbb{R})$ and show that it decreases monotonically on $\mathbb{R}$. Assuming the contrary, there exist points $x_1, x_2 \in \mathbb{R}$ such that $x_1 < x_2$, but $u(x_1) < u(x_2)$. Choose a positive integer n such that

$$|c_0(x_1+n)| < \delta/2, \quad |c_0(x_2+n)| < \delta/2, \quad \text{where } \delta = u(x_2) - u(x_1).$$

Observing that the function $\arg B(\exp\{i\theta(x)\}) = u(x) + c_0(x)$ is monotonically decreasing on $\mathbb{R}$, we have

$$\begin{aligned} 0 &< u(x_1+n) + c_0(x_1+n) - u(x_2+n) - c_0(x_2+n) \\ &= u(x_1) - u(x_2) + c_0(x_1+n) - c_0(x_2+n) \\ &= c_0(x_1+n) - c_0(x_2+n) - \delta < 0. \end{aligned}$$

This proves the monotonicity of $u(x)$. But then the function $y = u(x)$ is invertible on $\mathbb{R}$; its inverse function has the form

$$u^{-1}(y) = -y\pi^{-1} + \beta(y), \tag{5.16}$$

where $\beta(y)$ is continuous and π-periodic on $\mathbb{R}$. Indeed, $u(x+1) = u(x) - \pi$, $x \in \mathbb{R}$, thanks to the periodicity of $\alpha(x)$, and consequently $x+1 = u^{-1}(y-\pi) = u^{-1}(y)+1$, $y \in \mathbb{R}$. But then

$$\beta(y-\pi) = u^{-1}(y-\pi) + (y-\pi)\pi^{-1} = u^{-1}(y) + y\pi^{-1} = \beta(y), \quad y \in \mathbb{R},$$

i.e., the function $\beta(y)$ is periodic and hence uniformly continuous on $\mathbb{R}$.

Applying the map u^{-1} to both sides of equality (5.15), we obtain

$$u^{-1}(\arg B(\exp\{i\theta(x)\})) = x + c_1(x), \quad x \in \mathbb{R}\setminus\{0\}, \tag{5.17}$$

where the function

$$c_1(x) = u^{-1}(-\pi x - \alpha(x) + c_0(x)) - u^{-1}(-\pi x - \alpha(x))$$

is continuous on $\mathbb{R} \setminus \{0\}$ and, due to the uniform continuity of $u^{-1}(y)$, vanishes at $\pm\infty$. Recalling that $x = (\operatorname{sign}\theta) \log |\theta|^{-1}$, we conclude from (5.17) that

$$\exp\{-\pi i u^{-1}(-i \log B(t))\} = \exp\{-\pi i (\operatorname{sign}\theta) \log |\theta|^{-1}\} c_2(t),$$

where $t = \exp\{i\theta\}$, $c_2(t) \in GC(\Gamma_0)$. Using (5.16) and (5.12), we obtain

$$B(t) \exp\{-\pi i \beta(-i \log B(t))\} = \exp\{\arg a(t)\} c(t), \tag{5.18}$$

where $c(t) \in C(\Gamma_0)$. But the function

$$V(\tau) = \exp\{-\pi i \beta(-i \log \tau)\}$$

is continuous on Γ_0, because $\beta(y)$ has the period π, while $\log \tau$ increases by $2\pi i$ when τ runs once through Γ_0 counter-clockwise. From the periodicity of the function $\beta(y)$ it also follows that the topological index of the function $V(\tau)$ is equal to zero. By Remark 5.1 and Theorem 5.8, the function $V(B(t))$ admits the p-factorization

$$V(B(t)) = V^-(B(t)) V^+(B(t))$$

simultaneously for all p, $1 < p < \infty$. Hence, by Proposition 5.4, $V(B(t))$ is a MNAF. But then from (5.18) it follows that

$$a(t) = b(t) B(t),$$

where $b(t) = V(B(t))|a(t)|[c(t)]^{-1} \in \operatorname{fact}(p, 1)$ for all p, $1 < p < \infty$, because it is a product of MNAFs. □

The generalized factorization obtained above for functions $a(t)$ that have a logarithmic whirl point allows us to apply Theorem 2.6 to the corresponding operators $T(a)$ and to construct a normal solvability theory for them. Here the subspace $\ker T(a)$ can be described in terms of bases of the model subspace $\mathcal{M}(B)$ which consist either of rational or of meromorphic functions:

$$\tau_k(t) = \frac{1}{1 - x_k t}, \quad k \in \mathbb{N}, \qquad \omega_k(t) = \frac{B(t)}{B'(x_k)(t - x_k)}, \quad k \in \mathbb{N}.$$

To that end it suffices to use the Γ_0-versions of the relevant theorems of Section 3.9, which are obtained by means of fractional linear isomorphic transformations of the form (4.30). In this connection let us mention here only that the zeros

$$x_k = \frac{2 - e^{-k+1/2}}{2 + e^{-k+1/2}}, \quad k \in \mathbb{N},$$

of the function $B(z)$ satisfy the Carleson δ-condition for the circle, which now reads

$$\inf_k \prod_{j\neq k} \left| \frac{x_k - x_j}{1 - x_k x_j} \right| \geq \Delta > 0. \tag{5.19}$$

Indeed,

$$\begin{aligned} \prod_{j\neq k} \left| \frac{x_k - x_j}{1 - x_k x_j} \right| &= \prod_{j\neq k} \left| \frac{e^{-j} - e^{-k}}{e^{-j} + e^{-k}} \right| = \prod_{j=1}^{k-1} \frac{1 - e^{j-k}}{1 + e^{j-k}} \cdot \prod_{j=k+1}^{\infty} \frac{1 - e^{k-j}}{1 + e^{k-j}} \\ &> \left(\prod_{n=1}^{\infty} \frac{1 - e^{-n}}{1 + e^{-n}} \right)^2 = \Delta > 0, \end{aligned}$$

which proves (5.19).

5.5 Infinite index of arbitrary order

In this section, which is also based on the factorization theory of u-periodic functions, we consider a very general case when the index is infinite. This case includes the cases of power, logarithmic, exponential, etc. growth of the argument of the symbol near a discontinuity point. Here, as in the preceding section, we will construct a Blaschke product $B(t)$ corresponding to the given symbol and carry out the (B,p)-factorization of this symbol.

Suppose, as above, that $a(t)$ has one point of discontinuity $t_0 = 1$, that $\inf_{t\in\Gamma_0} |a(t)| > 0$, and that $a(t)$ admits the representation

$$a(\exp(i\theta)) = |a(\exp(i\theta))| \exp\{i2\pi f(\theta)\}, \tag{5.20}$$

where the function $f(\theta)$ is continuous and monotonically decreasing on the intervals $(-\pi, 0)$ and $(0, \pi)$, and satisfies

$$\lim_{\theta\to 0\pm 0} f(\theta) = \mp\infty.$$

With no loss of generality we can take $f(-\pi + 0) = f(\pi - 0) = 0$. Now let us introduce the function

$$\theta(x) = f^{-1}(-x) \tag{5.21}$$

on $\mathbb{R} \setminus \{0\}$. Then $\theta(x)$ is monotonically decreasing on the rays $(-\infty, 0)$ and $(0, \infty)$, and

$$\theta(\pm\infty) = 0, \quad \theta(0 \pm 0) \stackrel{\text{def}}{=} \theta(\pm 0) = \pm\pi.$$

Further, let

$$\Delta(n) = \begin{cases} \theta(n) - \theta(n+1), & n = +0, 1, 2, \ldots, \\ \theta(n-1) - \theta(n), & n = -0, -1, -2, \ldots. \end{cases}$$

On the segment $I = [-1/2, 1/2]$, consider the sequence of functions

$$\psi_n(s) = \frac{\theta(n) - \theta(n+s)}{\Delta(n)}, \quad s \in I.$$

We will assume that this sequence converges monotonically on I and that

$$\begin{aligned} \lim_{n\to+\infty} \psi_n(s) &= \psi(s), \\ \lim_{n\to-\infty} \psi_n(s) &= -\psi(-s), \end{aligned} \tag{5.22}$$

where the function $\psi(s)$ is monotonically increasing and continuous on I. Let us emphasize that conditions (5.22) play a crucial role. Specifically, they allow us to speak of a property that may be referred to as *generalized periodicity*. We finally put

$$\xi(n) = \begin{cases} \theta(n+1)/\theta(n), & n = +0, 1, 2, \dots, \\ \theta(n-1)/\theta(n), & n = -0, -1, -2, \dots, \end{cases} \tag{5.23}$$

$$\alpha(n) = 1 - \xi(n). \tag{5.24}$$

We first need several auxiliary results.

Proposition 5.5 *The following assertions hold for a function $\theta(x)$ of the form* (5.21) *and the positive numbers $\alpha(n)$ given by* (5.24).

(1) *The series $\sum_{j=1}^{\infty} \alpha(j)$ and $\sum_{j=-\infty}^{-1} \alpha(j)$ diverge.*

(2) *For integers k and n such that $kn > 0$ and $0 < |k| < |n|$ one has the estimate*

$$\sum_{j=k}^{n} \alpha_j < \log \frac{\theta(k)}{\theta(n+\sigma)}, \quad \sigma = \operatorname{sign} n. \tag{5.25}$$

Proof. Let us consider the case of positive k and n. Since

$$\frac{\theta(1)}{\theta(n)} = \prod_{j=1}^{n-1} \frac{1}{\xi(j)},$$

we have

$$\lim_{n\to\infty} \left(\prod_{j=1}^{n-1} \frac{1}{\xi(j)} \right) = \infty$$

and therefore

$$\lim_{n\to\infty} \sum_{j=1}^{n-1} (-\log \xi(j)) = \lim_{n\to\infty} \sum_{j=1}^{n-1} (-\log(1-\alpha(j))) = \infty.$$

Next, since $0 < \alpha(j) < 1$,one has $-\log(1-\alpha(j)) < 2$ for j large enough, and hence assertion (1) is established. To prove (2) we use the inequality

$$1 - x < \log x^{-1}, \quad 0 < x < 1.$$

Then

$$\begin{aligned}\sum_{j=k}^{n} \alpha(j) &= \sum_{j=k}^{n} \left(1 - \frac{\theta(j+1)}{\theta(j)}\right) < \sum_{j=k}^{n} \log \frac{\theta(j)}{\theta(j+1)} \\ &= \log \prod_{j=k}^{n} \frac{\theta(j)}{\theta(j+1)} = \log \frac{\theta(k)}{\theta(n+1)}.\end{aligned}$$

The case when the numbers k and n are negative is dealt with in a similar manner. □

Proposition 5.6 *Suppose that the function $\theta(x)$ has the form* (5.21) *and satisfies condition* (5.22). *Then*

$$\lim_{n\to\pm\infty} \frac{\Delta(n \pm 1)}{\Delta(n)} = d, \qquad 0 \le d \le 1. \tag{5.26}$$

Moreover,

$$\lim_{n\to\infty} \xi(n) = d, \qquad \lim_{n\to\infty} \alpha(n) = 1 - d, \tag{5.27}$$

$$\lim_{n\to\infty} \frac{\alpha(n+1)}{\alpha(n)} = 1, \tag{5.28}$$

and

$$\psi(1/2) - d\psi(-1/2) = 1. \tag{5.29}$$

Proof. Consider the case $n > 0$. By the definition of the functions $\psi_n(s)$ and $\psi_{n+1}(s)$, we can write

$$\theta(n + 1/2) = \theta(n) - \psi_n(1/2)\Delta(n)$$

and

$$\theta(n + 1/2) = \theta(n+1) - \psi_{n+1}(-1/2)\Delta(n+1).$$

From these equalities it follows that

$$\frac{\Delta(n+1)}{\Delta(n)} = -\frac{1 - \psi_n(1/2)}{\psi_{n+1}(-1/2)}.$$

Further, in view of conditions (5.22), this implies that

$$\lim_{n\to\infty} \frac{\Delta(n+1)}{\Delta(n)} = -\frac{1 - \psi(1/2)}{\psi(-1/2)} \stackrel{\text{def}}{=} d. \tag{5.30}$$

Since $-\psi(-1/2) > 0$ and $0 < \psi(1/2) \leq 1$, we see that $0 \leq d$. The quantity d cannot be larger than 1, because otherwise the sum $\sum_{j=1}^{\infty} \Delta(j)$ would diverge, while we know that

$$\lim_{n\to\infty} \sum_{j=1}^{n} \Delta(j) = \lim_{n\to\infty} (\theta(1) - \theta(n+1)) = \theta(1) < \infty.$$

Thus, (5.26) is proved.

Now let us prove relation (5.27). Suppose it is false. Then there exists a limit point d_0 of a bounded sequence $\{\xi_j\}_{j=0}^{\infty}$, such that $d_0 \neq d$. Denote

$$d_j \stackrel{\text{def}}{=} \frac{\Delta(j+1)}{\Delta(j)} = \frac{\theta(j+2) - \theta(j+1)}{\theta(j+1) - \theta(j)} = \xi(j)\frac{1 - \xi(j+1)}{1 - \xi(j)}$$

and rewrite this relation in the form

$$\xi(j+1) - d_j = (\xi(j) - d_j)\xi^{-1}(j). \tag{5.31}$$

Suppose that $d_0 < d$. Then there exists a number j_0 such that $\xi(j_0) < d_j$ for all $j \geq j_0$. Putting in (5.31) successively $j = j_0$, $j = j_0 + 1$, $j = j_0 + 2$, ..., and using the fact that $\xi^{-1}(j) > 1$, we conclude that the sequence of positive numbers $\{\xi(j)\}_{j=j_0}^{\infty}$ is monotonically decreasing, and hence has a limit when $j \to \infty$, equal to d_0.

Passing to the limit in (5.31), we obtain

$$d_0 - d = (d_0 - d)d_0^{-1}.$$

Since $d_0 \neq d$, this equality yields $d_0 = 1$, which is impossible, since we assumed that $d_0 < d \leq 1$.

Now let $d_0 > d$. In much the same way as above, one can show using (5.31) that the sequence $\{\xi(j)\}_{j=0}^{\infty}$ is monotonically increasing. Passage to the limit in (5.31) yields again $d_0 = 1$, which is impossible, since in this case (see the representation of the numbers d_j) one can find a $0 < \widetilde{d} < 1$ such that $1-\xi(j+1) < \widetilde{d}(1-\xi(j))$ for j large enough. Hence, there exists a number $c > 0$ such that $(1 - \xi(j)) < c\widetilde{d}^j$. This yields $\theta(j+1) > (1 - c\widetilde{d}^j)\theta(j)$, and hence that $\theta(j+1) > \prod_{k=1}^{j}(1 - c\widetilde{d}^k)\theta(1)$. Therefore, $\lim_{j\to\infty} \theta(j+1) > 0$, contrary to the properties of the function $\theta(x)$ given by (5.21).

Thus, the unique limit point of the sequence $\{\xi(j)\}_{j=1}^{\infty}$ is d, and (5.27) is proved.

To prove (5.28) we use the fact that the numbers d_j can be written as

$$d_j = \xi(j)\frac{\alpha(j+1)}{\alpha(j)}.$$

Letting here $j \to \infty$, we obtain (5.28). Finally, (5.29) follows from (5.30).

The case of negative n is dealt with in a similar manner. □

Proposition 5.7 *Let $u \in [-1/2, 1/2]$. Then*

$$\arctan u + \sum_{j=1}^{\infty} \left(\arctan \frac{1}{j+u} - \arctan \frac{1}{j-u} \right) = \arctan \left(\frac{\tan \pi u}{\tanh 1} \right), \tag{5.32}$$

where the series on the left-hand side converges uniformly on the indicated segment.

Formula (5.32) is essentially contained in Prudnikov, Brychkov, and Marychev [1], pp. 749–750.

Before we formulate the main theorem of this section let us introduce two auxiliary functions,

$$A(\theta) = \sum_{\theta(j) > \theta} \arctan(\alpha(j)) - \sum_{\theta(j) < -\theta} \arctan(\alpha(j)), \quad \theta \in (0, \pi), \tag{5.33}$$

and

$$C(n) = \sum_{j=m(n)}^{n-\sigma} \arctan \frac{\Delta(j)}{\theta(j) - \theta(n)} - \sum_{j=n+\sigma}^{M(n)} \arctan \frac{\Delta(j)}{\theta(n) - \theta(j)}, \tag{5.34}$$

where $\sigma = \operatorname{sign} n$, the number $m = m(n)$ is the j of smallest modulus for which $|\theta(j)| \le \frac{3}{2}|\theta(n)|$, while $M = M(n)$ is the j of largest modulus for which $|\theta(j)| \ge \frac{1}{2}|\theta(n)|$.

Notice that $A(\theta)$ characterizes the connection between the behaviors of the function $\theta(x)$ in a right and a left semi-neighborhood of zero; in other words, it connects the behavior of the function $f(\theta)$ (which has infinite index) when $x \to +\infty$ and $x \to -\infty$ (recall that $x = -f(\theta)$). The quantity $C(n)$ characterizes the behavior of $\theta(x)$ near the point $x = n$.

Now we are ready to formulate the main result of this section. Notice that here we consider the case $d = 1$ (see (5.26)).

Theorem 5.10 (On general oscillations) *Assume that the function $a(t) \in GL_\infty(\Gamma_0)$ is continuous on $\Gamma_0 \setminus \{1\}$ and has the form (5.20) with a function $f(\theta)$ that is monotonically increasing on $(-\pi, 0)$ and $(0, \pi)$ and satisfies the condition*

$$\lim_{\theta \to 0 \pm 0} f(\theta) = \mp\infty.$$

In addition, assume that conditions (5.22) are satisfied, that in (5.26) $d = 1$, and that the limits

$$\lim_{\theta \to 0 \pm 0} A(\theta) = a, \quad a \in \mathbb{R}, \tag{5.35}$$

$$\lim_{n \to \pm\infty} C(n) = 0 \tag{5.36}$$

exist, where the functions $A(\theta)$ and $C(n)$ are defined by formulas (5.33) and (5.34).

Then the function $a(t)$ admits the representation

$$a(t) = B(t)g(B(t))d(t)|a(t)|, \tag{5.37}$$

which is a (B,p)-factorization, with $g(t), d(t) \in C(\Gamma_0)$. Moreover, the topological index of the function $g(t)$ is equal to zero, the Blaschke product $B(t)$ is constructed from the zeros $z_j = r_j \exp\{i\theta(j)\}$, where $r_j = (1 - \Delta(j)/2)/(1 + \Delta(j)/2)$, $j = \pm 1, \pm 2, \ldots$, and the product

$$b(t) \stackrel{\text{def}}{=} g(B(t))d(t)|a(t)|$$

admits a $(p,1)$-factorization of the form (2.7) for any $1 < p < \infty$.

Proof. First of all, let us observe that the zeros z_j satisfy the Blaschke condition, so that the corresponding product $B(t)$ converges. Indeed,

$$\begin{aligned}
\sum_{j\in\mathbb{Z}\setminus\{0\}} (1 - |z_j|) &\le \text{const} \sum_{j\in\mathbb{Z}\setminus\{0\}} \Delta(j) \\
&= \text{const} \lim_{n\to\infty} \sum_{j=-n,\, j\neq 0}^{n} (\theta(j) - \theta(j+1)) \\
&= \text{const} \lim_{n\to\infty} [(\theta(1) - \theta(n+1)) - (\theta(-1) - \theta(1-n))] \\
&= \text{const}\,(\theta(1) - \theta(-1)) < \infty.
\end{aligned}$$

Now denote $f_B(x) = \arg B(\exp\{i\theta(x)\})$, where $x = -f(\theta)$. We will first consider the case where $x = n + s$ is positive, $n \in \mathbb{N}$, $s \in [-1/2, 1/2]$. Then $\theta = \theta(x) > 0$ and, by formula (2.43), we have

$$\begin{aligned}
f_B(x) &= -2\pi(n-1) - 2\sum_{k=1}^{n-1} \varphi_k(\theta(x)) - 2\widetilde{\varphi}_n(\theta(x)) \\
&\quad -2\sum_{k=n+1}^{\infty} \varphi_k(\theta(x)) - 2\sum_{k=-\infty}^{-1} \varphi_k(\theta(x)),
\end{aligned} \tag{5.38}$$

where

$$\varphi_j(\theta(x)) = \arctan\left(\frac{\Delta(j)}{2}\cot\left(\frac{\theta(x) - \theta(j)}{2}\right)\right),$$

$$\widetilde{\varphi_n}(\theta(x)) = \begin{cases} \varphi_n(\theta(x)), & \text{if } s \le 0, \\ \pi + \varphi_n(\theta(x)), & \text{if } s > 0. \end{cases}$$

Our aim is to show that under the assumptions of the theorem the function $f_B(x)$ has a form analogous to representation (5.15). The proof will be carried out in several stages, showing that a certain group of terms in (5.38), with indices lying "near" $j = n$, generate a function of the type $\alpha(x)$ in (5.15), while the remaining terms generate a function which is continuous on the closed real line.

In the first stage, let us isolate the following group of terms:

$$A_1(\theta(x)) = -2\left(\sum_{\theta(j)\geq\delta_1(x)} \varphi_j(\theta(x)) + \sum_{\theta(j)\leq-\delta_1(x)} \varphi_j(\theta(x))\right),$$

where $\delta_1 = \delta_1(x) > 0$ is an arbitrary function such that $\delta_1(x) \to 0$ when $x \to \infty$ and

$$\lim_{x\to\infty} \frac{\theta(x)}{\delta_1(x)} = 0. \tag{5.39}$$

Let us show that, thanks to (5.35),

$$\lim_{\theta\to 0+0} A_1(\theta) = a_1, \quad a_1 \in \mathbb{R}. \tag{5.40}$$

To that end we introduce the function

$$A_2(\theta(x)) = A_1(\theta(x)) + 2A(\delta_1(x)) = A_2^+(\theta) - A_2^-(\theta),$$

where

$$\begin{aligned} A_2^+(\theta) &= 2\sum_{\theta(j)\geq\delta_1(x)} (\arctan\alpha(j) + \varphi_j(\theta(x))), \\ A_2^-(\theta) &= 2\sum_{\theta(j)\leq-\delta_1(x)} (\arctan\alpha(j) - \varphi_j(\theta(x))). \end{aligned}$$

Consider the term $A_2^+(\theta)$ and rewrite it as

$$\begin{aligned} A_2^+(\theta) &= 2\sum_{\theta(j)\geq\delta_1(x)} \left[\left(\arctan\alpha(j) - \arctan\left(\frac{\Delta(j)}{2}\cot\frac{\theta(j)}{2}\right)\right)\right. \\ &\quad \left. + \left(\arctan\left(\frac{\Delta(j)}{2}\cot\frac{\theta(j)}{2}\right) + \varphi_j(\theta(x))\right)\right] \stackrel{\text{def}}{=} A_{2,1}^+(\theta) + A_{2,2}^+(\theta). \end{aligned}$$

To estimate $A_{2,2}^+(\theta)$, we use the elementary inequality

$$|\arctan u - \arctan v| \leq |\arctan(u-v)|, \quad u, v > 0. \tag{5.41}$$

We have

$$\begin{aligned} |A_{2,2}^+(\theta)| &\leq 2\left|\sum_{\theta(j)\geq\delta_1(x)} \arctan\left(\frac{\Delta(j)}{2}\left(\cot\left(\frac{\theta(j)-\theta(x)}{2}\right) - \cot\left(\frac{\theta(j)}{2}\right)\right)\right)\right| \\ &\leq \text{const} \sum_{\theta(j)\geq\delta_1(x)} \Delta(j)\frac{\sin\left(\frac{\theta(x)}{2}\right)}{\sin\left(\frac{\theta(j)-\theta(x))}{2}\right)\sin\left(\frac{\theta(j)}{2}\right)} \\ &\leq \text{const}\,\theta(x) \sum_{\theta(j)\geq\delta_1(x)} \frac{\Delta(j)}{(\theta(j)-\theta(x))\theta(j)} \\ &\leq \text{const}\,\theta(x) \sum_{\theta(j)\geq\delta_1(x)} \left(\frac{\Delta(j)}{\theta(j)\theta(j+1)}\right)\xi(j)\frac{1}{1-\frac{\theta(x)}{\theta(j)}}. \end{aligned}$$

Using the fact that $\theta(j) \geq \delta_1(x)$ and taking into account condition (5.39) and also the limit (5.27), we obtain the estimate

$$\begin{aligned} |A^+_{2,2}(\theta)| &\leq \text{const}\ \theta(x) \sum_{\theta(j)\geq\delta_1(x)} \left(\frac{1}{\theta(j+1)} - \frac{1}{\theta(j)}\right) \\ &= \text{const}\ \theta(x) \left(\frac{1}{\theta(j_*+1)} - \frac{1}{\theta(1)}\right), \end{aligned}$$

where j_* denotes the largest number for which $\theta(j_*) \geq \delta_1(x)$. Therefore, (5.39) and (5.27) yield

$$\lim_{\theta\to 0} A^+_{2,2}(\theta) = 0.$$

Next, notice that, for any $\theta > 0$, $A^+_{2,1}(\theta)$ is a partial sum of the convergent series

$$S = 2\sum_{j=1}^{\infty} \left(\arctan \alpha(j) - \arctan\left(\frac{\Delta(j)}{2} \cot \frac{\theta(j)}{2}\right)\right).$$

Convergence is verified using the inequality

$$\left|\cot u - \frac{1}{u}\right| \leq \text{const} \cdot |u|, \quad |u| \leq u_0 < \pi/2. \tag{5.42}$$

Indeed, combining (5.41) and (5.42) we have

$$\begin{aligned} S &\leq \text{const} \sum_{j=1}^{\infty} \arctan\left(\frac{\Delta(j)}{2}\left(\frac{2}{\theta(j)} - \cot\frac{\theta(j)}{2}\right)\right) \\ &\leq \text{const} \sum_{j=1}^{\infty} \Delta(j)\theta(j) < \infty. \end{aligned}$$

Therefore,

$$\lim_{\theta\to 0+0} A^+_{2,1}(\theta) = a^+_2 = \lim_{\theta\to 0+0} A^+_2(\theta), \quad a^+_2 \in \mathbb{R}.$$

In exactly the same manner one can show that an analogous inequality holds for the function $A^-_2(\theta)$. Thus, (5.40) is proven.

In the second stage we show that in the remaining terms of (5.38) (for which $|\theta(j)| < \delta_1(x)$) the function $\varphi_j(\theta(x))$ can be replaced, modulo quantities that are $o(1)$ as $x \to \infty$, by the function

$$\varphi^0_j(\theta(x)) = \arctan\left(\frac{\Delta(j)}{\theta(x) - \theta(j)}\right).$$

Indeed, let

$$O_1(x) \stackrel{\text{def}}{=} \sum_{\delta_1(x)>\theta(j)>\theta(n)} (\varphi_j(\theta(x)) - \varphi^0_j(\theta(x))).$$

Then using inequalities (5.41) and (5.42), we have

$$\begin{aligned}|O_1(x)| &\le \left|\sum_{\delta_1(x)>\theta(j)>\theta(n)} \arctan\left(\frac{\Delta(j)}{2}\left(\cot\frac{\theta(x)-\theta(j)}{2} - \frac{2}{\theta(x)-\theta(j)}\right)\right)\right| \\ &\le \text{const} \sum_{\delta_1(x)>\theta(x)>\theta(n)} \Delta(j)(\theta(j)-\theta(x)) \\ &\le \text{const}\, \delta_1(x) \sum_{\delta_1(x)>\theta(j)>\theta(n)} \Delta(j) \le \text{const}\, \delta_1(x),\end{aligned}$$

i.e.,

$$\lim_{x\to\infty} O_1(x) = 0. \tag{5.43}$$

Similarly, one can show that (5.43) holds for the functions

$$O_2(x) \stackrel{\text{def}}{=} \sum_{j=n+1}^{\infty} (\varphi_j(\theta(x)) - \varphi_j^0(\theta(x)))$$

and

$$O_3(x) \stackrel{\text{def}}{=} \sum_{0>\theta(j)>-\delta_1(x)} (\varphi_j(\theta(x)) - \varphi_j^0(\theta(x))),$$

as well as for

$$O_4(x) \stackrel{\text{def}}{=} \widetilde{\varphi}_n(\theta(x)) - \widetilde{\varphi}_n^0(\theta(x)),$$

where

$$\widetilde{\varphi}_n(\theta(x)) = \begin{cases} \varphi_n^0(\theta(x)), & \text{if } s \le 0; \\ \pi + \varphi_n^0(\theta(x)), & \text{if } s > 0. \end{cases} \tag{5.44}$$

Thus, starting with (5.38) we arrive at the equality

$$\begin{aligned}f_B(x) = -2\pi(n-1) - 2 &\sum_{\delta_1(x)>\theta(j)\ge\theta(n-1)} \varphi_j^0(\theta(x)) - 2\widetilde{\varphi}_n^0(\theta(x)) \\ -2\sum_{j=n+1}^{\infty} \varphi_j^0(\theta(x)) - 2 &\sum_{0>\theta(j)>-\delta_1(x)} \varphi_j^0(\theta(x)) + a_1 + O_5(x),\end{aligned} \tag{5.45}$$

where the number a_1 is taken from (5.40) and the continuous function $O_5(x)$ satisfies condition (5.43).

We reached the third stage of the proof, in which we will show that the group of central terms, with indices "not too far from n," gives a function $\xi(s)$ that is continuous and monotonically decreasing on the segment $[-1/2, 1/2]$ and satisfies

$$\xi(-1/2) - \xi(1/2) = 2\pi. \tag{5.46}$$

To do this we will use condition (5.22) and the limit (5.26) (with $d = 1$). We obtain the representations

$$\widetilde{\varphi_n^0}(\theta(x)) = \left\{ \begin{array}{ll} -\arctan\left(\frac{1}{\psi(s)}\right), & s \le 0 \\ \pi - \arctan\left(\frac{1}{\psi(s)}\right), & s > 0 \end{array} \right\} + O_n(x), \tag{5.47}$$

$$\begin{aligned} \varphi_{n+j}^0(\theta(x)) &= \arctan\left(\frac{1}{-\psi_n(s)\frac{\Delta(n)}{\Delta(n+j)} + \frac{\Delta(n+1)}{\Delta(n+j)} + \ldots + \frac{\Delta(n+j-1)}{\Delta(n+j)}}\right) \\ &= \arctan\left(\frac{1}{j - \psi(s)}\right) + O_{n+j}(x). \end{aligned}$$

Similarly,

$$\varphi_{n-j}^0(\theta(x)) = -\arctan\left(\frac{1}{j + \psi(s)}\right) + O_{n-j}(x),$$

where the continuous functions $O_n(x)$, $O_{n\pm j}(x)$, $j = 1, 2, \ldots$, satisfy condition (5.43). It follows that there exists a function $l = l(n)$ with values in the positive integers, which increases without bound (possibly, very slow) when $n \to \infty$, such that the function

$$\begin{aligned} O_6(x) \quad &= \quad \sum_{j=1}^{l(n)} \left(\varphi_{n+j}^0(\theta(x)) - \arctan\left(\frac{1}{j - \psi(s)}\right)\right) \\ &\quad + \sum_{j=1}^{l(n)} \left(\varphi_{n-j}^0(\theta(x)) + \arctan\left(\frac{1}{j + \psi(s)}\right)\right) \end{aligned} \tag{5.48}$$

also satisfies condition (5.43). Now let us introduce the function $\xi(s)$ by the formula

$$\xi(s) = 2\sum_{j=1}^{\infty}\left(\arctan\frac{1}{j - \psi(s)} - \arctan\frac{1}{j + \psi(s)}\right) - 2\xi_0(s), \tag{5.49}$$

where $\xi_0(s)$ denotes the function inside the braces in (5.47). Since the series in (5.49) converges uniformly for $s \in [-1/2, 1/2]$, (5.48) yields the following modification of (5.45):

$$\begin{aligned} f_B(x) = -2\pi(n-1) + \xi(s) - 2 &\sum_{\delta_1(x) > \theta(j) > \theta(n - l(n))} \varphi_j^0(\theta(x)) \\ - 2 \sum_{j=n+l(n)+1}^{\infty} \varphi_j^0(\theta_j(x)) - &\sum_{0 > \theta(j) > -\delta_1(x)} \varphi_j^0(\theta(x)) + a_1 + O_7(x), \end{aligned} \tag{5.50}$$

where the function $O_7(x)$ satisfies (5.43). Next, we use Proposition 5.7 to write the function $\psi(s)$ in a simpler form. Let us first consider the case where $\psi(-1/2) =$

$-1/2$, $\psi(1/2) = 1/2$. Then by using the identity

$$\arctan(\pm u) + \arctan\left(\pm\frac{1}{u}\right) = \pm\frac{\pi}{2}, \quad u > 0, \tag{5.51}$$

we obtain

$$\xi_0(s) = \frac{\pi}{2} + \arctan\psi(s), \quad -1/2 \le s \le 1/2.$$

Applying now (5.32) we see that in the case at hand

$$\xi(s) = -2\arctan\left(\frac{\tan(\pi\psi(s))}{\tanh 1}\right) - \pi. \tag{5.52}$$

Now let $\psi(1/2) > 1/2$, $\psi(-1/2) > -1/2$ (recall that, according to (5.29), $\psi(1/2) - \psi(-1/2) = 1$). Then exploiting the fact that the series in (5.32) represents a function analytic in the strip $-1 < \operatorname{Re} u < 1$, we conclude that

$$\xi(s) = \begin{cases} -2\arctan\left(\frac{\tan(\pi\psi(s))}{\tanh 1}\right) - \pi, & \text{if } \psi(s) \le 1/2, \\ -2\arctan\left(\frac{\tan(\pi\psi(s))}{\tanh 1}\right) - 3\pi, & \text{if } \psi(s) > 1/2. \end{cases}$$

Similarly, when $\psi(1/2) < 1/2$, $\psi(-1/2) < -1/2$,

$$\xi(s) = \begin{cases} -2\arctan\left(\frac{\tan(\pi\psi(s))}{\tanh 1}\right) + \pi, & \text{if } \psi(s) < -1/2, \\ -2\arctan\left(\frac{\tan(\pi\psi(s))}{\tanh 1}\right) - \pi, & \text{if } \psi(s) \ge -1/2. \end{cases}$$

Clearly, the function $\xi(s)$ is continuous and monotonically decreasing on the segment $[-1/2, 1/2]$ and satisfies condition (5.46).

In the fourth stage of the proof we show that the sum of all the remaining terms in (5.50) tends to a finite limit as $x \to \infty$. This will be done in several steps. The first step uses condition (5.36) to show that this is true for the sums in which the summation index ranges over the intervals $m(n) \le j < n - l(n)$ and $n + l(n) < j \le M(n)$, where $m(n)$ and $M(n)$ are the quantities figuring in (5.34). To do this we resort to the following obvious inequalities:

$$2\sum_{j=m(n)}^{n-l(n)-1} \arctan\frac{\Delta(j)}{\theta(j) - \theta(n+1)} < -2\sum_{j=m(n)}^{n-l(n)-1} \varphi_j^0(\theta(x))$$

$$< 2\sum_{j=m(n)}^{n-l(n)-1} \arctan\frac{\Delta(j)}{\theta(j) - \theta(n-1)};$$

$$-2\sum_{j=n+l(n)+1}^{M(n)} \arctan\frac{\Delta(j)}{\theta(j) - \theta(n-1)} < 2\sum_{j=n+l(n)+1}^{M(n)} \varphi_j^0(\theta(x))$$

$$< -2\sum_{j=n+l(n)+1}^{M(n)} \arctan\frac{\Delta(j)}{\theta(j) - \theta(n+1)}.$$

Indeed, subtracting the second inequality from the first, we see that the group of terms in question admits the two-sided estimate

$$C_1(n+1) \le -2\left(\sum_{j=m(n)}^{n-l(n)-1} \varphi_j^0(\theta(x)) + \sum_{j=n+l(n)+1}^{M(n)} \varphi_j^0(\theta(n))\right) < C_1(n-1),$$

where

$$\begin{aligned} C_1(n\pm 1) \ &\overset{\text{def}}{=}\ 2C(n\pm 1) - 2\sum_{j=1}^{l(n)}\left[\arctan\left(\frac{\Delta(n\pm 1-j)}{\theta(n\pm 1-j)-\theta(n\pm 1)}\right]\right. \\ &\quad + \arctan\left(\frac{\Delta(n\pm 1+j)}{\theta(n\pm 1+j)-\theta(n\pm 1)}\right)\Bigg), \end{aligned} \tag{5.53}$$

and $C(n)$ is the function defined by equality (5.34). By condition (5.36),

$$\lim_{n\to+\infty} C(n+1) = \lim_{n\to+\infty} C(n-1) = 0.$$

Moreover, in the third stage of the proof we have shown that the sum in the right-hand side of equality (5.53) converges to $(\xi(s)+2\xi_0(s))\big|_{s=0} = 0$ as $n\to\infty$. Hence, the group of terms under consideration,

$$-2\left(\sum_{j=m(n)}^{n-l(n)-1} \varphi_j^0(\theta(x)) + \sum_{j=n+l(n)+1}^{M(n)} \varphi_j(\theta(x))\right) \overset{\text{def}}{=} O_8(x), \tag{5.54}$$

satisfies condition (5.43).

To estimate the remaining group of terms in (5.50) we will show that the functions $\varphi_j^0(\theta(x))$ can be replaced, modulo $o(1)$ quantities, by the functions

$$\varphi_j^1(\theta(x)) = \sigma\log\left(1+\sigma\frac{\Delta(j)}{\theta(x)-\theta(j)}\right) = \sigma\log\left(\frac{\theta(x)-\theta(j+\sigma)}{\theta(x)-\theta(j)}\right),$$

where $\sigma = \operatorname{sign} j$.

Indeed, let

$$O_9(x) \overset{\text{def}}{=} \sum_{\delta_1(x)>\theta(j)>\frac{3}{2}\theta(x)} (\varphi_j^0(\theta(x)) - \varphi_j^1(\theta(x))).$$

Using the inequality

$$|\arctan u - \log(1+u)| < \text{const}\cdot u^2, \quad u \ge u_0 > -1, \tag{5.55}$$

we obtain the chain of estimates

$$\begin{aligned} |O_9(x)| \ &\le\ \text{const} \sum_{\delta_1(x)>\theta(j)>\frac{3}{2}\theta(x)} \left(\frac{\Delta(j)}{\theta(x)-\theta(j)}\right)^2 \\ &= \sum_{\delta_1(x)>\theta(j)>\frac{3}{2}\theta(x)} \alpha^2(j)\Big/\left(1-\frac{\theta(x)}{\theta(j)}\right)^2 \le 4 \sum_{\delta_1(x)>\theta(j)>\frac{3}{2}\theta(x)} \alpha^2(j). \end{aligned}$$

Now denoting

$$\delta_2(x) \stackrel{\text{def}}{=} \sup_{\delta_1(x)>\theta(j)>\frac{3}{2}\theta(x)} \alpha(j)$$

and using inequality (5.25), we obtain

$$|O_9(x)| \le 4\delta_2(x) \log \frac{\delta_1(x)}{\theta(x)}.$$

Notice that $\lim_{x\to\infty} \delta_2(x) = 0$, and that if $\delta_1(x)$ is replaced by a smaller quantity, then $\delta_2(x)$ does not increase. Since the function $\delta_1(x)$, which tends to zero as $x \to \infty$, can be chosen in an arbitrary manner, the only constraint being (5.39), we may assume that

$$\lim_{x\to\infty} \delta_2(x) \log \frac{\delta_1(x)}{\theta(x)} = 0. \tag{5.56}$$

Such a choice is made, for example, as follows. First pick $\widetilde{\delta}_1(x) = \theta^{1/2}(x)$. If it happens that $\widetilde{\delta}_2(x) \le \theta^{1/2}(x)$ for all sufficiently large x, then (5.56) holds. If this is not the case, then upon setting

$$\delta_1(x) = \max\left\{\theta^{3/4}(x); \min\left\{\widetilde{\delta}_2(x); \frac{\theta(x)}{\widetilde{\delta}_2(x)}\right\}\right\},$$

we have $\delta_1(x) \le \widetilde{\delta}_1(x)$, and condition (5.39) is fulfilled. Moreover, if $\delta_1(x) = \widetilde{\delta}_2(x)$ or $\delta_1(x) = \theta(x)/\widetilde{\delta}_2(x)$, then

$$\delta_2(x) \log \frac{\delta_1(x)}{\theta(x)} < \widetilde{\delta}_2(x) \log \theta^{-1/2}(x).$$

If, however, $\delta_1(x) = \theta^{3/4}(x)$, then

$$\delta_2(x) \log \frac{\delta_1(x)}{\theta(x)} \le \frac{\widetilde{\delta}_2(x)}{2} \log \theta^{-1/2}(x),$$

which proves (5.56). Thus, we conclude that

$$\begin{aligned}
\sum_{\delta_1(x)>\theta(j)>\frac{3}{2}\theta(x)} \varphi_j^0(\theta(x)) &= \sum_{\delta_j>\theta(j)>\frac{3}{2}\theta(x)} \varphi_j^1(\theta(x)) + 2O_9(x) \\
&= \log\left(\prod_{\delta_1(x)>\theta(j)>\frac{3}{2}\theta(x)} \frac{\theta(x)-\theta(j+1)}{\theta(x)-\theta(j)}\right) + 2O_9(x) \\
&= \log\left(\frac{\theta(x)-\theta(j_2)}{\theta(x)-\theta(j_1)}\right) + 2O_9(x),
\end{aligned} \tag{5.57}$$

where

$$\theta(j_1) < \delta_1(x) \le \theta(j_1-1), \quad \theta(j_2) < \frac{3}{2}\theta(x) < \theta(j_2-1),$$

and the function $O_9(x)$ satisfies condition (5.43). Now let us replace the functions $\varphi_j^0(\theta(x))$ by $\varphi_j^1(\theta(x))$ in the next group of terms. To this end, let us estimate the function

$$O_{10}(x) \stackrel{\text{def}}{=} \sum_{\frac{\theta(x)}{2}>\theta(j)>0} (\varphi_j^0(\theta(x)) - \varphi_j^1(\theta(x))).$$

Using inequality (5.55), we obtain

$$\begin{aligned}
|O_{10}(x)| &\leq \text{const} \sum_{\frac{\theta(x)}{2}>\theta(j)>0} \left(\frac{\Delta(j)}{\theta(x)-\theta(j)}\right)^2 \\
&\leq \frac{\text{const}}{\theta^2(x)} \sum_{\frac{\theta(x)}{2}>\theta(j)>0} \Delta^2(j) \\
&\leq \frac{\text{const}}{\theta^2(x)} \left(\sup_{\frac{\theta(x)}{2}>\theta(j)>0} \Delta(j)\right)\left(\sum_{\frac{\theta(x)}{2}>\theta(j)>0} \Delta(j)\right) \leq \text{const}\,\frac{\theta^2(j_3)}{\theta^2(x)}\delta_3(x),
\end{aligned}$$

where

$$\delta_3(x) = \sup_{\frac{\theta(x)}{2}>\theta(j)>0} \alpha(j)$$

and $\theta(j_3-1) \geq \theta(x)/2 > \theta(j_3)$. Since $(\theta(j_3)/\theta(x))^2 \leq 1/4$ and $\lim_{x\to\infty} \delta_3(x) = 0$, the function $O_{10}(x)$ satisfies condition (5.43). Therefore, we have

$$\begin{aligned}
\sum_{\frac{\theta(x)}{2}>\theta(j)>0} \varphi_j^0(\theta(x)) &= \sum_{\frac{\theta(x)}{2}>\theta(j)>0} \varphi_j^1(\theta(x)) + O_{10}(x) \\
&= \log\left(\prod_{\frac{\theta(x)}{2}>\theta(j)>0} \frac{\theta(x)-\theta(j+1)}{\theta(x)-\theta(j)}\right) + O_{10}(x) \\
&= \log\left(\frac{\theta(x)}{\theta(x)-\theta(j_3)}\right) + O_{10}(x). \qquad (5.58)
\end{aligned}$$

Finally, passing to the last remaining group of terms in (5.50), let us estimate the function

$$O_{11}(x) \stackrel{\text{def}}{=} \sum_{0>\theta(j)>-\delta_1(x)} (\varphi_j^0(\theta(x)) + \varphi_j^1(\theta(x))).$$

Again by inequality (5.55):

$$\begin{aligned}
|O_{11}(x)| &\leq \text{const} \sum_{0>\theta(j)>-\delta_1(x)} \left(\frac{\Delta(j)}{\theta(x)-\theta(j)}\right)^2 \\
&\leq \text{const}\left(\sum_{-\theta(x)>\theta(j)>-\delta_1(x)} \left(\frac{\Delta(j)}{\theta(j)}\right)^2 + \sum_{0>\theta(j)>-\theta(x)} \left(\frac{\Delta(j)}{\theta(j)}\right)^2\right).
\end{aligned}$$

The fact that the first (resp., second) of the sums obtained here tends to zero is shown exactly as we did for the function $O_9(x)$ (resp., $O_{10}(x)$). Therefore, $O_{11}(x)$ has property (5.43), and we deduce that

$$\begin{aligned}\sum_{0>\theta(j)>-\delta_1(x)} \varphi_j^0(\theta(x)) &= -\sum_{0>\theta(j)>-\delta_1(x)} \varphi_1(\theta(x)) + O_{11}(x)\\ &= -\log\left(\prod_{0>\theta(j)>-\delta_1(x)} \frac{\theta(x)-\theta(j-1)}{\theta(x)-\theta(j)}\right) + O_{11}(x)\\ &= \log\frac{\theta(x)-\theta(j_4)}{\theta(x)} + O_{11}(x), \qquad (5.59)\end{aligned}$$

where $\theta(j_4) > -\delta_1(x) > \theta(j_4-1)$.

Combining the relations (5.54), (5.57), (5.58) and (5.59) established above we are able to pass from the representation (5.50) to the following one:

$$\begin{aligned}f_B(x) &= -2\pi(n-1) + \xi(s) - 2\log\frac{\theta(j_2)-\theta(x)}{\theta(x)-\theta(j_3)}\\ &\quad -2\log\frac{\theta(j_4)-\theta(x)}{\theta(x)-\theta(j_1)} + a_1 + O_{12}(x), \qquad (5.60)\end{aligned}$$

where $\xi(s)$ is given by formulas (5.52) and a_1 is the constant figuring in (5.40). Here let us remark that, by the arguments used in the first stage of the proof, $a_1 = a_2 - 2a$, where a has the form (5.35) and

$$\begin{aligned}a_2 &= 2\sum_{j=1}^{\infty}\left[\arctan(\alpha(j)) - \arctan\left(\frac{\Delta(j)}{2}\cot\frac{\theta(j)}{2}\right)\right]\\ &\quad -2\sum_{-\infty}^{-1}\left[\arctan(\alpha(j)) + \arctan\left(\frac{\Delta(j)}{2}\cot\frac{\theta(j)}{2}\right)\right].\end{aligned}$$

Recalling the definition of the numbers j_2 and j_3 and relation (5.27) ($d=1$), we obtain

$$\lim_{x\to\infty}\frac{\theta(j_2)}{(3/2)\theta(x)} = 1 \quad\text{and}\quad \lim_{x\to\infty}\frac{\theta(j_3)}{(1/2)\theta(x)} = 1.$$

Similarly,

$$\lim_{x\to\infty}\frac{\theta(j_1)}{\delta_1(x)} = 1 \quad\text{and}\quad \lim_{x\to\infty}\frac{\theta(j_4)}{\delta_1(x)} = -1.$$

Consequently, the logarithmic terms in (5.60) tend to zero as $x\to\infty$. Now recalling that $x = n+s$ and $s = x - [x+1/2]$, we can rewrite (5.60) as

$$f_B(x) = -2\pi x + p(x - [x+1/2]) + \pi + a_1 + O_{13}(x), \qquad (5.61)$$

where

$$p(s) = \xi(s) + 2\pi s + \pi, \quad s \in [-1/2, 1/2],$$

and $O_{13}(x)$ has property (5.43). Observe that $p(-1/2) = p(0) = p(1/2) = 0$, and so the function $p(x - [x + 1/2])$ is continuous and 1-periodic.

Representation (5.61) is crucial for our analysis, but for the moment it is proved only for $x > 0$. To show that it holds for $x < 0$ as well, let us consider the Blaschke product $B_1(t)$ connected with $B(t)$ by the relation $B_1(t) = \overline{B(t^{-1})}$. Clearly,

$$\arg B_1(\exp(i\theta(x)) = -\arg B(\exp(-i\theta(x))), \tag{5.62}$$

and the zeros of $B_1(t)$ have the form

$$z_j^{(1)} = \overline{z}_j = r_j \exp(-i\theta(j)).$$

By (5.61), the function $B_1(t)$ admits the representation

$$f_{B_1}(x) = -2\pi x + p(-(x - [x + 1/2])) + \pi - a_1 + O_{14}(x) \tag{5.63}$$

for $x > 0$. Indeed, by formulas (5.22),

$$\begin{aligned}\psi_1(x) &= \lim_{n\to\infty} \frac{\theta_1(n) - \theta_1(n+s)}{\Delta_1(n)} = \lim_{n\to\infty} \frac{-\theta(-n) + \theta(-n-s)}{-\theta(-n) + \theta(-n-1)} \\ &= -\lim_{j\to-\infty} \frac{\theta(j) - \theta(j+s)}{\Delta(j)} = \psi(-s),\end{aligned}$$

which explains why the function $p(s)$ in (5.61) was replaced by $p(-s)$ in (5.63). The fact that the constant a_1 is replaced by $-a_1$ follows from the relation $\theta_1(j) = -\theta(-j)$. Next, since $-x - [-x+1/2] = -(x - [x+1/2])$, relation (5.62) shows that (5.61) holds for all $x \in \mathbb{R}$. (That π is replaced by $-\pi$ when we pass to (5.63) in the case of negative x is not important, because the resulting difference of arguments is 2π.) Finally let us mention that, with no loss of generality, the constants appearing in (5.61) can be omitted and one can use a representation of the form

$$f_B(x) = -2\pi x + p(x - [x + 1/2]) + O(x), \tag{5.64}$$

where the function $O(x)$ is continuous and tends to zero at infinity. From this point on one can argue in much the same way as in the proof of Theorem 5.9. Indeed, let us introduce the function

$$u(x) = -2\pi x + p(x - [x + 1/2])$$

and notice that it decreases monotonically on the real line. Since the function $p(x - [x + 1/2])$ is 1-periodic, the inverse function has the form

$$u^{-1}(y) = -\frac{y}{2\pi} + \beta(y), \tag{5.65}$$

where $\beta(y)$ is 2π-periodic on $\mathbb{R}$. Applying the map u^{-1} to both sides of equality (5.64) we obtain

$$u^{-1}(\arg B(\exp(i\theta(x)))) = x + d_1(x), \tag{5.66}$$

where the function

$$\begin{aligned} d_1(x) &= u^{-1}(-2\pi x + p(x - [x+1/2]) + O(x)) \\ &\quad - u^{-1}(2\pi x + p(x - [x+1/2])) \end{aligned}$$

is continuous on $\mathbb{R}$ and, thanks to the uniform continuity of the function $u^{-1}(y)$, vanishes at infinity. Recalling that $x = -f(\theta)$, (5.66) yields

$$\exp(-2\pi i u^{-1}(\arg B(\exp(i\theta)))) = \exp(-2\pi i f(\theta))\exp(-2\pi i d_1(-f(\theta))).$$

According to (5.65), this inequality implies that

$$B(t)\exp(-2\pi i\beta(-i\log B(t))) = \frac{a(t)}{|a(t)|d(t)},$$

where $d(t) \stackrel{\text{def}}{=} \exp(-2\pi i d_1(-f(\theta)))$ is continuous on Γ_0, $\beta(y)$ is 2π-periodic, and $\log\tau$ acquires an increment of $2\pi i$ after a complete circuit of Γ_0 in the positive direction. Since $\operatorname{ind} g(t) = \beta(2\pi) - \beta(0) = 0$, representation (5.37) is proved.

Let us emphasize that (5.37) is indeed a $(B(t), p)$-factorization. This follows from Proposition 5.4, because a continuous function is an MNAF. Since the positive function $|a(t)|$ is also an MNAF, the proof of the theorem is complete. □

5.6 Sufficient conditions for the theorem on general oscillations. Examples

Conditions (5.22), (5.35), (5.36), under which Theorem 5.10 holds, cover a very large class of symbols whose argument increases monotonically near a point of discontinuity. However, sometimes they are not easy to verify directly. In this section we establish readily verifiable conditions that are sufficient for the indicated relations to be fulfilled. The main efforts will be focussed on deriving conditions in terms of the function $f(\theta)$ itself, and not of the function $\theta(x) = f^{-1}(-x)$, as was the case in Theorem 5.10. Throughout this section we will assume (like in the preceding one) that the function $f(\theta)$ is monotonically increasing on $[-\pi, 0)$ and $(0, \pi]$, and satisfies $\lim_{\theta\to 0\pm 0} f(\theta) = \mp\infty$.

As a preliminary step, let us mention the following well-known relations between the derivatives of the function $f(x)$ and those of the function $\theta(x) = f^{-1}(-x)$. Assuming that $f(\theta)$ has a first and a second derivative on the set $[-\pi, \pi] \setminus \{0\}$, we have

$$f'(\theta) = -\frac{1}{\theta'(x)} \tag{5.67}$$

and

$$f''(\theta) = -\frac{\theta''(x)}{(\theta'(x))^3}. \tag{5.68}$$

Since conditions (5.22) play the crucial role for us, we will deal with them first.

Proposition 5.8 *Let the function $f(\theta)$ be twice continuously differentiable on $[-\pi, \pi] \setminus \{0\}$ and let $f'(\theta)$ be monotonically decreasing (resp., increasing) on $(0, \pi)$ (resp., $(-\pi, 0)$) and satisfy*

$$\lim_{\theta \to 0} \frac{f''(\theta)}{(f'(\theta))^2} = 0. \tag{5.69}$$

Then (5.22) holds with the function $\psi(s) \equiv s$.

Proof. We consider here the case $\theta > 0$ ($n > 0$). Applying Lagrange's theorem three times, we get

$$\begin{aligned}\psi_n(s) &= \frac{\theta(n) - \theta(n+s)}{\theta(n) - \theta(n+1)} = \frac{s\theta'(n_1)}{\theta'(n_2)} \\ &= s\left(1 + \frac{\theta'(n_1) - \theta'(n_2)}{\theta'(n_2)}\right) = s\left(1 + \frac{\theta''(n_3)}{\theta'(n_2)}(n_1 - n_2)\right),\end{aligned}$$

where $n_1 \in [n, n+s]$, $n_2 \in [n, n+1]$, $n_3 \in [\min(n_1, n_2), \max(n_1, n_2)]$.

If $n_1 \geq n_2$, then $n_3 \geq n_2$ and, thanks to the fact that the function $|\theta'(x)|$ is monotonically decreasing, equalities (5.67) and (5.68) yield

$$\left|\frac{\theta''(n_3)}{\theta'(n_2)}\right| < \left|\frac{\theta''(n_3)}{\theta'(n_3)}\right| = \left|\frac{f''(\theta_3)}{(f'(\theta_3))^2}\right|,$$

where $\theta_3 = \theta(n_3)$.

If now $n_1 \leq n_2$, then $n_1 \leq n_3$, and modifying slightly the representation of the function $\psi_n(s)$, we have

$$\psi_n(s) = s\left(\frac{\theta'(n_2)}{\theta'(n_1)}\right)^{-1} = s\left(1 + \frac{\theta''(n_3)}{\theta'(n_1)}(n_2 - n_1)\right)^{-1}.$$

As above,

$$\left|\frac{\theta''(n_3)}{\theta'(n_1)}\right| \leq \left|\frac{\theta''(n_3)}{\theta'(n_3)}\right| = \left|\frac{f''(\theta_3)}{(f'(\theta_3))^2}\right|$$

and $\lim_{\theta \to 0} \theta_3 = 0$. Hence, by (5.69), the assertion of the proposition is proved.

The case $\theta < 0$ is dealt with in a similar manner. □

Remark 5.5 Under the assumptions of Proposition 5.8, the function $f'(\theta)$ is monotone on $[-\pi, 0)$ and $(0, \pi]$, and consequently

$$\lim_{\theta \to \pm 0} f'(\theta) = +\infty.$$

Indeed, otherwise the function

$$f(\theta) = \begin{cases} -\int_\theta^\pi f'(s)ds, & \theta > 0, \\ \int_{-\pi}^\theta f'(s)ds, & \theta < 0, \end{cases}$$

would be bounded.

Furthermore, from condition (5.69) it follows that

$$\lim_{\theta\to 0} \frac{1}{\theta f'(\theta)} = 0. \tag{5.70}$$

Indeed, considering first the case $\theta > 0$, we have

$$\int_0^\theta \left(-\frac{f''(s)}{(f'(s))^2}\right) ds = \frac{1}{f'(\theta)},$$

whence, by the mean value theorem for integrals,

$$-\frac{f''(\theta_1)}{(f'(\theta_1))^2} = \frac{1}{\theta f'(\theta)}, \quad \theta_1 \in (0, \theta],$$

which yields (5.70). The case $\theta < 0$ is dealt with similarly.

Now let us address condition (5.36).

Proposition 5.9 *Suppose that* $f(\theta)$ *is twice continuously differentiable on* $[-\pi, \pi] \setminus \{0\}$ *and that* $f'(\theta)$ *is monotonically nonincreasing (resp., nondecreasing) on* $(0, \pi)$ *(resp.,* $(-\pi, 0)$*) and satisfies* (5.70) *and*

$$\lim_{\theta\to 0} \frac{f''(\theta)|\theta|^{1/2}}{(f'(\theta))^{3/2}} = 0. \tag{5.71}$$

Then (5.36) *holds.*

Proof. Before we address the proof itself, let us observe that the hypotheses of the proposition guarantee that (5.69) holds. Indeed, for θ sufficiently small (5.70) yields

$$\frac{f''(\theta)}{(f'(\theta))^2} = \frac{f''(\theta)|\theta|^{1/2}}{(f'(\theta))^{3/2}} \cdot \frac{1}{(|\theta| f'(\theta))^{1/2}} < \frac{f''(\theta)|\theta|^{1/2}}{(f'(\theta))^{3/2}}.$$

Hence, from (5.71) one obtains (5.69).

Further, by (5.34), the function $C(n)$ has the form

$$C(n) = \sum_{j=m(n)}^{n-1} \arctan \frac{\Delta(j)}{\theta(j) - \theta(n)} - \sum_{j=n+1}^{M(n)} \arctan \frac{\Delta(j)}{\theta(n) - \theta(j)}.$$

Now let us use the fact that the function $O_6(n)$ given by (5.48) tends to zero as $n \to \infty$. Since for $x = n$ we have $s = 0$ and $\psi(0) = 0$,

$$O_6(n) = \sum_{j=1}^{l(n)} \left[\arctan\left(\frac{\Delta(n+j)}{\theta(n-j) - \theta(n)}\right) - \arctan\left(\frac{\Delta(n+j)}{\theta(n) - \theta(n+j)}\right)\right],$$

where, we recall, $\lim_{n\to\infty} l(n) = \infty$. Therefore, instead of $C(n)$ we can consider the truncated expression

$$C_1(n) \stackrel{\text{def}}{=} \sum_{j=m(n)}^{n-l(n)-1} \arctan\frac{\Delta(j)}{\theta(j) - \theta(n)} - \sum_{j=n+l(n)+1}^{M(n)} \arctan\frac{\Delta(j)}{\theta(n) - \theta(j)}.$$

As in the proof of Theorem 5.10, let us "isolate the logarithms," i.e., write $C_1(n)$ in the form

$$\begin{aligned} C_1(n) &= \sum_{j=m(n)}^{n-l(n)-1} \left[\arctan\frac{\theta(j) - \theta(j+1)}{\theta(j) - \theta(n)} + \log\left(1 - \frac{\theta(j) - \theta(j+1)}{\theta(j) - \theta(n)}\right)\right] \\ &\quad - \sum_{j=n+l(n)+1}^{M(n)} \left[\arctan\frac{\theta(j) - \theta(j+1)}{\theta(n) - \theta(j)} - \log\left(1 + \frac{\theta(j) - \theta(j+1)}{\theta(n) - \theta(j)}\right)\right] \\ &\quad - \log\frac{\theta(n) - \theta(M(n)+1)}{\theta(m(n)) - \theta(n)} \cdot \frac{\theta(n - l(n)) - \theta(n)}{\theta(n) - \theta(n + l(n) + 1)}. \end{aligned} \tag{5.72}$$

By the choice of the numbers $m(n)$ and $M(n)$ (see the proof of the fact that the logarithmic term in (5.60) tends to zero),

$$\lim_{n\to\infty} \frac{\theta(n) - \theta(M(n)+1)}{\theta(m(n)) - \theta(n)} = 1.$$

Next, by the choice of $l(n)$ (see the reasoning after (5.48)),

$$\theta(n) - \theta(n + l(n) + 1) = (l(n) + 1)\Delta(n) + o(1)$$

and

$$\theta(n - l(n)) - \theta(n) = l(n)\Delta(n) + o(1).$$

Therefore,

$$\lim_{n\to\infty} \frac{\theta(n - l(n)) - \theta(n)}{\theta(n) - \theta(n + l(n) + 1)} = 1.$$

We see that the last logarithmic term in (5.72) tends to zero, and since

$$|\arctan u \mp \log(1 \pm u)| \le \text{const} \cdot u^2, \quad u \in (u_0, 1), \quad u_0 > 0,$$

it suffices to show that the sums

$$C_2(n) = \sum_{j=m(n)}^{n-l(n)-1} \left(\frac{\Delta(j)}{\theta(j)-\theta(n)}\right)^2$$

and

$$C_3(n) = \sum_{j=n+l(n)+1}^{M(n)} \left(\frac{\Delta(j)}{\theta(n)-\theta(j)}\right)^2$$

tend to zero.

Let us estimate $C_3(n)$, using Lagrange's theorem and the fact that the function $|\theta'(x)|$ tends to zero monotonically when $x \to \infty$. What results is

$$\begin{aligned} |C_3(n)| &= \sum_{j=n+l(n)+1}^{M(n)} \left(\frac{\theta'(j_1)}{\theta'(j_2)}\right)^2 \frac{1}{(n-j)^2} \leq \sum_{j=n+l(n)+1}^{M(n)} \frac{1}{(j-n-1)(j-n)} \\ &= \sum_{j=n+l(n)+1}^{M(n)} \left(\frac{1}{j-n-1} - \frac{1}{j-n}\right) = \frac{1}{l(n)} - \frac{1}{M(n)-n}, \end{aligned}$$

where $j_1 \in [j, j+1]$, $j_2 \in [n, j]$.

Since the functions $l(n)$ and $(M(n) - n)$ tend to infinity as $n \to \infty$,

$$\lim_{n\to\infty} C_3(n) = 0.$$

Now let us estimate the function $C_2(n)$. Denote by $C_{2,1}(n)$ the sum of all its terms (with indices no smaller than $m_1(n)$) for which

$$\Delta(j) \leq g \cdot \Delta(n),$$

where $g > 1$ is an arbitrarily fixed number. Then

$$\begin{aligned} C_{2,1}(n) &\stackrel{\text{def}}{=} \sum_{j=m_1(n)}^{n-l(n)-1} \frac{\Delta^2(j)}{(\theta(j)-\theta(n))^2} \\ &\leq g \cdot \Delta(n) \sum_{j=m_1(n)}^{n-l(n)-1} \left(\frac{1}{\theta(j+1)-\theta(n)} - \frac{1}{\theta(j)-\theta(n)}\right) \\ &= g\left(\frac{\theta(n)-\theta(n+1)}{\theta(n-l(n))-\theta(n)} - \frac{\theta(n)-\theta(n+1)}{\theta(m_1(n))-\theta(n)}\right). \end{aligned}$$

Since both $l(n)$ and $n - m_1(n)$ tend to infinity as $n \to \infty$ ($m(n) \leq m_1(n) \leq n - l(n) - 1$), we obtain

$$\lim_{n\to\infty} C_{2,1}(n) = 0.$$

Thus, it remains to examine the sum

$$C_{2,2}(n) \stackrel{\text{def}}{=} \sum_{j=m(n)}^{m_1(n)-1} \left(\frac{\Delta(j)}{\theta(j) - \theta(n)} \right)^2 .$$

We estimate it by the integral

$$I(n) = \int_{\frac{3}{2}\theta_n \le \theta(s) \le \theta_n + \delta} \left(\frac{\theta'(u)}{\theta(u) - \theta(n)} \right)^2 du,$$

where $\theta_n \stackrel{\text{def}}{=} \theta(n)$ and $\delta > 0$ is chosen so that if $\theta(u_g) = \theta_n + \delta$, then $\theta'(u_g) = g \cdot \theta'(n)$. Indeed, since

$$|\theta'(j+1)| \le \Delta(j) \le |\theta'(j)| \quad \text{and} \quad \lim_{j \to \infty} \frac{\theta'(j+1)}{\theta'(j)} = 1,$$

which was essentially shown in the proof of Proposition 5.8 (whose assumption, as we have pointed out above, are satisfied here), we have

$$\left(\frac{\Delta(j)}{\theta(j) - \theta(n)} \right)^2 \le \text{const} \int_j^{j+1} \left(\frac{\theta'(u)}{\theta(u) - \theta(n)} \right)^2 du.$$

Now changing the variables in the integral $I(n)$ to $s = \theta(u)$, $u = -f(s)$, we get

$$I(n) = \int_{\theta_n + \delta}^{\frac{3}{2}\theta_n} \frac{1}{(s - \theta_n)^2} \cdot \frac{ds}{f'(s)}. \tag{5.73}$$

Let us calculate the quantity δ. By Lagrange's theorem,

$$f'(\theta_n + \delta) - f'(\theta_n) = \delta f''(\theta_n^*), \qquad \theta_n^* \in (\theta_n + \delta, \theta_n).$$

Since $f'(\theta_n + \delta) = f'(\theta_n)/g$, we obtain

$$\delta = \gamma \left| \frac{f'(\theta_n)}{f''(\theta_n^*)} \right|, \quad \gamma = (g-1)/g.$$

Now let us estimate the maximum of the integrand in $I(n)$. To this end, denoting the denominator in (5.73) by $F(s)$, let us find its derivative:

$$F'(s) = (s - \theta_n)(2f'(s) + (s - \theta_n)f''(s)).$$

If $F'(s_0) = 0$, then

$$s_0 - \theta_n = -\frac{2f'(s_0)}{f''(s_0)}. \tag{5.74}$$

Therefore, the maximum of the integrand can only be attained at a point of the form (5.74) or at the endpoints of the interval $(\theta_n + \delta, \frac{3}{2}\theta)$. If the maximum is attained at points of the form (5.74), then

$$\frac{1}{F(s_0)} = \frac{(f''(s_0))^2}{4(f'(s_0))^3}. \tag{5.75}$$

If the maximum is attained at the left end-point of the integration interval, then

$$\frac{1}{F(\theta_n+\delta)} = \frac{1}{\delta^2 f'(\theta_n+\delta)} = \frac{g(f''(\theta_n^*))^2}{\gamma^2(f'(\theta_n))^3} \le \frac{g}{\gamma^2} \cdot \frac{(f''(\theta_n^*))^2}{(f'(\theta_n^*))^3}. \tag{5.76}$$

In the case of the right end-point,

$$\frac{1}{F(\frac{3}{2}\theta_n)} = \frac{4}{\theta_n^2 f'(\frac{3}{2}\theta_n)}. \tag{5.77}$$

Thus, in the cases (5.75), (5.76),

$$I(n) \le \text{const}\; \theta_n \sup_{s\in(\theta_n+\delta, \frac{3}{2}\theta_n)} \frac{(f''(s))^2}{(f'(s))^3} \le \text{const} \sup_{s\in(\theta_n+\delta,\frac{3}{2}\theta_n)} \left(\frac{f''(s)s^{1/2}}{f^{3/2}(s)}\right)^2,$$

while in the case (5.77),

$$I(n) \le \text{const}\, \frac{1}{\left(\frac{3}{2}\theta_n\right) f'\left(\frac{3}{2}\theta_n\right)}.$$

According to the assumptions of the proposition, $\lim_{\theta\to 0} I(\theta) = 0$. □

Let us give yet another assertion which guarantees that (5.36) is satisfied, and which in some situations is more convenient than Proposition 5.9.

Proposition 5.10 *Suppose that all the assumptions of Proposition* 5.8 *are satisfied and that*

$$\left|\frac{f'(\theta)}{f'\left(\frac{3}{2}\theta\right)}\right| \le g, \tag{5.78}$$

where $g > 1$ *does not depend on* $\theta \in (-\pi, \pi) \setminus \{0\}$. *Then* (5.36) *holds.*

Proof. Essentially, this was shown to be true in the proof of Proposition 5.8. Indeed, if there exists a $g > 1$ such that (5.78) is fulfilled, then in the sum $C_{2,1}(\theta)$ the lower limit is $m_1(n) = m(n)$ and the sum $C_{2,2}(\theta)$ is identically 0. Hence, here there is no need to estimate the integral $I(n)$. □

Now let us pass to condition (5.35).

First of all, let us mention that (5.35) holds in a case that is often encountered, namely, then when the function $f(\theta)$ behaves symmetrically relative to the point of discontinuity.

Proposition 5.11 *Suppose that the function $f(\theta)$ is odd:*

$$f(\theta) = -f(-\theta). \tag{5.79}$$

Then (5.35) *holds.*

Proposition 5.12 *Suppose that $f(\theta)$ is continuously differentiable on $(-\pi, \pi) \setminus \{0\}$ and the function $\psi(\theta) = (\theta f'(\theta))^{-1}$ tends monotonically to zero when $\theta \to 0$. Then* (5.35) *holds whenever one of the following conditions is satisfied:*

(a)

$$I \stackrel{\text{def}}{=} \int_0^\pi \left[f'(s) \arctan \frac{1}{sf'(s)} - f'(-s) \arctan \frac{1}{sf'(-s)} \right] ds < \infty; \tag{5.80}$$

(b)

$$I_1 \stackrel{\text{def}}{=} \int_0^\pi \left| \frac{1}{(sf'(s))^2} - \frac{1}{(sf'(-s))^2} \right| \frac{ds}{s} < \infty; \tag{5.81}$$

(c)

$$I_2 \stackrel{\text{def}}{=} \int_0^\pi \frac{1}{(sf'(s))^2} \cdot \frac{ds}{s} < \infty \quad \textit{and} \quad I_3 \stackrel{\text{def}}{=} \int_0^\pi \frac{1}{(sf'(-s))^2} \cdot \frac{ds}{s} < \infty. \tag{5.82}$$

Proof. For the sake of definiteness we will consider the case $\theta > 0$. By (5.34), the function $A(\theta)$ can be written as the difference $A(\theta) = A_+(\theta) - A_-(\theta)$, where

$$\begin{aligned} A_+(\theta) &= \sum_{j=1}^{N_+} \arctan \frac{\theta(j) - \theta(j+1)}{\theta(j)}, \\ A_-(\theta) &= \sum_{j=1}^{N_-} \arctan \frac{\theta(-j) - \theta(-j-1)}{\theta(-j)}, \end{aligned}$$

with N_+ (resp., N_-) denoting the largest number for which $\theta(N_+) < \theta$ (resp., $\theta(-N_-) > -\theta$). We claim that the sum $A_+(\theta)$ can be replaced, up to a term that tends to a finite limit when $\theta \to 0$, by an integral of the form

$$I_+(\theta) = \int_0^{x_+} \arctan \frac{\theta'(x)}{\theta(x)} dx, \quad \theta(x_+) = \theta.$$

(Recall that $\theta(x) = f^{-1}(-x)$.) Indeed, using Lagrange's theorem and the mean value theorem for integrals we obtain

$$\begin{aligned} I_+(\theta) - A_+(\theta) &= \sum_{j=0}^{N_+-1} \left(\arctan \frac{\theta'(j_1)}{\theta(j_1)} - \arctan \frac{\theta'((j+1)_2)}{\theta(j+1)} \right) \\ &\quad + \int_{N_+-1}^{x_+} \arctan \frac{\theta'(x)}{\theta(x)} dx, \end{aligned} \tag{5.83}$$

where $j_1 \in [j, j+1]$, $(j+1)_2 \in (j+1, j+2)$. The last term in (5.83) tends to zero when $\theta \to 0$ (because $x_+ - (N_+ - 1) < 2$ and the function $\psi(\theta)$ tends to zero). Consequently, it suffices to examine the sum appearing in (5.83). Thanks to the monotonicity of the functions $\psi(\theta)$ and $\theta'(x)$, the terms of this sum are nonnegative, and hence to show that the sum tends to a finite limit it suffices to show that it is uniformly bounded for $\theta \in (0, \pi]$.

Using the monotonicity of the functions $\theta'(x)$ and $\psi(\theta)$ one more time, we get

$$\begin{aligned}
&\sum_{j=0}^{N_+-1}\left(\arctan\frac{\theta'(j_1)}{\theta(j_1)} - \arctan\frac{\theta'((j+1)_2)}{\theta(j+1)}\right)\\
&\leq \sum_{j=0}^{N_+-1}\left(\arctan\frac{\theta'(j_1)}{\theta(j_1)} - \arctan\frac{\theta'((j+1)_2)}{\theta((j+1)_2)}\right)\\
&\leq \sum_{j=0}^{N_+-1}\left(\arctan\frac{\theta'(j)}{\theta(j)} - \arctan\frac{\theta'(j+2)}{\theta(j+2)}\right)\\
&= \arctan\frac{\theta'(0)}{\theta(0)} + \arctan\frac{\theta'(1)}{\theta(1)} - \arctan\frac{\theta'(N_+)}{\theta(N_+)} - \arctan\frac{\theta'(N_++1)}{\theta(N_++1)}\\
&\leq \arctan\frac{\theta'(0)}{\theta(1)} + \arctan\frac{\theta'(1)}{\theta(1)}.
\end{aligned}$$

We see that instead of the sum $A_+(\theta)$ once can consider the integral $I_+(\theta)$. Making the change of variables $\theta(x) = s$, $x = -f(s)$ in it, we obtain

$$I_+(\theta) = -\int_\theta^\pi f'(s)\arctan\frac{1}{sf'(s)}ds.$$

Similarly, one can show that the sum $A_-(\theta)$ can be replaced by the integral

$$I_-(\theta) = -\int_\theta^\pi f'(-s)\arctan\frac{1}{sf'(-s)}ds,$$

and therefore $A(\theta)$ can be replaced by the integral

$$I(\theta) = -\int_\theta^\pi \left[f'(s)\arctan\frac{1}{sf'(s)} - f'(-s)\arctan\frac{1}{sf'(-s)}\right]ds.$$

Now it is obvious that $A(\theta)$ tends to a finite limit as $\theta \to 0$ if only the integral I defined by (5.80) converges.

Now let us prove (b). To this end we will show that (b) $\Longrightarrow$ (a). Consider the function

$$F(v) = \frac{1}{v}\arctan v.$$

Denoting $v_+ = 1/(sf'(s))$ and $v_- = 1/(sf'(-s))$, we obtain

$$|I| \le \int_0^\pi |F(v_+) - F(v_-)| \frac{ds}{s} = \int_0^\pi |v_+ - v_-||F'(v^*)| \frac{ds}{s},$$

where

$$v^* \in [v_+, v_-], \quad F'(v) = \left[v - (1+v^2)\arctan v\right] / \left[v^2(1+v^2)\right].$$

Clearly, $|F'(v)| \le \text{const} \cdot v$. Therefore, (5.81) yields

$$\begin{aligned} |I| &\le \text{const} \int_0^\pi |v_+ - v_-| v^* \frac{ds}{s} \\ &\le \text{const} \int_0^\pi |v_+ - v_-||v_+ + v_-| \frac{ds}{s} \\ &= \text{const} \int_0^\pi |v_+^2 - v_-^2| \frac{ds}{s} < \infty. \end{aligned}$$

To complete the proof, we note that the convergence of I_2 and I_3 in (5.82) implies the convergence of I_1 in (5.81). □

Remark 5.6 As already mentioned above, conditions (5.22) and (5.36), respectively (5.69) and (5.71), impose restrictions on the behavior of the function $f(\theta)$ at the point θ itself and also in some neighborhood of it. The conditions of Proposition 5.9 are sufficient for the assertions of Proposition 5.8 to hold. Let us point out that conditions (5.69) and (5.71) are very close to one another, though not equivalent. One can give an example showing this, but it is quite tedious and will be omitted. Condition (5.35) (resp., (5.80)) connects the behaviors of the functions $f(\theta)$ and $f(-\theta)$ near zero. This connection is however rather weak. Thus, conditions (5.82), which cover a very wide spectrum of cases, allow (within these limitations) the function $f(\theta)$ to behave "independently" to the left and to the right of the point $\theta_0 = 0$.

Remark 5.7 The conditions imposed on the function $f(\theta)$ in Propositions 5.8–5.12 have local character (in a neighborhood of the point $\theta = 0$). What we mean by this is that what really matters is only the behavior of $f(\theta)$ in any (two-sided) neighborhood of $\theta = 0$. In particular, in Propositions 5.8–5.12 the set $(-\pi, 0) \cup (0, \pi)$ may be replaced by $(-\delta, 0) \cup (0, \delta)$ for any $\delta \in (0, \pi)$.

Let us give several examples in which the conditions of Theorem 5.10 are satisfied.

Example 5.1 *Power whirls.*
As a first example consider the function

$$f(\theta) = \begin{cases} -c_+\theta^{-\lambda_+}, & \text{if } \theta > 0, \\ c_-|\theta|^{-\lambda_-}, & \text{if } \theta < 0, \end{cases}$$

where $c_\pm > 0$ and $\lambda_\pm \in (0, \infty)$. It obviously satisfies the conditions of Propositions 5.8 and 5.9 (or Proposition 5.10), as well as condition (5.82), and consequently all the conclusions of Theorem 5.10 are valid for $f(\theta)$. Notice that this example is more general than the case studied in Section 4.10, where the numbers $\lambda_\pm$ were assumed to be in $(0, 1)$.

Example 5.2 *Power-logarithmic whirls.*
Now let

$$f(\theta) = \begin{cases} -c_+\theta^{-\lambda_+}(\log|\theta|^{-1})^{\beta_+}, & \text{if } \theta > 0, \\ c_-|\theta|^{-\lambda_-}(\log|\theta|^{-1})^{\beta_-}, & \text{if } \theta < 0, \end{cases}$$

where $c_\pm > 0$ and $\lambda_\pm \in (0, \infty)$, $\beta_\pm \in \mathbb{R}$. The applicability of Theorem 5.10 in this case is verified in the same way as in Example 5.1.

Example 5.3 *Exponential and superexponential growth of the argument.*
Let

$$f(\theta) = \begin{cases} -c_+\exp\{d_+\theta^{-\lambda_+}\}, & \text{if } \theta > 0, \\ c_-\exp\{d_-|\theta|^{-\lambda_-}\}, & \text{if } \theta < 0, \end{cases},$$

where $c_\pm > 0$, $d_\pm > 0$ and $\lambda_\pm \in (0, \infty)$ or let

$$f(\theta) = \begin{cases} -c_+\exp\{g_+\exp(d_+\theta^{-\lambda_+})\}, & \text{if } \theta > 0, \\ c_-\exp\{g_-\exp(d_-|\theta|^{-\lambda_-})\}, & \text{if } \theta < 0, \end{cases} \tag{5.84}$$

where $c_\pm > 0$, $d_\pm > 0$, $g_\pm > 0$, and $\lambda_\pm \in (0, \infty)$. The conditions of Theorem 5.10 are verified as in the preceding cases. Let us mention only that here Proposition 5.10 does not hold, whereas Proposition 5.9 does.

This example shows that the conditions of Theorem 5.10 are well fitted for rapidly growing arguments $f(\theta)$. In particular, it is readily verified that a function $f(\theta)$ constructed via a superposition of any finite number of exponents by analogy with (5.84) also satisfies (5.69), (5.71), (5.82).

Now let us consider the case of slowly growing functions $f(\theta)$.

Example 5.4 *Logarithmic whirls.*
Let

$$f(\theta) = \begin{cases} -c(\log\theta^{-1})^{\beta}, & \text{if } \theta > 0, \\ c(\log|\theta|^{-1})^{\beta}, & \text{if } \theta < 0, \end{cases} \tag{5.85}$$

where $\beta \in (1, \infty)$, $c > 0$. This function, too, satisfies the conditions of Theorem 5.10, which can be verified by calculating the limits (5.69), (5.71) and applying Proposition 5.11.

Let us observe that the case $\beta = 1$ ("pure" logarithmic whirls), and the more so the cases with $0 < \beta < 1$, already fail to satisfy the basic condition (5.22). Indeed, for $\beta = 1$ and $\theta > 0$,

$$\alpha(x) = \exp\left\{-\frac{x}{c}\right\},$$

and so

$$\lim_{n\to\infty}\frac{\Delta(n+1)}{\Delta(n)}=\lim_{n\to\infty}\frac{\exp\{-(n+1)/c\}-\exp\{-(n+2)/c\}}{\exp\{-n/c\}-\exp\{-(n+1)/c\}}=e^{-1}.$$

Therefore, $d=e^{-1}<1$, which contradicts the conditions of Theorem 5.10. We see that the choice $\beta=1$ in (5.85) is a limit case for Theorem 5.10. This observation is confirmed by the following example, which is examined in much the same way as (5.85):

$$f(\theta)=\begin{cases} c\log(\theta^{-1})(\log\log(\theta^{-1}))^{\beta}, & \text{if } \theta>0,\\ -c\log(|\theta|^{-1})(\log\log(|\theta|^{-1}))^{\beta}, & \text{if } \theta<0,\end{cases} \tag{5.86}$$

where $\beta>0$, $c>0$.

The reader should notice that the functions $f(\theta)$ considered in the cases (5.85), (5.86) behave symmetrically to the left and to the right of the discontinuity point. This is connected with the fact that condition (5.35) is not satisfied for all (different) exponents β in (5.85) for $\theta>0$ and $\theta<0$. More precisely, if $\beta_+>3/2$ and $\beta_->3/2$, then using condition (5.82) it is not difficult to verify that the following function satisfies (5.35) and all the conditions of Theorem 5.10:

$$f(\theta)=\begin{cases} c_+(\log\theta^{-1})^{\beta_+}, & \text{if } \theta>0,\\ -c_-(\log\theta^{-1})^{\beta_-}, & \text{if } \theta<0,\end{cases}\qquad c_\pm>0. \tag{5.87}$$

At the same time, it is readily seen that when $\beta_+\neq\beta_-$ and at least one of these numbers is not larger than 3/2 relations (5.80)–(5.82) and condition (5.35) are not satisfied. In this case for Theorem 5.10 to be applicable it is necessary that the behaviors of the function $f(\theta)$ to the left and to the right of $\theta=0$ be compatible. For example, the function

$$f(\theta)=\begin{cases} c\log^{\beta}(\theta^{-1}), & \text{if } \theta>0\\ -c\log^{\beta}(|\theta|^{-1})+c_1\log^{\beta_1}(|\theta|^{-1}), & \text{if } \theta<0,\end{cases}$$

where $c>0$, $c_1\in\mathbb{R}$, $1<\beta<3/2$, and $\beta_1<3\beta-3$, satisfies conditions (5.81). Indeed,

$$\begin{aligned} I_1 &= \int_0^{\pi}\left|\tfrac{1}{(c\beta\log^{\beta-1}(\theta^{-1}))^2}-\tfrac{1}{(c\beta\log^{\beta-1}(\theta^{-1})+c_1\beta_1\log^{\beta_1-1}(\theta^{-1}))^2}\right|\tfrac{ds}{s}\\ &\le \text{const}\int_0^{\pi}\left|\tfrac{(1+(c_1\beta_1/(c\beta))\log^{\beta_1-\beta}(\theta^{-1}))^2-1}{\log^{2\beta-2}(\theta^{-1})(1+(c_1\beta_1/(c\beta))\log^{\beta_1-\beta}(\theta^{-1}))^2}\right|\tfrac{ds}{s}\\ &\le \text{const}\int_0^{\pi}\tfrac{1}{\log^{3\beta-\beta_1-2}(\theta^{-1})}\cdot\tfrac{ds}{s}<\infty, \end{aligned}$$

because $3\beta-\beta_1-2>1$.

Example 5.5 *Asymmetric whirls.*
In Examples 5.1–5.3 and in the case (5.87) (when $\beta_\pm>3/2$), condition (5.36) can be verified separately for left and right semi-neighborhoods of the point $\theta=0$

with the help of (5.82). In connection with this, new examples that satisfy the conditions of Theorem 5.10 can be constructed from the ones mentioned above by combining different types of whirling to the left and to the right. For instance, take

$$f(\theta) = \begin{cases} -c_+ \exp(g \exp(\theta^{-\lambda})), & \text{if } \theta > 0, \\ c_- \log^\beta(|\theta|^{-1}), & \text{if } \theta < 0, \end{cases} \tag{5.88}$$

where $c_\pm > 0$, $g > 0$, $\lambda > 0$, and $\beta > 3/2$. The function (5.88) combines very fast oscillations to the right of the point $\theta = 0$ with very slow oscillations to its left.

Example 5.6 *Perturbation by a nonmonotone term.*
Up to this point our examples were "regular" functions, for which, in particular, all the derivatives are monotone. Of course, Theorem 5.10 also holds for wide classes of functions that do not enjoy such nice properties. Consider, for example, the function

$$f(\theta) = -\sigma c|\theta|^{-\lambda} + c_1 \sin|\theta|^{-\beta}, \quad \sigma = \operatorname{sign}\theta,$$

where $c > 0$, $c_1 \in \mathbb{R}$, and $\lambda > 2\beta > 0$. The last constraint ensures that the derivative $f'(\theta)$ is monotone. Here condition (5.22) is verified by means of (5.69), (5.35) is guaranteed by Proposition 5.11, while (5.36) is guaranteed by Proposition 5.10.

Example 5.7 *Infinite index of power type.*
In the theory of the Riemann boundary value problem with infinite index one often encounters symbols with argument of the form

$$f(\theta) = \begin{cases} -c_+(\theta)\theta^{-\lambda_+}, & \text{if } \theta > 0, \\ c_-(\theta)|\theta|^{-\lambda_-}, & \text{if } \theta < 0, \end{cases} \tag{5.89}$$

where $\lambda_\pm > 0$, and the functions $c_\pm(\theta)$ are continuous on the segments $[0, \pi]$ and $[-\pi, 0]$. Let us assume that $c_\pm(\theta)$ are twice continuously differentiable on the semi-intervals $[-\pi, 0)$ and $(0, \pi]$ and that

$$\lim_{\theta \to 0} c'_\pm(\theta)\theta = 0, \qquad \lim_{\theta \to 0} c''_\pm(\theta)\theta^2 = 0. \tag{5.90}$$

Theorem 5.10 covers the cases (5.89)–(5.90), which is again readily verified by means of Propositions 5.8, 5.9, and 5.12.

Let us remark that the device of replacing the constants $c_\pm$ by continuous functions works not only in Example 5.1, as we just did, but also in Examples 5.2–5.4.

5.7 Slow oscillations

As already mentioned in the examples given above, Proposition 5.4 is well fitted for rapidly growing functions $f(\theta)$. The natural boundary of the domain of applicability of this proposition is the case of pure logarithmic whirls, which was already considered in Section 5.4 and was shown to violate the conditions of Theorem 5.10.

The following result completes Theorem 5.10 and is designed to deal with the case of slowly growing functions $f(\theta)$.

Theorem 5.11 *Let the function* $a(t) \in GL_\infty(\Gamma_0)$, *be continuous on* $\Gamma_0 \setminus \{1\}$ *and have the form* (5.20) *with a function* $f(\theta)$ *that is monotonically increasing on* $(-\pi, 0) \cup (0, \pi)$ *and satisfies the condition*

$$\lim_{\theta \to \pm} f(\theta) = \mp\infty.$$

Assume, in addition, that condition (5.22) *is satisfied, that* (5.27) *holds with some number* $0 < d < 1$, *and that*

$$\lim_{n \to +\infty} \left(-\frac{\theta(n)}{\theta(-n)} \right) = 1.$$

Then the function $a(t)$ *admits the following representation, which is a* $(B(t), p)$*-factorization simultaneously for all* $1 < p < \infty$:

$$a(t) = B(t) g(B(t)) d(t) |a(t)|,$$

where $g(t), d(t) \in C(\Gamma_0)$. *Moreover, the topological index of the function* $g(t)$ *is equal to zero and the Blaschke product* $B(t)$ *is constructed from the zeros* $z_j = r_j \exp\{i\theta(j)\}$, *where*

$$r_j = (1 - \Delta(j)/2)/(1 + \Delta(j)/2), \quad j = \pm 1, \pm 2, \ldots$$

The proof of this result is based on the same ideas as the proof of Theorem 5.10 and will be omitted.

By (5.27), for j sufficiently large,

$$d - \delta < \frac{\theta(j+1)}{\theta(j)} < d + \delta,$$

where δ is an arbitrarily small fixed number. It follows that

$$\text{const} \cdot \exp(-n \log(\Delta - \delta)^{-1}) \le \theta(n) \le \text{const} \cdot \exp(-n \log(\Delta + \delta)^{-1}),$$

and so the function $f(\theta)$ itself obeys the two-sided estimate

$$c_1 \log \theta^{-1} \le |f(\theta)| \le c_2 \log \theta^{-1},$$

which shows that Theorem 5.7 is indeed devoted to the case of slowly growing arguments.

Example 5.8 *Purely logarithmic whirls.*
The simplest case in which the conditions of Theorem 5.11 are satisfied is the case

$$f(\theta) = -c\sigma \log |\theta|^{-1}, \quad c > 0, \quad \sigma = \operatorname{sign} \theta,$$

which was already considered in Section 5.4. Indeed, here the function $\theta(x)$ has the form

$$\theta(x) = \exp(-x/c), \quad x > 0.$$

Therefore,

$$\psi_n(s) \equiv \psi(s) = (1 - \Delta^s)/(1 - \Delta), \quad s \in [-1/2, 1/2],$$

where $\Delta = \exp(-c^{-1}) < 1$.

Needless to say, the conditions of Theorem 5.11 are also satisfied for functions $f(\theta)$ of a more general form.

5.8 Modelling of oscillations by means of inner functions

The aim of this section is to construct an inner function $U(t)$ on the unit circle Γ_0 such that the unimodular function $a(t)$ given by

$$a(\exp(i\theta)) = \exp(2i\pi f(\theta)) \tag{5.91}$$

admits the representation

$$a(t) = U(t)c(t), \tag{5.92}$$

where $c(t)$ is a continuous function on Γ_0. As before, we will assume that the real-valued function $f(\theta)$ is continuous and monotonically increasing on the semi-intervals $[-\pi, 0)$ and $(0, \pi]$, and that

$$\lim_{\theta \to \pm} f(\theta) = \mp\infty.$$

From (5.92) it follows that

$$\lim_{t \to 1} (\arg a(t) - \arg U(t)) = 0,$$

since with no loss of generality one can take $c(1) = 0$.

Representation (5.92), interesting in its own right, is very important for the theory treated here. It clearly is a particular case of (U, p)-factorization for any $p \in (1, \infty)$, and as such it enables us to construct a theory of normal solvability for the operator $T(a)$. Furthermore, (5.92) means that $a(t) \in H^\infty(\Gamma_0) + C(\Gamma_0)$. The last remark allows us to consider operators whose symbols have generalized periodic and almost periodic discontinuities. We will embark on this topic in the next two sections.

Before stating the main result of this section, let us recall that the investigation of an operator $T(a)$ with nondegenerate symbol $a(t)$ can always be reduced to that of the operator with the unimodular symbol $a(t)/|a(t)|$.

Theorem 5.12 *Suppose that the function* $f(\theta)$ *satisfies all the conditions of Theorem* 5.10 *and that in condition* (5.22)

$$\psi(s) = s. \tag{5.93}$$

Then the unitary valued function $a(t)$ *given by* (5.91) *admits a representation* (5.92) *in which the inner function* $U(t)$ *has the form*

$$U(t) = \frac{r_0 + B(t)}{1 + r_0 B(t)}, \quad r_0 = e^{-2},$$

and $B(t)$ *is the Blaschke product figuring in* (5.37).

Proof. We will use the final representation (5.64) for the argument of a Blaschke product, obtained in the proof of Theorem 5.10 (see also (5.61) and (5.49)):

$$f_B(x) = -2\pi n - 2\arctan\left(\frac{\tan \pi s}{\tanh 1}\right) + O(x), \tag{5.94}$$

where $x = n+s$, $s \in (-1/2, 1/2]$, and the function $O(x)$ is continuous and vanishes at infinity. Recall that, by definition, $x = -f(\theta)$, $f_B(x) = \arg B(\exp(i\theta(x)))$, $\theta(x) = f^{-1}(-x)$.

Notice that the continuous branch of $\arg U(t)$ can be selected so that

$$|\arg U(t) - \arg B(t)| < \pi. \tag{5.95}$$

Now let us use formula (2.44). Observing that $(1-r_0)/(1+r_0) = \tanh 1$, we have

$$\begin{aligned}\arg \frac{r_0 + \exp(i\varphi)}{1 + r_0 \exp(i\varphi)} &= -2\arctan\left(\tanh 1 \cot\left(\frac{\varphi - \pi}{2}\right)\right) \\ &= 2\arctan\left(\tanh 1 \tan \frac{\varphi}{2}\right),\end{aligned}$$

where the branch of the argument presented is continuous on the interval $\varphi \in (-\pi, \pi)$. Now using (5.94) and (5.95), the continuous branch of $\arg U(t)$ can be calculated as follows:

$$\begin{aligned}\arg U(\exp(i\theta(x))) &= -2\pi n + 2\arctan\left(\tanh 1 \left[\tan\left(-\arctan\frac{\tan \pi s}{\tanh 1}\right) + O(x)\right]\right) \\ &= -2\pi n - 2\pi s + \widetilde{O}(x) = -2\pi x + \widetilde{O}(x),\end{aligned}$$

where $\lim_{x\to\infty} \widetilde{O}(x) = 0$. This proves (5.92). Indeed,

$$U(t) = U(\exp(i\theta)) = \exp(i2\pi f(\theta))\exp(-i\widetilde{O}(x)) = a(t)c(t). \qquad \square$$

Let us mention here that all the Examples 5.1–5.7 given in Section 5.6 satisfy the conditions of Theorem 5.12 because, by Proposition 5.8, condition (5.93) is fulfilled for them. Theorem 5.12 can be generalized to a sufficiently wide class of functions $\psi(s)$, different from s, but we will not discuss this further.

5.9 Generalized almost periodic discontinuities

Let us return to the case of almost periodic discontinuities, considered in Chapter 4. According to Theorem 4.2, the argument of an arbitrary nondegenerate continuous almost periodic function $p(x)$, $x \in \mathbb{R}$, can be written as

$$\arg p(x) = \sigma x + \psi(x), \tag{5.96}$$

where $\sigma \in \mathbb{R}$, $\psi(x) \in \Pi_c(\mathbb{R})$; the classes $\Pi_c(\mathbb{R})$ and $\Pi_w(\mathbb{R})$ were defined in Section 4.1. Therefore, $p(x)$ is a power-type whirl (near the point at infinity) with exponent 1. Notice, however, that in general the conditions on the argument of the symbol given in the preceding sections do not cover the case of almost periodic discontinuities. In particular, the presence of a bounded oscillating component in (5.96) may result in a function $\arg p(x)$ failing even to be monotone.

The aim of the present section is to extend the conditions imposed on the behavior of the function $a(t)$ near its discontinuity points so that analogues of almost periodic discontinuities with arbitrary growth of the argument (in the sense of Sections 5.6 and 5.8) become amenable to study.

Let us give precise statements. Assume that the function $\alpha(x)$ is continuous and monotonically increasing on the whole real line $\mathbb{R}$ and that

$$\lim_{x \to \pm\infty} \alpha(x) = \pm\infty.$$

Denote by $\Pi_{c,\alpha}(\Gamma_0)$ (resp., $\Pi_{w,\alpha}(\Gamma_0)$) the set of all functions $a(t)$ representable in the form

$$a(t) = p(\alpha(\xi(t))), \quad t \in \Gamma_0, \quad \xi(t) = -i\frac{t+1}{t-1},$$

where $p(x)$ belongs to the class $\Pi_c(\mathbb{R})$ (resp., $\Pi_w(\mathbb{R})$). Let us observe that the fractional linear transformation $x = \xi(t)$ maps the interior (resp., exterior) of Γ_0 conformally onto the upper (resp., lower) complex half-plane and that it maps Γ_0 itself onto $\mathbb{R}$.

The main result of this section is a theory of normal solvability of the operator $T(a)$ with a symbol $a(t) \in \Pi_{c,\alpha}(\Gamma_0)$. Notice that in the case $\alpha(x) = x$, $\Pi_{c,\alpha}(\Gamma_0) \equiv \Pi_c(\Gamma_0)$ and $\Pi_{w,\alpha}(\Gamma_0) \equiv \Pi_w(\Gamma_0)$.

The basic requirement imposed in this section on the function $\alpha(x)$ is that for every $\sigma > 0$ it admit a representation

$$\exp\{i\sigma\alpha(\xi(t))\} = U_\sigma(t)c_\sigma(t), \tag{5.97}$$

where $U_\sigma(t)$ is an inner function and $c_\sigma(t) \in C(\Gamma_0)$. Theorem 5.12 gives sufficient conditions for such a representation to hold when $\sigma = 1$. Let us point out here that in the examples given in Section 5.6 and, more generally, in the sufficient conditions formulated in Propositions 5.8–5.12, one is allowed to multiply the argument by an arbitrary positive constant σ.

Theorem 5.13 *Let $a(t) = p(\alpha(\xi(t)))$ be a function in $\Pi_{c,\alpha}(\Gamma_0)$. Then for the operator $T(a)$ to be normally solvable in the space $L_p^+(\Gamma_0)$, $1 < p < \infty$, it is necessary and sufficient that*

$$\inf_{t\in\Gamma_0} |a(t)| > 0.$$

When this condition is satisfied, the almost periodic index $\sigma(p)$ of the function $p(x)$ is well defined (see (4.4)) and the following assertions hold true.

(1) *If $\sigma(p) = 0$, then $T(a)$ is a Φ-operator in $L_p^+(\Gamma_0)$.*

(2) *If $\sigma(p) > 0$, then $T(a)$ is left invertible in $L_p^+(\Gamma_0)$ and* $\dim\operatorname{coker} T(a) = \infty$.

(3) *If $\sigma(p) < 0$, then $T(a)$ is right invertible in $L_p^+(\Gamma_0)$, and* $\dim\ker T(a) = \infty$.

Proof. The condition $\inf_{t\in\Gamma_0} |a(t)| > 0$ is necessary for the normal solvability of the operator $T(a)$ for any function $a(t) \in L_\infty(\Gamma_0)$ (see Proposition 1.6), and for that reason we will henceforth assume that it is satisfied.

Suppose first that $a(t) \in \Pi_{w,\alpha}(\Gamma_0)$. In this case $p(x) \in \Pi_w(\Gamma_0)$, and hence, by Theorem 4.5,

$$p(x) = p_+(x)\exp(i\sigma(p)x)p_-(x). \tag{5.98}$$

The functions $p_+^{\pm1}(x) \in \Pi_w^+(\mathbb{R})$ and $p_-^{\pm1}(x) \in \Pi_w^-(\mathbb{R})$ admit absolutely convergent series expansions

$$p_+^{\pm1}(x) = \sum_{j=0}^{\infty} c_j^{\pm}\exp(i\sigma_j^{\pm}), \quad \sigma_j^{\pm} \geq 0, \quad c_j^{\pm} \in \mathbb{C}, \tag{5.99}$$

$$p_-^{\pm1}(x) = \sum_{j=0}^{\infty} d_j^{\pm}\exp(\nu_j^{\pm}), \quad \nu_j^{\pm} \leq 0, \quad d_j^{\pm} \in \mathbb{C}. \tag{5.100}$$

Changing the variables in (5.98), we obtain

$$a(t) = p_+(\alpha(\xi(t)))\exp(i\sigma(p)\alpha(\xi(t)))p_-(\alpha(\xi(t))).$$

In view of (5.97), (5.99), and (5.100), Theorem 2.5 shows that

$$p_+^{\pm1}(\alpha(\xi(t))) \in L_\infty^+(\Gamma_0) + C(\Gamma_0)$$

and

$$p_-^{\pm1}(\alpha(\xi(t))) \in L_\infty^-(\Gamma_0) + C(\Gamma_0).$$

Therefore, the functions $p_\pm^{\pm1}(\alpha(\xi(t)))$ admit a p-factorization for every $p \in (1,\infty)$ and are MNAFs (see Theorem 4.6). We conclude that Theorem 2.1 implies assertion (1) in case $\sigma(p) = 0$.

Further, if $\sigma(p) \neq 0$, then by (5.97),

$$\exp\{i\sigma(p)\alpha(\xi(t))\} = U_{|\sigma(p)|}^{\varepsilon}(t)c_{\sigma(p)}(t),$$

where $\varepsilon = \operatorname{sign}\sigma(p)$ and $c_{\sigma(p)}(t)$ is a continuous function that does not vanish on Γ_0. Hence, for $\sigma(p) \neq 0$ the function $a(t)$ admits a $(U_{|\sigma(p)|}, p)$-factorization, and then Theorem 2.6 yields assertions (2) and (3). The case where $a(t) \in \Pi_{\mathrm{w},\alpha}(\Gamma_0)$ can be regarded as already dealt with.

Now let us pass to the general case, where

$$a(t) \in \Pi_{\mathrm{c},\alpha}(\Gamma_0) \qquad \text{and} \qquad \inf_{t\in\Gamma_0\setminus\{1\}} |a(t)| > 0.$$

Let $\{a_n(t)\} \in \Pi_{\mathrm{w},\alpha}(\Gamma_0)$ be a sequence of functions which converges uniformly to $a(t)$. Using the stability of the almost periodic index we may, with no loss of generality, assume that

$$\inf_{t\in\Gamma_0\setminus\{1\}} |a_n(t)| > 0 \qquad \text{and} \qquad \sigma(p_n) = \sigma(p),$$

where $a_n(t) = p_n(\alpha(\xi(t)))$. It then follows that the function

$$a_0(t) = \exp\{-i\sigma(p)\alpha(\xi(t))\}a(t)$$

is a uniform limit of MNAFs. By Theorem 4.6, $a_0(t)$ itself is an MNAF. Using again representation (5.97), we conclude that the function $a(t)$ admits a $(U_{|\sigma(p)|}, p)$-factorization of the form

$$a(t) = U^{\varepsilon}_{|\sigma(p)|}(t)c_{\sigma(p)}(t)a_0(t)$$

for every $p \in (1, \infty)$, which completes the proof of the theorem. □

This result is completely analogous to the theorem concerning the classical almost periodic case. Let us mention that the usual technique of change of variables $x \mapsto \alpha(x)$ does not work here (see Theorem 4.11), because in general the homeomorphism $u = \alpha(x)$ of $\mathbb{R}$ onto itself is not a diffeomorphism of the closed real line.

5.10 Generalized matrix periodic discontinuities

Consider a matrix function $A(t) = \{a_{k,j}(t)\}_{k,j=1}^n$ with elements of the form

$$a_{k,j}(t) = c_{k,j}(t)g_{k,j}(\exp(2\pi i f(t))), \tag{5.101}$$

where $c_{k,j}(t), g_{k,j}(t)$ are functions in $C_0(\Gamma)$, the function $f(t)$ is real-valued on $\Gamma_0 \setminus \{1\}$ and satisfies $\lim_{t\to 1\pm 0} f(t) = \mp\infty$, and the function

$$a(t) = \exp(2\pi i f(t)) \tag{5.102}$$

admits a representation (5.92). With no loss of generality we may assume that the continuous function figuring in (5.92) satisfies $c(1) = 1$.

It is clear that the stated restrictions allow each element of the matrix function $A(t)$ to have a second-kind discontinuity of oscillating type. Thus, in the simplest case, if $c_{k,j}(t) \equiv 1$, $g_{k,j}(t) = t^{n_{k,j}}$, $n_{k,j} \in \mathbb{Z}$, then $\arg a_{k,j} = 2\pi n_{k,j} f(t)$. As Theorem 5.12 and Examples 5.1–5.7 show, the class of admissible functions $f(t)$ is quite broad. In the case where $f(t) \equiv \xi(t)\ (= -i(t+1)/(t-1))$ and all $c_{k,j}(t) \equiv 1$, we are dealing with *periodic discontinuities*. This means that the matrix function $\widetilde{A}(x)$ defined via $\widetilde{A}(\xi(t)) \equiv A(t)$ is periodic on the real line $\mathbb{R}$.

To formulate the main result of this section, let us associate with $A(t)$ the continuous matrix function on the circle Γ_0 given by

$$A_0(t) = \{c_{k,j}(1) g_{k,j}(t)\}_{k,j=1}^{n}. \tag{5.103}$$

We will assume that the composition $A_0(u(t))$, where $u(t)$ is form (5.92), admits a $(u^{|\kappa_j|}, p)$-factorization (for example, this is the case when $A(t)$ satisfies a Hölder condition).

Theorem 5.14 *Suppose that the matrix function* $A(t)$ *has the form* (5.101), (5.102) *and that*

$$\inf_{t \in \Gamma_0 \setminus \{1\}} |\det A(t)| > 0 \quad \text{and} \quad \det A_0(t) \neq 0.$$

Let κ_j, $1 \leq j \leq n$, *denote the partial indices of the matrix function* $A_0(t)$. *Consider* $T(A)$ *as an operator on* $L_p^{+(n)}(\Gamma_0)$, *where* $1 < p < \infty$. *Then* $T(a)$ *is*

(1) *a* Φ_+*-operator if all* $\kappa_j \geq 0$;

(2) *a* Φ_-*-operator if all* $\kappa_j \leq 0$;

(3) *a* Φ*-operator if all* $\kappa_j = 0$.

Proof. Using representation (5.92) for the function $a(t)$ given by (5.102), we write $A(t)$ in the form

$$A(t) = A_0(u(t)) + C(t), \tag{5.104}$$

where $C(t) \in \mathbb{C}^{n \times n}(\Gamma_0)$ and $C(1) = \emptyset$ (here the symbol $\emptyset$ denotes the zero matrix). By (5.103), the matrix function $A_0(u(t))$ admits a (u,p)-factorization with diagonal factor

$$D(u(t)) = \operatorname{diag}\{u^{\kappa_1}(t), u^{\kappa_2}(t), \ldots, u^{\kappa_n}(t)\}.$$

Suppose first that all $\kappa_j \geq 0$. Then, by Proposition 5.2, the operator $T(A_0(u(t)))$ is left invertible. Consequently, for any arbitrarily small $\delta > 0$ there exists a rational matrix function $R_\delta(t)$ such that

$$\|A_0(t) - R_\delta(t)\|_{L_\infty^{n \times n}(\Gamma_0)} < \delta. \tag{5.105}$$

If δ is sufficiently small, then the matrix function $R_\delta(t)$ admits a standard factorization with rational elements

$$R_\delta(t) = R_{\delta,-}(t) D(t) R_{\delta,+}(t),$$

where $R^{\pm 1}_{\delta,\pm}(t)$ are rational matrix functions. Notice that the left inverse operator

$$T^{-1}(R_\delta(u(t)) = P^+ R^{-1}_{\delta,+}(u(t)) D^{-1}(u(t)) P^+ R^{-1}_{\delta,-}(u(t)) P^+$$

is bounded uniformly in δ in the norm of $L_p^{+(n\times n)}(\Gamma_0)$. Indeed, from (5.105) it follows that the same estimate is satisfied by the function

$$O_\delta(u(t)) = A_0(u(t)) - R_\delta(u(t)),$$

and consequently

$$\lim_{\delta\to 0} \|T^{-1}(a_0(u(t))) - T^{-1}(R_\delta(u(t)))\|_{L_p^{+(n)}(\Gamma_0)} = 0.$$

Now letting the operator $T^{-1}(R_\delta(u(t)))$ act from the left on $T(A)$ and using identity (5.104) we obtain

$$\begin{aligned} T^{-1}(R_\delta(u(t)))T(a) = I &+ T^{-1}(R_\delta(u(t)))T(O_\delta(u(t))) \\ &+ P^+ R_\delta^{-1}(u(t)) C(t) P^+ \\ &- P^+ R^{-1}_{\delta,+}(u(t)) D^{-1}(u(t)) P^- R^{-1}_{\delta,-}(u(t)) C(t) P^+. \end{aligned} \tag{5.106}$$

The first three terms on the right-hand side of this equality tend to the Toeplitz operator $T(B) \overset{\text{def}}{=} T(E_n + A_0^{-1}(t)C(t))$ as $\delta \to 0$, where E_n is the unit matrix of order n. Since the matrix function $B(t)$ is continuous (recall that $C(1) = \emptyset$) and, by the assumptions of the theorem, $\det B(t) \neq 0$, the operator $T(B)$ is Fredholm. Thus, for sufficiently small δ the sum of the first three terms on the right-hand side of (5.106) is also a Fredholm operator. Finally, let us mention that, thanks to the continuity of the matrix function $R^{-1}_{\delta,-}(u(t))C(t)$, the last term in (5.106) is a compact operator. We conclude that $T^{-1}(R_\delta(u(t)))T(a)$ is a Φ-operator, and, by a well-known theorem of the theory of normally solvable operators, $T(A)$ is a Φ_+-operator.

Assertion (2), and together with it (3), is established in much the same manner. □

Remark 5.8 In the case where all $\kappa_j = 0$, the above proof enables us also to calculate the index of the operator $T(A)$:

$$\operatorname{ind} T(A) = -\frac{1}{2\pi} \arg\det(E_n + A_0^{-1}(t)C(t))\big|_{\Gamma_0}.$$

Remark 5.9 Theorem 5.14 is valid also for the case of the real line $\mathbb{R}$ because the Toeplitz operators defined on $\mathbb{R}$ and Γ_0 are similar (see (4.30)). Here the fact that the similarity operator K acts from the space $L_p^+(\Gamma_0, |t-1|^{p-2})$ into the space $L_p^+(\mathbb{R})$ is inessential, because all the arguments remain valid in the case of weighted spaces as well.

References and comments

5.1 The theory of block Toeplitz operators is quite completely treated in Litvinchuk and Spitkovsky [1] (see also Böttcher and Silbermann [1], Böttcher, Karlovich, and Spitkovsky [3], and Clancey and Gohberg [1]). Propositions 5.1, 5.2 and Theorem 5.1 belong to Simonenko [4]. Theorem 5.2 originates in Gohberg's work [1].

5.2 In this section we follow Grudsky's works [5]–[7], [9]. Another sufficiently general approach to the factorization of matrix functions is worked out in Spitkovsky [7], [8].

5.3 The foundation of the proposed method for the factorization of u-periodic matrix functions (u-factorization) is Theorem 5.5. This result is well known in the theory of composition operators (see Nordgren [1], Shapiro [1]). The theory of the u-factorization appeared for the first time in Grudsky [5].

The following question, which is of great interest in connection with u-factorization, had been open for a long time (since about 1982): If the function $a(t) \in L_\infty(\Gamma_0)$ admits a p-factorization and $u(t)$ is an arbitrary function, is it true that $(\gamma_u a)(t) = a(u(t))$ also admits a (u,p)-factorization? Böttcher and Grudsky showed in [2] that this need not be the case.

5.4 Theorem 5.4 is due to Grudsky [5].

5.5–5.8 The results obtained in these sections are considerable generalizations and sharpenings of results of Grudsky [5]–[7], [9]–[11], Böttcher and Grudsky [1], and Nitievskiĭ [1], [2].

5.9 In this section we follow the paper of Böttcher and Grudsky [1]. For generalized semi-almost periodic symbols see also the recent paper by Böttcher, Grudsky, and Spitkovsky [3].

5.10 This section is based on Grudsky's papers [7] and [9].

Chapter 6

Toeplitz Operators Whose Symbols Have Zeros

Let X be a topological linear space and let the operator $A \in \mathcal{L}(X) = \mathcal{L}(X, X)$ have a nonclosed image ($\operatorname{im} A \neq \operatorname{clos} \operatorname{im} A$). Since a linear operator is in fact a triplet (correspondence rule, domain of definition, domain of values), its properties depend in an essential manner on each of the components of the triplet and there are reasons to hope that by appropriately perturbing these components one can achieve that the image of the "new" (perturbed) operator will be closed. This problem is intimately connected with problems that arise when solving the equation $Ax = y$. If we are interested in describing all its right-hand sides for which the solution lies in a given space, then we fix the domain of definition of the operator and try to restrict its domain of values, simultaneously modifying the topology in the latter. If, however, we are interested in describing all "reasonable" solutions for a given class of right-hand sides, then the domain of values of the operators is kept fixed, while the operator itself and its domain of definition are subject to a certain extension. The "new" operator with closed image, as well and the procedure for constructing it, are called a *normalization* of the operator A.

In this chapter we consider some aspects of normalization of the operator $T(a)$ paying attention, first and foremost, to the case where the symbol a has a locally-polynomial degeneracy on the curve Γ.

The normalization principle

6.1 Normally solvable operators

Let us list a few known results on the normal solvability of operators in Banach spaces. If X is a Banach space and $M \subset X$ (resp., $M \subset X^*$), we denote by $M^{\perp}$

the annihilator of the set M in the space X^* (resp., X), i.e., the set of all elements $y \in X^*$ (resp., $y \in X$) such that $\langle y, x \rangle = 0$ for all $x \in X$ (resp., $\langle x, y \rangle$ for all $x \in X^*$). Let X_1, X_2, X_3 be Banach spaces.

Theorem 6.1 *For the operator $A \in \mathcal{L}(X_1, X_2)$ to be normally solvable it is necessary and sufficient that*

$$\operatorname{im} A = (\ker A^*)^{\perp}.$$

Theorem 6.2 *Let $A = BC$, where $B \in \mathcal{L}(X_2, X_3)$, $C \in \mathcal{L}(X_1, X_2)$. If the operators B and C are normally solvable, then for the operator A to be normally solvable it is necessary and sufficient that the linear manifold $\ker B + \operatorname{im} C$ be closed in X_2. In particular, if $\dim \ker B < \infty$, then A is normally solvable.*

Theorem 6.3 *The operator $A \in \mathcal{L}(X_1, X_2)$ is normally solvable if and only if the operator $A^* \in \mathcal{L}(X_2^*, X_1^*)$ is normally solvable.*

Theorem 6.4 *If the operator $A \in \mathcal{L}(X_1, X_2)$ is normally solvable, then*

$$\operatorname{im} A^* = (\ker A)^{\perp}.$$

6.2 Normalization of linear operators

Throughout this section it will be assumed that X is a reflexive Banach space and that X^* is its dual space, equipped with the topology induced by the norm

$$\|f\| = \sup_{x \neq 0} \frac{|\langle f, x \rangle|}{\|x\|}, \quad f \in X^*. \tag{6.1}$$

It is known that the space X^* is complete and that an operator A lies in $\mathcal{L}(X)$ if and only if A^* lies in $\mathcal{L}(X^*)$. Moreover, $(A^*)^* = A$.

Let $A \in \mathcal{L}(X)$ have a nonclosed image.

We say that the operator A *admits a right normalization* if there exists a Banach space $D_1 X$ which contains X algebraically, such that A extends to an operator $\widetilde{A} \in \mathcal{L}(D_1 X, X)$ with closed image. In this case the operator $\widetilde{A}$ is called a *right normalization* of the operator A.

We say that the operator A *admits a left normalization* if there exists a Banach space $D_2 X$ which is algebraically contained in X, such that the operator $\widetilde{A}$ that coincides on X with A but takes values in $D_2 X$ belongs to $\mathcal{L}(X, D_2 X)$ and has closed image. In this case the operator $\widetilde{A}$ is called a *left normalization* of the operator A.

Let L be a subspace of X. A direct complement of L in X, whenever it exists, will be denoted by $\operatorname{add} L$. The subspace $\operatorname{add} L$ of X is in general not uniquely determined but, whenever it exists,

$$X = L \oplus \operatorname{add} L.$$

If $A \in \mathcal{L}(X)$ and the subspace $\operatorname{clos}\operatorname{im} A$ has a direct complement $\operatorname{add}\operatorname{clos}\operatorname{im} A$ in X, the latter will be denoted for simplicity by $\operatorname{add} A$.

Let $A \in \mathcal{L}(X)$ be such that the linear manifold $\operatorname{im} A$ is not closed in X, and let the subspaces $\ker A$ and $\operatorname{clos}\operatorname{im} A$ have direct complements in X, i.e., in X there exists subspaces $X_1 = \operatorname{add}\ker A$ and $X_2 = \operatorname{add} A$. The operator A induces an algebraic isomorphism of the subspace X_1 onto the linear manifold $\operatorname{im} A$. Let A^{-1} denote the inverse isomorphism (defined on $\operatorname{im} A$) and let P denote a projector of X onto the subspace $\operatorname{clos}\operatorname{im} A$. Now let us introduce the normed linear space

$$DX = \operatorname{im} A \oplus X_2,$$

equipped with the norm

$$\|x\|_1 = \|x\| + \|A^{-1}Px\|. \tag{6.2}$$

Theorem 6.5 (1) *The space DX is complete in the norm (6.2) and $DX \subset X$. The space $(DX)^*$ is complete in the norm (6.1) associated with the norm (6.2) and $X^* \subset (DX)^*$.*

(2) *The operator $\widetilde{A}$ which coincides on X with A belongs to $\mathcal{L}(X, DX)$ and is normally solvable, and hence is a left normalization of the operator A. Moreover, $\ker \widetilde{A} = \ker A$ and $\ker \widetilde{A}^* = \ker A^*$.*

(3) *The operator $\widetilde{A}^*$ belongs to $\mathcal{L}((DX)^*, X^*)$, is normally solvable, coincides on X^* with the operator A^*, and hence is a right normalization of the operator A^*. Moreover, $\operatorname{im} \widetilde{A}^* = (\ker A)^\perp$.*

Remark 6.1 The embedding $X^* \subset (DX)^*$ means that the restriction of a functional from X^* to $DX = \operatorname{im} A \oplus X_2$ yields a functional from $(DX)^*$. The functionals from $(DX)^*$ are defined not on the whole X, but only on the linear manifold DX, which is everywhere dense in X. In connection with this, the equality $\ker \widetilde{A}^* = \ker A^*$ means, first, that the restriction of any functional from $\ker A^*$ to DX belongs to $\ker \widetilde{A}^*$, and, secondly, that any functional from $\ker \widetilde{A}^*$ admits a unique extension to a functional from $\ker A^*$.

Proof. First, let us show that the subspace DX is complete in the norm (6.2). Let $\{x_n\}$ be a fundamental sequence in DX. Then the sequences $\{x_n\}$ and $\{A^{-1}Px_n\}$ are fundamental in X. Let $\widetilde{x}$ and $\widetilde{y}$ denote their limits in X. We claim that $\widetilde{x} \in DX$. Indeed, $x_n \to \widetilde{x}$ implies $Px_n \to P\widetilde{x}$, and so $A^{-1}Px_n \to \widetilde{y}$. This shows that $Px_n \to A\widetilde{y}$. By the uniqueness of the limit, $P\widetilde{x} = A\widetilde{y}$, and since

$$\|\widetilde{x}\|_1 = \|\widetilde{x}\| + \|A^{-1}P\widetilde{x}\| = \|\widetilde{x}\| + \|\widetilde{y}\| < \infty,$$

we conclude that $\widetilde{x} \in DX$. Moreover,

$$\|\widetilde{x} - x_n\|_1 = \|\widetilde{x} - x_n\| + \|A^{-1}P(\widetilde{x} - x_n)\| = \|\widetilde{x} - x_n\| + \|\widetilde{y} - A^{-1}Px_n\| \to 0,$$

and so $x_n \to \widetilde{x}$ in DX.

Further, the space $(DX)^*$ is complete in the norm (6.1) generated by the norm $\|\cdot\|_1$ given by (6.2), and since $\|x\| \leq \|x\|_1$ for all x, $X^* \subset (DX)^*$. The operator $\widetilde{P} = P|_{DX}$ is a projector of DX onto the linear manifold $\operatorname{im}\widetilde{A}$ and $\widetilde{P} \in \mathcal{L}(DX)$, because for any $x \in DX$,

$$\|\widetilde{P}x\|_1 = \|Px\| + \|A^{-1}Px\| \leq \max\{\|P\|, 1\} \cdot \|x\|_1.$$

Therefore, the linear manifold $\operatorname{im}\widetilde{A}$ is closed in DX, and the operator $\widetilde{A}$ belongs to $\mathcal{L}(X, DX)$ and is a left normalization of the operator A. Also, $\ker\widetilde{A} = \ker A$. Let us show that $\ker\widetilde{A}^* = \ker A^*$. To that end, let $f \in \ker A^*$. Then $f \in (DX)^*$ and $\langle A^*f, x\rangle = \langle f, Ax\rangle = 0$ for all $x \in X$. It follows that $\langle \widetilde{A}^*f, x\rangle = \langle f, \widetilde{A}x\rangle = \langle f, Ax\rangle = 0$ for all $x \in DX$, and hence $f \in \ker\widetilde{A}^*$. We conclude that $\ker A^* \subset \ker\widetilde{A}^*$. Conversely, let $f \in \ker\widetilde{A}^*$. Then $f \in (DX)^*$ and $\langle \widetilde{A}^*f, x\rangle = \langle f, \widetilde{A}x\rangle = 0$ for all $x \in DX$. But then $\langle f, Ax\rangle = 0$ for all $x \in DX$.

Let us consider the linear manifold $\mathcal{M} = A(DX) \oplus X_2$ and show that it is dense in X. Indeed, let $y \in \operatorname{im} A$. Then $y = Ax$ with $x \in X$. Since the linear manifold DX is dense in X, there exists a sequence $\{x_n\}$, $x_n \in DX$, such that $\|x - x_n\| \to 0$. This implies that $\|y - Ax_n\| = \|A(x - x_n)\| \to 0$. Therefore, the linear manifold $A(DX)$ is dense in $\operatorname{im} A$, which in turn implies that $A(DX)$ is dense in $\operatorname{clos}\operatorname{im} A$, and hence that the linear manifold $\mathcal{M}$ is dense in X. Further, let $y \in \mathcal{M}$. Then $y = Ax + x_2$, where $x \in DX$, $x_2 \in X_2$. Therefore,

$$\begin{aligned} |\langle f, y\rangle| &= |\langle f, Ax + x_2\rangle| = |\langle f, x_2\rangle| = |\langle f, (I-P)y\rangle| \\ &\leq c(f)\|(I-P)y\|_1 = c(f)\|(I-P)y\| \leq c(f)\|I-P\|\|y\|, \end{aligned}$$

i.e., the functional f is bounded on $\mathcal{M}$. But then f extends by continuity in a unique manner to a functional $\widetilde{f} \in X^*$. Moreover, $\widetilde{f}|_{\operatorname{im} A} = 0$, because $\widetilde{f}|_{A(DX)} = f|_{A(DX)} = 0$. Therefore, $\widetilde{f} \in \ker A^*$. We have shown that $\ker\widetilde{A}^* \subset \ker A^*$, and so $\ker\widetilde{A}^* = \ker A^*$, as claimed.

The operator $\widetilde{A}^*$ lies in $\mathcal{L}((DX)^*, X^*)$, has closed image, and $\operatorname{im}\widetilde{A}^* = (\ker A)^\perp$ (Theorems 6.3 and 6.4). Moreover, for any $x \in X$ and any $f \in X^*$,

$$\langle \widetilde{A}^*f, x\rangle = \langle f, \widetilde{A}x\rangle = \langle f, Ax\rangle = \langle A^*f, x\rangle.$$

Consequently, $\widetilde{A}^*$ is a right normalization of the operator A^*. □

Remark 6.2 The correspondence between the subspaces $\ker A^*$ and $\ker\widetilde{A}^*$ described above takes linearly independent systems of functionals into linear independent systems. Indeed, let $\{f_1, \ldots, f_n\}$ be a linearly independent system of functionals from $\ker A^*$. Then the restrictions of these functionals to DX yield a linearly independent system of functionals from $(DX)^*$. Assuming the contrary, one can find complex numbers c_k, not simultaneously equal to zero, such that

$$f = \sum_{k=1}^{n} c_k(f_k|_{DX}) = 0.$$

Thus, $\langle f, x\rangle = 0$ for all $x \in DX$. Since the functional f is bounded on the linear manifold DX, we have $\langle f, x\rangle = 0$ for all $x \in X$ by continuity, which contradicts the assumption that the system $\{f_1, \dots, f_n\}$ is linearly independent. The converse assertion is proved in much the same way. This shows that the normalization of the operator A described in Theorem 6.5 preserves the defect numbers and the index of the operator:

$$\dim\ker A = \dim\ker \widetilde{A}, \qquad \dim\ker A^* = \dim\ker \widetilde{A}^*, \qquad \operatorname{ind} A = \operatorname{ind} \widetilde{A}.$$

In applications Theorem 6.5 is not very effective, because the normalization of an operator in accordance with that theorem assumes that the main characteristics of the operator – kernel, direct complement of the kernel, image, and so on – are known beforehand. Incidentally, for a certain class of operators, described below and important in applications, the normalization procedure described in Theorem 6.5 becomes considerably simpler.

An operator $A \in \mathcal{L}(X)$ is said to be *unboundedly invertible* if there exist a linear submanifold X_0 of X that is dense in X and contains $\operatorname{im} A$ and a linear operator $B : X_0 \to X$ such that

(i) $BA = I$ on X;

(ii) $AB = I$ on X_0;

(iii) the operator B does not extend to a linear operator $\widetilde{B}$ on X that is bounded on X.

If the operator A is unboundedly invertible, then the operator B is uniquely determined and called the *unbounded inverse* of A.

Theorem 6.6 *For an operator $A \in \mathcal{L}(X)$ to be unboundedly invertible it is necessary and sufficient that*

(1) $\ker A = \{0\}$;

(2) $\operatorname{clos}\operatorname{im} A = X$ *and* $\operatorname{im} A \neq X$.

Proof. Let the operator A be unboundedly invertible. From condition (i) it follows that $\ker A = \{0\}$. We claim that $X_0 \subset \operatorname{im} A$. Indeed, if $y \in X_0$, then $By \in X$ and hence, by condition (ii) above, $ABy = y$. Therefore, $y \in \operatorname{im} A$, and so $X_0 \subset \operatorname{im} A$, as claimed. Since X_0 is dense in X, we conclude that $\operatorname{clos}\operatorname{im} A = X$. Finally, from condition (iii) in the definition of unbounded invertibility it follows that $\operatorname{im} A \neq X$.

Conversely, suppose that conditions (1) and (2) of the theorem are satisfied. Denote $X_0 = \operatorname{im} A$ and consider the operator $\widetilde{A} : X \to X_0$ given by $\widetilde{A}x = Ax$ for $x \in X$. Since $\ker \widetilde{A} = \{0\}$ and $\operatorname{im} \widetilde{A} = X_0$, the operator $\widetilde{A}$ is an algebraic isomorphism of the linear space X onto the linear manifold X_0. The inverse operator B satisfies conditions (i) and (ii) in the definition of the unbounded invertibility of A.

Condition (iii) of that definition is also satisfied. Indeed, assuming the contrary, the operator B admits an extension to a linear operator $\widetilde{B}$ on X that is bounded on X, and then the equality $AB = I$ extends to the equality $A\widetilde{B} = I$ on X. Hence, the operator A has a continuous right inverse and $\operatorname{im} A = X$, which contradicts assumption (2) of the theorem. □

Theorem 6.7 *Let $A \in \mathcal{L}(X)$. For the operator A to be unboundedly invertible in the space X it is necessary and sufficient that the operator A^* be unboundedly invertible in the space X^*.*

Proof. If $A \in \mathcal{L}(X)$, then $A^* \in \mathcal{L}(X^*)$. Notice also that $\operatorname{clos} \operatorname{im} A = (\ker A^*)^\perp$ and that, thanks to the reflexivity of X, $(A^*)^* = A$ and consequently $\operatorname{clos} \operatorname{im} A^* = (\ker A)^\perp$ (see, e.g., Kantorovich and Akilov [1], Chap. XII, §2).

If A is unboundedly invertible, then $\operatorname{clos} \operatorname{im} A = X$, i.e., $\ker A^* = \{0\}$. Simultaneously, $\ker A = \{0\}$, and hence $\operatorname{clos} \operatorname{im} A^* = X^*$. At the same time, $\operatorname{im} A^* \neq X^*$, because otherwise $A^* \in G\mathcal{L}(X^*)$ by Banach's theorem on the inverse operator, which is equivalent to $A \in G\mathcal{L}(X)$ (see, e.g., Dunford and Schwartz [1], Chap. VI). Therefore $\operatorname{im} A = X$, which contradicts the unbounded invertibility of the operator A. We conclude that the operator A^* is unboundedly invertible in X^*.

The converse is proved in much the same manner. □

For unboundedly invertible operators Theorem 6.5 takes on a simpler form. Let $A \in \mathcal{L}(X)$ be unboundedly invertible. Denote by DX the linear manifold $\operatorname{im} A$, equipped with the new norm

$$\|x\|_1 = \|x\| + \|A^{-1}x\|, \tag{6.3}$$

where A^{-1} is the unbounded inverse of A.

Theorem 6.8 (1) *The subspace DX is complete in the norm* (6.3) *and $DX \subset X$. The subspace $(DX)^*$ is complete in the norm* (6.1) *associated with the norm $\|\ \|_1$ given by* (6.3) *and $X^* \subset (DX)^*$.*

(2) *The operator $\widetilde{A}$ that coincides on X with A lies in $\mathcal{L}(X, DX)$, has a continuous inverse, and hence is a left normalization of the operator A.*

(3) *The operator $\widetilde{A}^*$ lies in $\mathcal{L}((DX)^*, X^*)$, has a continuous inverse, coincides on X^* with A^*, and hence is a right normalization of the operator A^*.*

Proof. This theorem is a direct corollary of Theorem 6.5. All the assertions concerning invertibility follow from Banach's theorem on the inverse operator and the reflexivity of the space X. □

The next result covers a situation that is often encountered in applications.

Theorem 6.9 *Let $A \in \mathcal{L}(X)$ and suppose $A = BC$, where $B, C \in \mathcal{L}(X)$, B is unboundedly invertible, and C is normally solvable. Further, let $DX = \operatorname{im} B$ be equipped with the norm* (6.3)*, and let $\widetilde{B}$ be the left normalization of the operator B furnished by Theorem* 6.8, (2).

Then the operator $\widetilde{A} = \widetilde{B}C$ belongs to $\mathcal{L}(X, DX)$ and is a left normalization of the operator A; also, $\ker \widetilde{A} = \ker A$, *and if* $\ker C^* \subset \operatorname{im} B^*$, *then* $\ker \widetilde{A}^* = \ker A^*$.

Furthermore, the operator $\widetilde{A}^* = C^* \widetilde{B}^*$ lies in $\mathcal{L}((DX)^*, X^*)$ and is a right normalization of the operator A^*, and $\operatorname{im} \widetilde{A}^* = (\ker A)^{\perp}$.

Proof. By Theorem 6.8, the operator $\widetilde{B}$ has a continuous inverse. Since C is normally solvable, so is the operator $\widetilde{A} = \widetilde{B}C$ (see Theorem 6.2). Clearly, $\ker \widetilde{A} = \ker A = \ker C$. Let us show that $\ker \widetilde{A}^* = \ker A^*$ (see Remark 6.1). The set $\ker A^*$ coincides with the set of all solutions of the equation $C^* B^* x = 0$ in X^*, while $\ker \widetilde{A}^*$ coincides with the set of all solutions of the equation $C^* \widetilde{B}^* x = 0$ in $(DX)^*$. Since $X^* \subset (DX)^*$ and $B^*|_{X^*} = \widetilde{B}^*|_{X^*}$, we have $\ker A^* \subset \ker \widetilde{A}^*$. But $\ker C^* \subset \operatorname{im} B^*$ by assumption. Therefore, in $\ker \widetilde{A}^*$ there are no functionals that lie in $(DX)^* \setminus X^*$. Indeed, $y \in \ker \widetilde{A}^*$ means that $C^* \widetilde{B}^* y = 0$, and so $\widetilde{B}^* y \in \ker C^*$. This implies that $\widetilde{B}^* y \in \operatorname{im} B^*$, i.e., $B^* y = B^* y_1$ for some $y_1 \in X^*$ and so, thanks to the invertibility of the operator $\widetilde{B}^*$, we have $y \in X^*$. Therefore, $\ker \widetilde{A}^* \subset \ker A^*$. We conclude that $\ker \widetilde{A}^* = \ker A^*$.

The operator $\widetilde{A}^* = C^* \widetilde{B}^*$ belongs to $\mathcal{L}((DX)^*, X^*)$ and is normally solvable (Theorem 6.3). By Theorem 6.4, $\operatorname{im} \widetilde{A}^* = (\ker \widetilde{A})^{\perp}$. Finally, on the space X^*,

$$\widetilde{A}^* = C^* \widetilde{B}^* = C^* B^* = A^*,$$

i.e., $\widetilde{A}^*$ is a right normalization of the operator A^*. □

Notice that Remark 6.2 remains valid in the case of Theorem 6.9.

Next we will describe a class of operators for which the conditions of Theorem 6.9 are satisfied. Namely, let $\mathcal{KL}(X)$ denote the set of all operators $A \in \mathcal{L}(X)$ with the following properties:

(1) every subspace of $\ker A$ has a direct complement in $\ker A$ and, in addition, has a direct complement in X which is isomorphic to X;

(2) the subspace $\operatorname{clos} \operatorname{im} A$ has a direct complement $\operatorname{add} A$ in X such that any subspace of $\operatorname{add} A$ has a direct complement in $\operatorname{add} A$ and, in addition, has a direct complement in X which is isomorphic to X;

(3) either $\ker A$ is isomorphic to a subspace lying in $\operatorname{add} A$ or $\operatorname{add} A$ is isomorphic to some subspace lying in $\ker A$.

Condition (3), for example, is satisfied by all operators for which either the kernel or the cokernel is trivial. If X is a Banach space in which every subspace of finite codimension is isomorphic to X, the class $\mathcal{KL}(X)$ contains all operators with finite-dimensional kernel and cokernel (M. I. Khaĭkin [1]). From the definition of the class $\mathcal{KL}(X)$ it follows that if A is in this class, then the subspace $\ker A$ has a direct complement in X isomorphic to X and $\operatorname{clos}\operatorname{im} A \cong X$.

Theorem 6.10 *Let $A \in \mathcal{KL}(X)$ and suppose A has a nonclosed image. Then there exist an unboundedly invertible operator $B \in \mathcal{L}(X)$ and a normally solvable operator $C \in \mathcal{L}(X)$ such that $A = BC$.*

Proof. Let $X_1 = \operatorname{add}\ker A$ be a direct complement of $\ker A$ in X which is isomorphic to X and let $\operatorname{add} A$ satisfy condition (2) in the definition of the class $\mathcal{KL}(X)$. Denote by Q the projector of X onto $\ker A$ parallel to X_1.

First, let us assume that $\ker A \cong \operatorname{add} A$, and let U be an isomorphism $\ker A \to \operatorname{add} A$. Consider the operator

$$B = A(I - Q) + UQ.$$

Clearly, $B \in \mathcal{L}(X)$ and $\operatorname{im} B = \operatorname{im} A \oplus \operatorname{add} A$, i.e., $\operatorname{add} B = \{0\}$. We claim that $\ker B = \{0\}$. Indeed, $Bx = 0$ means $A(I-Q)x + UQx = 0$, whence $A(I-Q)x = 0$ and $UQx = 0$. This implies that $(I-Q)x \in \ker A$ and $Qx = 0$, and hence that $x = 0$, because $(I-Q)x \in X_1$. Therefore, the operator B is unboundedly invertible. At the same time, B effects an isomorphism of $\ker A$ onto $\operatorname{add} A$ and an algebraic isomorphism of the subspace $X_1 = (I-Q)(X)$ onto the linear manifold $\operatorname{im} A$. Consequently, $B^{-1}A = I - Q$. Setting $C = B^{-1}A$, we have $A = BC$, where the operator B is unboundedly invertible and the operator $C = I - Q$, being a projector, is normally solvable.

Now assume that $\ker A \cong L_1$, where L_1 is a proper subspace of $\operatorname{add} A$, and let U be an isomorphism of $\ker A$ to L_1. Then $\operatorname{add} A = L_1 \oplus L_2$, where L_2 is a subspace in $\operatorname{add} A$. First, let us show that the subspace X_1 (a direct complement of $\ker A$ in X) can be written as $X_1 = L_3 \oplus X_2$, where L_3 and X_2 are subspaces of X_1, and $L_3 \cong L_2$, $X_2 \cong X_1$.

By the hypothesis of the theorem, the subspaces X_1 and $\operatorname{clos}\operatorname{im} A$ are isomorphic to X. Moreover, the linear manifold $\operatorname{clos}\operatorname{im} A \oplus L_2$ is a subspace of X (see Gohberg and Krupnik [2], Chap. II, Theorem 1.1), which coincides with a direct complement $\operatorname{add} L_1$ in the space X that is also isomorphic to X. Let U_1 denote an isomorphism of X_1 onto $\operatorname{clos}\operatorname{im} A \oplus L_2$. Since any isomorphism of Banach spaces takes a direct sum of subspaces into the direct sum of their images, $X_1 = L_3 \oplus X_2$, where $L_3 = U_1^{-1}(L_2)$ and $X_2 = U_1^{-1}(\operatorname{clos}\operatorname{im} A)$. Clearly, $X_2 \cong X_1$.

Let U_2 be an isomorphism of X_2 onto X_1, and let Q_1 denote the projector of X onto the subspace L_3 parallel to the subspace $\ker A \oplus X_2$. Consider the operator

$$B = AU_2(I - Q - Q_1) + U_1 Q_1 + UQ.$$

Clearly, $B \in \mathcal{L}(X)$. Moreover,

$$\operatorname{im} B = \operatorname{im} A \oplus L_2 \oplus L_1 = \operatorname{im} A \oplus \operatorname{add} A,$$

and so add $B = \{0\}$. We claim that $\ker B = \{0\}$. Indeed, let $Bx = 0$. Then

$$AU_2(I - Q - Q_1)x = U_1Q_1x = UQx = 0,$$

whence

$$(I - Q - Q_1)x = Q_1x = Qx = 0,$$

which implies that $x = 0$. Therefore, the operator B is unboundedly invertible.

Further, since $QQ_1 = Q_1Q = 0$, the operator $Q_2 = I - Q - Q_1$ is the projector of X onto the subspace X_2 parallel to the subspace $\ker A \oplus L_3$. Let P_1 denote the projector of X onto X_1 parallel to $\ker A$, and consider the operator $C = U_2^{-1}P_1$.

First, note that $C \in \mathcal{L}(X)$. Second, $C(X) = U_2^{-1}(P_1(X)) = U_2^{-1}(X_1) = X_2$, that is, the operator C is normally solvable. And third,

$$\begin{aligned} BC &= (AU_2Q_2 + U_1Q_1 + UQ)U_2^{-1}P_1 \\ &= AU_2Q_2U_2^{-1}P_1 = AU_2U_2^{-1}P_1 = AP_1 = A. \end{aligned}$$

This proves our theorem in the case where $\ker A$ is isomorphic to a proper subspace of add A.

Finally, suppose that add $A \cong L_1$, where L_1 is a subspace of $\ker A$. Then $\ker A = L_1 \oplus L_2$, where L_2 is a subspace of X. In this case

$$B = AU_1Q_2 + UQ_1, \quad C = U_1^{-1}P_1,$$

where U is an isomorphism of L_1 onto add A, U_1 is an isomorphism of $L_2 \oplus X_1$ onto X_1, Q_1 is the projector of X onto L_1 parallel to $L_2 \oplus X_1$, $Q_2 = I - Q_1$, P_1 is the projector of X onto X_1 parallel to $\ker A$, and X_1 is, as above, a direct complement of $\ker A$ in X. As in the preceding case, one can verify that the operator B is unboundedly invertible, the operator C is normally solvable, and $A = BC$. □

Let us assume again that $A \in \mathcal{L}(X)$ and that its image is not closed. An operator N defined on a dense (in X) linear manifold X_0 such that $\operatorname{im} A \subset X_0$ and acting in X is called a *left normalizer* of the operator A if the operator NA belongs to $\mathcal{L}(X)$ and is normally solvable.

It follows from Theorem 6.10 that every operator A in the class $\mathcal{KL}(X)$ has a left normalizer $N = B^{-1}$. It is clear that the operator B, and together with it the left normalizer N, is not uniquely determined. In the investigation of concrete operators with nonclosed image the construction of a left normalizer of maximal simplicity is a main task, because such a normalizer is necessary for defining the norm and hence for describing convergence in the space DX.

One can show that the operators A of class $\mathcal{KL}(X)$ admit a representation $A = BC$, where $B, C \in \mathcal{L}(X)$, the operator B is normally solvable and the operator C is unboundedly invertible, and one can also establish the analogues of Theorems 6.8 and 6.9 for right normalization. We will not do this here, because the class of operators studied below is considered in reflexive Banach spaces and is "closed" under passage to the adjoint (conjugate) operator. For that reason, Theorems 6.7–6.10 provide a sufficient basis for the normalization of these operators.

Normalization of Toeplitz operators

6.3 Symbols with polynomial degeneracy

In this section it will be assumed that the symbol a of the Toeplitz operator $T(a)$ can be written in the form $a = wb$, where w is a polynomial whose roots lie on the curve Γ and $b \in \mathrm{fact}_\Gamma(\infty, p, \varrho)$. Our plan for investigating the operator $T(a)$ goes as follows.

Let $\{\tau_1, \dots, \tau_m\}$ be the roots of the polynomial w and let n_k be the multiplicity of the root τ_k, $k = 1, \dots, m$. For simplicity, we will assume that $0 \in \mathcal{D}_\Gamma^+$. Then the function a can be written in the form

$$a = \omega^- a_1, \tag{6.4}$$

where

$$\omega^-(t) = \prod_{k=1}^{m} (1 - \tau_k t^{-1})^{n_k} \in L_\infty^-(\Gamma), \tag{6.5}$$

and

$$a_1(t) \in \mathrm{fact}_\Gamma(\infty, p, \varrho).$$

The operator $T(a)$, considered in the space $L_p^+(\Gamma, \varrho)$, can be decomposed as $T(a) = T(\omega^-)T(a_1)$, where the operator $T(\omega^-)$ is unboundedly invertible and the operator $T(a_1)$ has a continuous inverse in $L_p^+(\Gamma, \varrho)$, at least from one side. Hence, finding ourselves under the conditions of Theorem 6.9, we can carry out a left normalization of the operator $T(a)$.

Let us implement this plan.

We say that a function $f \in L_p(\Gamma, \varrho)$ has a *first-order derivative in the mean that is summable in the pth power with weight* ϱ (or *that belongs to the space* $L_p(\Gamma, \varrho)$) at the point $t_0 \in \Gamma$ if t_0 is a Lebesgue point for f and

$$f_{t_0}^{(1)}(t) = \frac{f(t) - f(t_0)}{t - t_0} \in L_p(\Gamma, \varrho).$$

Higher-order derivatives in the mean are defined inductively as follows. We say that the function $f \in L_p(\Gamma, \varrho)$ has a kth *derivative in the mean that belongs to the space* $L_p(\Gamma, \varrho)$ at the point $t_0 \in \Gamma$ if all the derivatives in the mean $f_{t_0}^{(j)}(t)$, $j = 1, 2, \dots, k-1$, exist at t_0, if they belong to $L_p(\Gamma, \varrho)$, and if t_0 is a Lebesgue point for the function $f_{t_0}^{(k-1)}(t)$ and

$$f_{t_0}^{(k)}(t) = \frac{f_{t_0}^{(k-1)}(t) - f_{t_0}^{(k-1)}(t_0)}{t - t_0} \in L_p(\Gamma, \varrho).$$

Let us remark that a $k-1$ times continuously differentiable function on Γ whose $(k-1)$th derivative is absolutely continuous and whose kth derivative lies in

$L_p(\Gamma, \varrho)$ has a kth derivative in the mean that belongs to $L_p(\Gamma, \varrho)$ at any point $t_0 \in \Gamma$.

We let $DL_p(\Gamma, \varrho)$ denote the linear manifold of all functions from $L_p(\Gamma, \varrho)$ which have derivatives in the mean of order up to and including n_k, $k = 1, 2, \ldots, m$, that belong to $L_p(\Gamma, \varrho)$ at the points τ_k. Notice that multiplication by sufficiently smooth functions, in particular, by functions in $C^\infty(\Gamma)$, leaves $DL_p(\Gamma, \varrho)$ invariant. In $DL_p(\Gamma, \varrho)$ we introduce the finite-dimensional operator $\mathcal{Y}_{n-1}$ that sends a function f into its Hermite interpolation polynomial $\mathcal{Y}_{n-1}(f)$ with interpolation nodes τ_k of respective multiplicities n_k:

$$\mathcal{Y}_{n-1}(f) = \sum_{k=1}^{m} \sum_{j=0}^{n_k-1} f_{\tau_k}^{(j)}(\tau_k) e_{kj}(t), \tag{6.6}$$

where

$$e_{kj}(t) = \sum_{l=j}^{n_k-1} \frac{1}{j!(l-j)!} \left[\frac{(t-\tau_k)^{n_k}}{w(t)} \right]^{(l-j)} \Bigg|_{t=\tau_k} \frac{w(t)}{(t-\tau_k)^{n_k-l}}, \tag{6.7}$$

$w(t) = t^n \omega^-(t)$ and $n = \sum_{k=1}^{m} n_k$.

Notice that $\deg \mathcal{Y}_{n-1}(f) < n$ and $\mathcal{Y}_{n-1}(u) = u$ for every polynomial u such that $\deg u < n$. Let us also introduce the operator N_1 which act according to the rule

$$N_1 = \frac{1}{w}[I - \mathcal{Y}_{n-1}], \tag{6.8}$$

defined at least on $C^\infty(\Gamma)$.

Here are some properties of the space $DL_p(\Gamma, \varrho)$.

Proposition 6.1 *The following conditions are equivalent:*

(a) $f \in DL_p(\Gamma, \varrho)$;

(b) *the derivatives in the mean* $f_{\tau_k}^{(j)}(t)$, $j = 0, 1, \ldots, n_k - 1$, $k = 1, 2, \ldots, m$, *exist and* $N_1(f) \in L_p(\Gamma, \varrho)$;

(c) $f = wg + u$, *where* $w(t) = \prod_{k=1}^{m} (t - \tau_k)^{n_k}$, $g \in L_p(\Gamma, \varrho)$ *and* u *is a polynomial with complex coefficients of degree* $\deg u < n$.

Proof. (a) $\Longrightarrow$ (c). By the definition of the derivative in the mean, $f = (t - \tau_1)^{n_1} h_1 + u_1$, where $h_1 \in L_p(\Gamma, \varrho)$ and u_1 is a polynomial with $\deg u_1 < n_1$. Let $\chi_1 + \chi_2 \equiv 1$ be a sufficiently smooth partition of unity such that $\chi_j = 0$ in a sufficiently small neighborhood of τ_j, $j = 1, 2$. Then $\chi_j(t - \tau_j)^{-n_j}$ are smooth functions, multiplication by which leaves $DL_p(\Gamma, \varrho)$ invariant. Therefore,

$$\chi_1 h_1 = \chi_1 (t - \tau_1)^{-n_1} (f - u_1) \in DL_p(\Gamma, \varrho),$$

and hence

$$\chi_1 h_1 = (t - \tau_2)^{n_2} h_2 + u_2,$$

where $h_2 \in L_p(\Gamma, \varrho)$ and u_2 is a polynomial with $\deg u_2 < n_2$. Using this relation, we obtain

$$\begin{aligned} f &= (t-\tau_1)^{n_1}\chi_1 h_1 + (t-\tau_1)^{n_1}\chi_2 h_1 + u_1 \\ &= (t-\tau_1)^{n_1}(t-\tau_2)^{n_2}h_2 + (t-\tau_1)^{n_1}(t-\tau_2)^{n_2}[\chi_2 h_1 (t-\tau_2)^{-n_2}] \\ &\quad +(t-\tau_1)^{n_1}u_2 + u_1 = (t-\tau_1)^{n_1}(t-\tau_2)^{n_2} g + u, \end{aligned}$$

where $g \in L_p(\Gamma, \varrho)$ and $\deg u < n_1 + n_2$. After that one proceeds by induction.

(c) $\Longrightarrow$ (a). It suffices to show that $f = wg \in DL_p(\Gamma, \varrho)$ whenever $g \in L_p(\Gamma, \varrho)$. We have $f = (t - \tau_k)h$, where $h \in L_p(\Gamma, \varrho)$. Therefore, $t = \tau_k$ is a Lebesgue point for the function f and $f(\tau_k) = 0$ (see Section 1.7). Similarly, $f^{(j)}_{\tau_k}(\tau_k) = 0$, $j = 1, 2, \ldots, n_k - 1$. By the definition of the derivative of order n_k in the mean, we obtain that $f^{(n_k)}_{\tau_k}(t) = (t - \tau_k)^{-n_k} wg \in L_p(\Gamma, \varrho)$, $k = 1, 2, \ldots m$.

(b) $\Longleftrightarrow$ (c). If $g = N_1(f) \in L_p(\Gamma, \varrho)$, then $f = wg + \mathcal{Y}_{n-1}(f)$, i.e., the implication (b) $\Longrightarrow$ (c) is true. Conversely, let $f = wg + u$, where $g \in L_p(\Gamma, \varrho)$ and the degree of the polynomial u is smaller than n. Since all such polynomials are annihilated by the operator N_1, we conclude that $N_1(f) = N_1(wg) = g \in L_p(\Gamma, \varrho)$, i.e., the implication (c) $\Longrightarrow$ (b) is true as well. □

Remark 6.3 From Proposition 6.1 it follows that any $f \in DL_p(\Gamma, \varrho)$ can be written in the form $f = wN_1(f) + \mathcal{Y}_{n-1}(f)$.

Proposition 6.2 *The space $DL_p(\Gamma, \varrho)$ is invariant under the operator S_Γ. Moreover, the operator N_1 given by* (6.8) *commutes on $DL_p(\Gamma, \varrho)$ with the operators S_Γ and $P_\Gamma^\pm$:*

$$N_1 S_\Gamma = S_\Gamma N_1, \quad N_1 P_\Gamma^\pm = P_\Gamma^\pm N_1.$$

Proof. Let $f \in DL_p(\Gamma, \varrho)$. Then, by Proposition 6.1, $f = wg + u$, where $g \in L_p(\Gamma, \varrho)$ and u is a polynomial of degree $\deg u < n$. Hence, since $S_\Gamma(u) = u$, we obtain

$$S_\Gamma(f) = S_\Gamma(wg) + u = wS_\Gamma(g) + u + u_1,$$

where

$$u_1(t) = \frac{1}{\pi i}\int_\Gamma \frac{w(\tau) - w(t)}{\tau - t} g(\tau) d\tau$$

is also a polynomial and $\deg u_1 < n$. But then, by Proposition 6.1, $S_\Gamma(f) \in DL_p(\Gamma, \varrho)$. Moreover,

$$N_1(S_\Gamma(f)) = S_\Gamma(g) = S_\Gamma(N_1(f)).$$

We conclude that for any $f \in DL_p(\Gamma, \varrho)$,

$$P_\Gamma^\pm(N_1(f)) = N_1(P_\Gamma^\pm(f)).$$ □

Now let us turn to the operator $T(a)$ with a symbol a of the form (6.4), (6.5).

Proposition 6.3 *The operator $T(\omega^-)$ with the symbol ω^- given by* (6.5) *is unboundedly invertible in the space $L_p^+(\Gamma, \varrho)$ and $\operatorname{im} T(\omega^-) = P^+(DL_p(\Gamma, \varrho))$. The corresponding unbounded inverse operator $N = [T(\omega^-)]^{-1}$ acts according to the rule*

$$N = \frac{1}{\omega^-}[I - t^{-n}\mathcal{Y}_{n-1}(t^n I)]. \tag{6.9}$$

Proof. First let us show that the kernel of the operator $T(\omega^-)$ is trivial. Let $f \in \ker T(\omega^-)$. Then $[T(\omega^-)]f = 0$, i.e., $P^+\omega^- f = \omega^- f - P^-\omega^- f = 0$. Denote $g = wf = t^n P^-\omega^- f$. Then $g = P^+ g = P^+ t^n P^- \omega^- f$ is a polynomial satisfying the following conditions: $\deg g < n$,

$$g_{\tau_k}^{(j)}(\tau_k) = g^{(j)}(\tau_k) = 0, \quad k = 1, 2, \ldots, m, \quad j = 0, 1, \ldots, n_k - 1.$$

It follows that $g \equiv 0$, which in turn implies that $f = 0$ almost everywhere on Γ. This shows that, indeed, $\ker T(\omega^-) = \{0\}$.

Now let us describe the image of the operator $T(\omega^-)$. Let $f \in \operatorname{im} T(\omega^-)$, i.e., $f = P^+\omega^- g$ for some function $g \in L_p^+(\Gamma, \varrho)$. Then,

$$f = \omega^- g - P^-\omega^- g = wP^+ t^{-n} g + wP^- t^{-n} g - P^-\omega^- g.$$

Applying the projector P^+ to both sides of this equality, we obtain that

$$f = wh + u, \quad \text{where } h = P^+ t^{-n} g = N_1(f) \in L_p^+(\Gamma, \varrho),$$

and that $u = P^+ w P^- t^{-n} g$ is a polynomial of degree $\deg u < n$. By Proposition 6.1, $\operatorname{im} T(\omega^-) \subset P^+(DL_p(\Gamma, \varrho))$.

Conversely, let $f \in P^+(DL_p(\Gamma, \varrho))$. Then $N_1(f) \in L_p^+(\Gamma, \varrho)$ (by Propositions 6.1 and 6.2), whence $N(f) \in L_p^+(\Gamma, \varrho)$, because for every $f \in DL_p(\Gamma, \varrho)$ we have

$$N(f) = t^n N_1(f) + K_n(f); \tag{6.10}$$

here the finite-dimensional operator K_n acts according to the rule

$$\begin{aligned} K_n(f) &= N(f) - t^n N_1(f) = w^{-1}[t^n \mathcal{Y}_{n-1}(f) - \mathcal{Y}_{n-1}(t^n f)] \\ &= N_1[t^n \mathcal{Y}_{n-1}(f)] = \sum_{j=0}^{n-1} c_j(f) t^j, \end{aligned}$$

where $c_j(f)$, $j = 1, 2, \ldots, n-1$, are linear functionals of f. Since

$$(T(\omega^-)N)(f) = P^+\omega N(f) = P^+[f - t^{-n}\mathcal{Y}_{n-1}(t^n f)] = f,$$

we conclude that $f \in \operatorname{im} T(\omega^-)$. Thus, we have shown that

$$\operatorname{im} T(\omega^-) = P^+(DL_p(\Gamma, \varrho)).$$

It remains to show that $\operatorname{clos}\operatorname{im} T(\omega^-) = L_p^+(\Gamma, \varrho)$. It is known that

$$\operatorname{clos} P^+(C^\infty(\Gamma)) = L_p^+(\Gamma, \varrho).$$

At the same time, if $f \in C^\infty(\Gamma)$, then

$$N_1(f) = w^{-1}[f - \mathcal{Y}_{n-1}(f)] \in C^\infty(\Gamma).$$

By Propositions 6.1 and 6.2,

$$P^+(C^\infty(\Gamma)) \subset P^+(DL_p(\Gamma, \varrho)).$$

Therefore,

$$\operatorname{clos} P^+(C^\infty(\Gamma)) = \operatorname{clos} P^+(DL_p(\Gamma, \varrho)) = L_p^+(\Gamma, \varrho).$$

By Theorem 6.6, the operator $T(\omega^-)$ is unboundedly invertible in the space $L_p^+(\Gamma, \varrho)$. The above arguments also show that the operator N given by (6.9) is its unbounded right inverse and hence also its unbounded inverse. □

Let us denote by $DL_p^+(\Gamma, \varrho)$ the space of all functions $f \in L_p^+(\Gamma, \varrho)$ that can be written in the form $f = wg + u$, where $g \in L_p^+(\Gamma, \varrho)$ and u is a polynomial of degree $\deg u < n$. Notice that

$$P^+(DL_p(\Gamma, \varrho)) = DL_p^+(\Gamma, \varrho).$$

Indeed, let $f \in DL_p(\Gamma, \varrho)$ and write $f = wg+u$ with $g \in L_p(\Gamma, \varrho)$ and a polynomial u of degree $\deg u < n$. Then $P^+f = wP^+g + P^+wP^-g + u$, where P^+wP^-g+u is a polynomial of degree smaller than n. Therefore, $P^+(DL_p(\Gamma, \varrho)) \subset DL_p^+(\Gamma, \varrho)$. The opposite inclusion is obvious. From this one easily obtains the following analogue of Proposition 6.1 for the space $DL_p^+(\Gamma, \varrho)$.

Proposition 6.4 *The following conditions are equivalent:*

(a) $f \in DL_p^+(\Gamma, \varrho)$;

(b) *the derivatives in the mean* $f_{\tau_k}^{(j)}(t)$, $j = 0, 1, \ldots, n_k - 1$, $k = 1, 2, \ldots, m$, *exist and* $N_1(f) \in L_p^+(\Gamma, \varrho)$;

(c) $f = wg + u$, *where* $g \in L_p^+(\Gamma, \varrho)$ *and* u *is a polynomial with complex coefficients of degree* $\deg u < n$.

Let us equip the space $DL_p^+(\Gamma, \varrho)$ with the norm

$$\|f\|_2 = \|N_1(f)\| + \alpha(f), \tag{6.11}$$

where $\|\cdot\| = \|\cdot\|_{p,\varrho}$ and $\alpha(f) = \sum_{k=1}^m \sum_{j=0}^{n_k-1} |f_{\tau_k}^{(j)}(\tau_k)|$. The norm $\|f\|_2$ given by (6.11) is equivalent to the norm $\|f\|_1 = \|f\| + \|N(f)\|$ given by (6.3). Indeed, let $\mathfrak{L}_n = \operatorname{span}\{1, t, \ldots, t^{n-1}\}$ and let $f \in \mathfrak{L}_n$. The functional α is a norm on

$\mathfrak{L}_n$, and since in a finite-dimensional space all norms are equivalent, $\alpha(f) \asymp \|f\|$ whenever $f \in \mathfrak{L}_n$. Observing that $f \in DL_p^+(\Gamma, \varrho)$ if and only if $f = wg + u$, where $g \in L_p^+(\Gamma, \varrho)$, $u \in \mathfrak{L}_n$, $\alpha(f) = \alpha(u)$, $N(f) = t^n g$, and $N_1(f) = g$, we have

$$\begin{aligned} \|f\|_1 &= \|f\| + \|N(f)\| = \|wg + u\| + \|t^n g\| \le c_1\|g\| + \|u\| \\ &\le c_1\|N_1(f)\| + c_2\alpha(u) \le c_3\|f\|_2. \end{aligned}$$

Furthermore,

$$\begin{aligned} \|f\|_2 &= \|N_1(f)\| + \alpha(f) \le \|g\| + c_4\|u\| = \|g\| + c_4\|f - wg\| \\ &\le c_4\|f\| + c_5\|t^n g\| \le c_6\|f\|_1. \end{aligned}$$

The next result describes a left normalization of Toeplitz operators with polynomially degenerate symbol.

Theorem 6.11 *Suppose that the function* $a \in L_\infty(\Gamma)$ *is representable in the form* (6.4), (6.5), *where* $a_1 \in \mathrm{fact}(\infty, p, \varrho)$, h *satisfies condition* (2.31) *and* $\mathrm{ind}\, b = 0$ ($a_1 = bh^{-1}$ *or* $a_1 = bh$, *where* $h \in L_\infty^+(\Gamma) \cap GL_\infty(\Gamma)$, $b = b^+b^- \in \mathrm{fact}(p, \varrho)$).

Then $T(a) \in \mathcal{L}(L_p^+(\Gamma, \varrho), DL_p^+(\Gamma, \varrho))$.

If $a_1 = bh^{-1}$, *then the operator* $T(a)$ *is right invertible,* $\dim\ker T(a) = \infty$, *and the operator*

$$[T(a)]^{-1} = \frac{h}{b^+}P^+\frac{1}{b^-}N \in \mathcal{L}(DL_p^+(\Gamma, \varrho), L_p^+(\Gamma, \varrho)),$$

where N *is given by* (6.9), *is a right inverse of* $T(a)$. *Moreover,* $\ker T(a) = \ker T(a_1)$.

If $a_1 = bh$, *then the operator* $T(a)$ *is left invertible,* $\dim \mathrm{coker}\, T(a) = \infty$, *and the operator*

$$[T(a)]^{-1} = P^+\frac{1}{b^+h}P^+\frac{1}{b^-}N \in \mathcal{L}(DL_p^+(\Gamma, \varrho), L_p^+(\Gamma, \varrho))$$

and is a left inverse of $T(a)$. *Moreover, for a function* f *to belong to* $\mathrm{im}\, T(a)$ *it is necessary and sufficient that*

$$\int_\Gamma \psi_j(t)(Nf)(t)dt = 0, \quad j = 1, 2, \ldots,$$

where the functions ψ_j *are the same as in Theorem* 2.6.

Proof. Since in the present case $T(a)$ can be written as

$$T(a) = T(\omega^-)T(a_1), \tag{6.12}$$

where the operator $T(\omega^-)$ is unboundedly invertible and the operator $T(a_1)$ has a continuous left or right inverse in $L_p^+(\Gamma, \varrho)$, Theorem 6.11 is a corollary of Theorem 2.6, Proposition 6.3, and Theorem 6.9. It remains to observe that the condition $f \in \mathrm{im}\, T(a)$ is equivalent to $N(f) \in \mathrm{im}\, T(a_1)$, because $N = [T(\omega^-)]^{-1}$. □

One of the peculiarities of Toeplitz operators whose symbols degenerate on the contour Γ is that the corresponding homogeneous equation $[T(a)](f) = 0$ may have generalized solutions. To show this, we will examine the simplest case of the operator $T(\omega^-)$ in the space $(DL_p^+(\Gamma_0, \varrho))^*$.

First let us describe the general form of a linear functional from the space $(DL_p^+(\Gamma_0, \varrho))^*$.

Proposition 6.5 *Let $f \in (DL_p^+(\Gamma_0, \varrho))^*$. Then one can find complex numbers c_{kj}, $j = 0, 1, \ldots, n_k - 1$, $k = 1, 2, \ldots, m$ and functions $g \in L_q^+(\Gamma_0, \varrho^{1-q})$, where $q = p(p-1)^{-1}$, such that for any $\varphi \in DL_p^+(\Gamma, \varrho)$,*

$$\langle f, \varphi \rangle = \sum_{k=1}^{m} c_{kj} \overline{\varphi_{\tau_k}^{(j)}(\tau_k)} + \langle g, N_1(\varphi) \rangle, \tag{6.13}$$

where the operator N_1 has the form (6.8).

Proof. First of all let us remark that any functional of the form (6.13) lies in the space $(DL_p^+(\Gamma_0, \varrho))^*$, because

$$\begin{aligned} |\langle f, \varphi \rangle| &\le \sum_{k=1}^{m} \sum_{j=0}^{n_k - 1} |c_{kj}| |\varphi_{\tau_k}^{(j)}(\tau_k)| + |\langle g, N_1(\varphi) \rangle| \\ &\le c_1 \alpha(\varphi) + \|g\|_{q, \varrho^{q-1}} \|N_1(\varphi)\| \le c_2 \|\varphi\|_2. \end{aligned}$$

Further, by Theorem 6.8, the operator $T(\omega^-)$ lies in the space

$$\mathcal{L}(L_p^+(\Gamma_0, \varrho), DL_p^+(\Gamma_0, \varrho))$$

and has a continuous inverse. It follows that the operator

$$T\left(\overline{\omega^-}\right) = [T(\omega^-)]^* \in \mathcal{L}((DL_p^+(\Gamma_0, \varrho))^*, (L_p^+(\Gamma_0, \varrho))^*)$$

also has a continuous inverse. Since $(L_p^+(\Gamma_0, \varrho))^* \cong L_q^+(\Gamma_0, \varrho^{1-q})$, this shows that if $f \in (DL_p^+(\Gamma_0, \varrho))^*$, then $\overline{\omega^-} f = \left[T\left(\overline{\omega^-}\right)\right](f) \in L_q^+(\Gamma_0, \varrho^{1-q})$.

Now let $f \in (DL_p^+(\Gamma_0, \varrho))^*$ and $\varphi \in DL_p^+(\Gamma_0, \varrho)$. By Proposition 6.4, Remark 6.3 and equalities (6.6) and (6.7),

$$\begin{aligned} \langle f, \varphi \rangle &= \langle f, w N_1(\varphi) + \mathcal{Y}_{n-1}(\varphi) \rangle \\ &= \sum_{k=1}^{m} \sum_{j=0}^{n_k - 1} \langle f, e_{kj} \rangle \overline{\varphi_{\tau_k}^{(j)}(\tau_k)} + \langle P^+ \overline{w} f, N_1(\varphi) \rangle, \end{aligned}$$

where

$$g = P^+ \overline{w} f = P^+ t^{-n} \overline{\omega^-} f = P^+ t^{-n} P^+ \overline{\omega^-} f \in L_q^+(\Gamma_0, \varrho^{1-q}).$$

This proves the validity of representation (6.13). $\square$

Remark 6.4 In distribution theory the functionals $\delta^{(j)+}(t-\tau_k)$ are defined via the equality

$$\langle \delta^{(j)+}(t-\tau_k), \varphi(t)\rangle = \langle \delta^{(j}(t-\tau_k), (P^+\varphi)(t)\rangle = (-1)^j \overline{(P^+\varphi)^{(j)}(\tau_k)}.$$

Thus the linear functional from $(DL_p^+(\Gamma_0, \varrho))^*$ can also be written as

$$f = \sum_{k=1}^{m} \sum_{j=0}^{n_k-1} c_{kj}\delta^{(j)+}(t-\tau_k) + N_1^*(g),$$

where $c_{kj} \in \mathbb{C}$, $g \in L_q^+(\Gamma_0, \varrho^{1-q})$, and $\langle N_1^*(g), \varphi\rangle = \langle g, N_1(\varphi)\rangle$.

We are now ready to describe the action of the operator $T(\omega^-)$ in $(DL_p^+(\Gamma_0, \varrho))^*$.

Proposition 6.6 *Let ω^- be the symbol* (6.5). *The operator $T(\omega^-)$ defined by*

$$\langle [T(\omega^-)](f), \varphi\rangle = \langle f, [T(\overline{\omega^-}](\varphi)\rangle$$

belongs to $\mathcal{L}((DL_p^+(\Gamma_0, \varrho))^*, L_q^+(\Gamma_0, \varrho^{1-q}))$ and is right invertible. One of its right inverses is given by the formula

$$[T(\omega^-)]^{-1} = \overline{\mu} N^* t^n I \in \mathcal{L}(L_q^+(\Gamma_0, \varrho^{1-q}), (DL_p^+(\Gamma_0, \varrho)^*),$$

where $\mu = \prod_{k=1}^m (-\tau_k)^{n_k}$, $\langle N^*(f), \varphi\rangle = \langle f, N(\varphi)\rangle$, and the operator N is given by (6.9). *Moreover,* $\dim\ker T(\omega^-) = n$ *and* $\ker T(\omega^-) = L_n$, *where*

$$L_n \stackrel{\text{def}}{=} \operatorname{span}\{\delta^{(j)+}(t-\tau_k), j = 0, 1, \dots, n_k - 1; k = 1, 2, \dots, m\}.$$

Proof. Since

$$\overline{\omega^-(t)} = \prod_{k=1}^{m} (1 - \overline{\tau}_k t)^{n_k} = \prod_{k=1}^{m} (-\overline{\tau_k})^{n_k} (t-\tau_k)^{n_k} = \overline{\mu} t^n \omega^-(t),$$

Theorem 6.9 guarantees that

$$T\left(\overline{\omega^-}\right) = T(\omega^-) T(\overline{\mu} t^n) \in \mathcal{L}(L_p^+(\Gamma_0, \varrho), DL_p^+(\Gamma_0, \varrho)).$$

Next, by Theorem 6.8 and Proposition 6.3,

$$T(\omega^-) = \left[T\left(\overline{\omega^-}\right)\right]^* \in \mathcal{L}((DL_p^+(\Gamma_0, \varrho))^*, L_q^+(\Gamma_0, \varrho^{1-q}))$$

and

$$\overline{\mu} N^* t^n I \in \mathcal{L}(L_q^+(\Gamma_0, \varrho^{1-q}), (DL_p^+(\Gamma_0, \varrho))^*).$$

Moreover, the operator $\overline{\mu}N^*t^nI$ is a right inverse for the operator $T(\omega^-)$. Indeed, for any $f \in L_q^+(\Gamma_0, \varrho^{1-q})$ and $\varphi \in L_p^+(\Gamma_0, \varrho)$,

$$\begin{aligned}\langle T(\omega^-)(\overline{\mu}N^*t^n)f, \varphi\rangle &= \langle f, T(t^{-n})(\mu N)T(\overline{\omega^-})\varphi\rangle \\ &= \langle f, T(t^{-n})(\mu N)T(\omega^-)T(\overline{\mu}t^n)\varphi\rangle = \langle f, \varphi\rangle.\end{aligned}$$

Further, for any j, $0 \le j \le n_k - 1$, and $k = 1, \dots, m$

$$\begin{aligned}\langle T(\omega^-)\delta^{(j)+}(t-\tau_k), \varphi\rangle &= \langle \delta^{(j)}(t-\tau_k), \overline{\omega^-}\varphi\rangle \\ &= (-1)^j(\overline{\mu w(t)\varphi(t)})^{(j)}|_{t=\tau_k} = 0,\end{aligned}$$

i.e., $L_n \subset \ker T(\omega^-)$.

Conversely, if $f \in \ker T(\omega^-)$, then, for any $\varphi \in L_p^+(\Gamma_0, \varrho)$,

$$\langle [T(\omega^-)](f), \varphi\rangle = \langle f, \overline{\mu}w\varphi\rangle = 0.$$

Using Proposition 6.4, Remark 6.3, and equalities (6.6) and (6.7), we conclude that for any function $\varphi \in DL_p^+(\Gamma_0, \varrho)$,

$$\begin{aligned}\langle f, \varphi\rangle &= \langle f, wN_1(\varphi) + \mathcal{Y}_{n-1}(\varphi)\rangle = \langle f, \mathcal{Y}_{n-1}(\varphi)\rangle \\ &= \sum_{k=1}^{m}\sum_{j=0}^{n_k-1}\langle f, e_{kj}\rangle\overline{\varphi_{\tau_k}^{(j)}(\tau_k)} = \langle \sum_{k=1}^{m}\sum_{j=0}^{n_k-1} c_{kj}\delta^{(j)+}(t-\tau_k), \varphi\rangle,\end{aligned}$$

i.e., $f \in L_n$. Therefore, $\ker T(\omega^-) = L_n$. □

6.4 Symbols with locally-polynomial degeneracy

In this section we weaken the assumptions on the function ω^- figuring in the representation (6.4) of the symbol a of the operator $T(a)$, which will now be studied in the space $L_p^+(\mathbb{R})$, $1 < p < \infty$. We exhibit a class of symbols (a priori they are infinitely differentiable outer functions belonging to the algebra $H^\infty(\Pi_\pm)$ which degenerate on separated real sets) for which the space $DL_p^+(\mathbb{R})$ admits, like in Section 6.3, two equivalent descriptions: a structural description in the spirit of Proposition 6.4, (c) and one in terms of the local behavior of functions in a neighborhood of the degeneracy set of the symbol. After that we will prove analogues of Proposition 6.3 and Theorem 6.11 for Toeplitz operators of the class alluded to above.

A function $h^\pm \in H^\infty(\Pi_\pm)$ is called a *model function of class* $\mathcal{U}^\pm(\mathbb{R})$ if $(h^\pm)^{-1} \in H^\infty(\Pi_\mp)$. Accordingly, the subspace

$$\mathcal{M}(h^\pm) = P^\pm h^\pm P^\mp(L_p(\mathbb{R}))$$

is called the *model subspace* in $L_p(\mathbb{R})$, $1 < p < \infty$, generated by the model function $h^\pm$.

The following assertion is a simple corollary of Propositions 3.1–3.3.

Proposition 6.7 *Let* $h^{\pm} \in \mathcal{U}^{\pm}(\mathbb{R})$. *Then*

(1) $\mathcal{M}(h^{\pm}) = \ker P^{\pm}\overline{h^{\pm}}I$ *in* $L_p^{\pm}(\mathbb{R})$;

(2) $f^{\pm} \in \mathcal{M}(h^{\pm})$ *if and only if*

$$(f^{\pm})(x) = \frac{1}{2\pi i}\int_{\mathbb{R}} \frac{h^{\pm}(t) - h^{\pm}(x)}{t - x} g^{\mp}(t)dt,$$

where $g^{\mp} \in L_p^{\mp}(\mathbb{R})$;

(3) $L_p^{\pm}(\mathbb{R}) = \mathcal{M}(h^{\pm}) \oplus (h^{\pm}I)(L_p^{\pm}(\mathbb{R}))$.

Notice also that $h^{\pm} \in \mathcal{U}^{\pm}(\mathbb{R})$ if and only if $\overline{h^{\pm}} \in \mathcal{U}^{\mp}(\mathbb{R})$, and consequently $\mathcal{M}(\overline{h^{\pm}}) \subset L_p^{\mp}(\mathbb{R})$. Moreover, $h^{\pm} \in \mathcal{U}^{\pm}(\mathbb{R})$ if and only if $(h^{\pm})^{-1} \in \mathcal{U}^{\mp}(\mathbb{R})$, and consequently $\mathcal{M}((h^{\pm})^{-1}) \subset L_p^{\mp}(\mathbb{R})$.

Let $C_B^{\infty}(\mathbb{R})$ denote the algebra of infinitely differentiable functions on the real line $\mathbb{R}$ that are bounded together with all their derivatives, and let $H_{\infty}^{\infty}(\Pi_{\pm}) \overset{\text{def}}{=} H^{\infty}(\Pi_{\pm}) \cap C_B^{\infty}(\mathbb{R})$.

We are now ready to describe the classes $\mathcal{K}^{\pm}(\mathbb{R})$ of what we call *generators of zeros*, i.e., functions $\omega^{\pm} \in H_{\infty}^{\infty}(\Pi_{\pm})$ whose sets of zeros, denoted below by $\zeta(\omega^{\pm})$, may serve as degeneracy sets of the symbols of the singular integral operators under study. Namely, let $\mathcal{K}^{\pm}(\mathbb{R})$ denote the set of functions $\omega^{\pm}$ on $\mathbb{R}$ with the following properties:

(i) $\omega^{\pm} \in H_{\infty}^{\infty}(\Pi_{\pm})$;

(ii) there exists a model function $h^{\pm} \in \mathcal{U}^{\pm}(\mathbb{R}) \cap C_B^{\infty}(\mathbb{R})$ such that $u^{\pm} = h^{\pm}\omega^{\mp} \in H_{\infty}^{\infty}(\Pi_{\pm})$;

(iii) the set $\zeta(\omega^{\pm}) = \{x_j\}$ of the zeros of $\omega^{\pm}$ is a subset of $\mathbb{R}$, $(\omega^{\pm})'(x_j) \neq 0$ for all j, and $\zeta(\omega^{\pm})$ is an *interpolation set* in the subspace $\mathcal{M}(h^{\pm})$, i.e.,

   (a) if $f^{\pm} \in \mathcal{M}(h^{\pm})$, then $\zeta(\omega^{\pm})$ consists of Lebesgue points of the function $f^{\pm}$ and $\{f^{\pm}(x_j)\} \in l_p(\mathbb{Z})$;

   (b) $f^{\pm} \in \mathcal{M}(h^{\pm})$ if and only if

$$f^{\pm}(x) = [\mathcal{Y}_{\omega^{\pm}}(f^{\pm})](x) = \sum_{j\in\mathbb{Z}} f^{\pm}(x_j)\frac{\omega^{\pm}(x)}{(\omega^{\pm})'(x_j)(x - x_j)},$$

   where the series converges in $L_p(\mathbb{R})$;

   (c) the mapping defined in items (a) and (b) is an isomorphism $l_p(\mathbb{Z}) \cong \mathcal{M}(h^{\pm})$, and

$$\|\{f^{\pm}(x_j)\}\|_{l_p(\mathbb{Z})} \asymp \|f^{\pm}\|_{L_p(\mathbb{R})}.$$

Next, let us introduce the spaces $X(\omega^{\pm})$ and the operators $N_{\omega^{\pm}}$ by

$$X(\omega^{\pm}) = \{f \in L_p(\mathbb{R}) : f = \omega^{\pm} f_1 + f_2^{\pm}, \text{ where } f_1 \in L_p(\mathbb{R}), f_2^{\pm} \in \mathcal{M}(h^{\pm})\}, \quad (6.14)$$

$$N_{\omega^{\pm}} = (\omega^{\pm})^{-1}(I - \mathcal{Y}_{\omega^{\pm}}). \quad (6.15)$$

The operators $N_{\omega^{\pm}}$ are defined only on those functions $f \in L_p(\mathbb{R})$ for which the points of the set $\zeta = \zeta(\omega^{\pm})$ are Lebesgue points and $f|_{\zeta} \in l_p(\mathbb{Z})$. The next proposition gives a complete description of the domain of the operator $N_{\omega^{\pm}}$.

Proposition 6.8 *Let $f \in L_p(\mathbb{R})$. Then for f to lie in $X(\omega^{\pm})$ it is necessary and sufficient that the set $\zeta = \zeta(\omega^{\pm})$ consist of Lebesgue points of the function f, that $f|_{\zeta} \in l_p(\mathbb{Z})$, and that $N_{\omega^{\pm}}(f) \in L_p(\mathbb{R})$.*

Proof. Let $f = \omega^{\pm} f_1 + f_2^{\pm}$, where $f_1 \in L_p(\mathbb{R})$, $f_2^{\pm} \in \mathcal{M}(h^{\pm})$, and $\zeta = \{x_j\}_{j \in \mathbb{Z}}$. As in the proof of Theorem 6.1, we deduce that the x_j's are Lebesgue points of the function $\omega^{\pm} f_1$ and $(\omega^{\pm} f_1)(x_j) = 0$. At the same time, ζ is an interpolation set in the subspace $\mathcal{M}(h^{\pm})$, and consequently the points x_j are Lebesgue points of the function $f_2^{\pm}$, and $f_2^{\pm}|_{\zeta} \in l_p(\mathbb{Z})$. Therefore, the operator $N_{\omega^{\pm}}$ is defined on $X(\omega^{\pm})$, and since $N_{\omega^{\pm}}(f_2^{\pm}) = 0$, it follows that $N_{\omega^{\pm}}(f) = N_{\omega^{\pm}}(\omega^{\pm} f_1) = f_1 \in L_p(\mathbb{R})$.

Conversely, if the set ζ consists of Lebesgue points of a function $f \in L_p(\mathbb{R})$ with $f|_{\zeta} \in l_p(\mathbb{Z})$, and if $f_1 = N_{\omega^{\pm}}(f) \in L_p(\mathbb{R})$, then, by formula (6.15), $f = \omega^{\pm} f_1 + \mathcal{Y}_{\omega^{\pm}}(f) \in X(\omega^{\pm})$, because $f_2^{\pm} = \mathcal{Y}_{\omega^{\pm}}(f) \in \mathcal{M}(h^{\pm})$. □

Remark 6.5 From Proposition 6.8 it follows that

$$\mathcal{M}(h^{\pm}) = \ker N_{\omega^{\pm}}.$$

Indeed, by condition (iii)-item (b) in the definition of the class $\mathcal{K}^{\pm}(\mathbb{R})$, $f \in \mathcal{M}(h^{\pm})$ if and only if $f = \mathcal{Y}_{\omega^{\pm}}(f)$, which in turn is equivalent to the equality $(\omega^{\pm})^{-1}(f - \mathcal{Y}_{\omega^{\pm}} f) = 0$. But this means precisely that $N_{\omega^{\pm}}(f) = 0$.

Proposition 6.9 *Suppose that, under the conditions defining the class $\mathcal{K}^{\pm}(\mathbb{R})$,*

$$h^- = (h^+)^{-1}, \quad \omega^- = h^- \omega^+.$$

Then the following equalities hold on $X(\omega^{\pm})$:

(1) $P^+ N_{\omega^{\pm}} = N_{\omega^{\pm}} P^+$, $P^- N_{\omega^{\pm}} = N_{\omega^{\pm}} P^-$;

(2) $h^{\pm} \mathcal{Y}_{\omega^{\mp}} = \mathcal{Y}_{\omega^{\pm}} h^{\pm} I$.

Proof. (1) Consider the operator N_{ω^+} and let $f \in X(\omega^+)$. Then $f = \omega^+ f_1^+ + \omega^+ f_1^- + f_2^+$, where $f_1^{\pm} \in L_p^{\pm}(\mathbb{R})$ and $f_2^+ \in \mathcal{M}(h^+)$. It follows that $P^+ f = \omega^+ f_1^+ + P^+ \omega^+ f_1^- + f_2^+$, where $f_3^+ = P^+ \omega^+ f_1^- = P^+ h^+ (\omega^- f_1^-) \in \mathcal{M}(h^+)$ (see the definition of $\mathcal{M}(h^+)$). Consequently, $(N_{\omega^+}P^+)(f) = f_1^+ = (P^+N_{\omega^+})(f)$, because $N_{\omega^+}(f_2^+ + f_3^+) = 0$. Since f is arbitrary, we conclude that

$$N_{\omega^+}P^+ = P^+N_{\omega^+}.$$

Now let $f \in X(\omega^+)$. Then $P^-f = P^-\omega^+f_1^- = \omega^+f_1^- - f_3^+$, whence $(N_{\omega^+}P^-)(f) = f_1^- = (P^-N_{\omega^+})(f)$, because $N_{\omega^+}(f_3^+) = 0$. This shows that $N_{\omega^+}P^- = P^-N_{\omega^+}$, as claimed. The case of the operator N_{ω^-} is dealt with is a similar manner.

(2) For the upper signs,

$$\begin{aligned}
\mathcal{Y}_{\omega^+}(h^+f) &= \sum_{j\in\mathbb{Z}} h^+(x_j)f(x_j)\frac{\omega^+(x)}{(\omega^+)'(x_j)(x-x_j)} \\
&= h^+(x)\sum_{j\in\mathbb{Z}} f(x_j)\frac{\omega^-(x)}{(\omega^-)'(x_j)(x-x_j)} = h^+\mathcal{Y}_{\omega^-}(f),
\end{aligned}$$

since $\omega^+ = h^+\omega^-$ and $(\omega^+)'(x_j) = h^+(x_j)(\omega^-)'(x_j)$. The case of the lower signs is similar. □

Proposition 6.10 *The norm*

$$\|f\| = \|f\|_p + \|N_{\omega^\pm}(f)\|_p \tag{6.16}$$

makes the space $X(\omega^\pm)$ into a Banach space.

Proof. Let $\{f_n\}$ be a fundamental sequence in $X(\omega^+)$. Then the sequences $\{f_n\}$ and $\{N_{\omega^+}(f_n)\}$ are fundamental in $L_p(\mathbb{R})$, and since this space is complete, there exist elements $\widetilde{f}, \widehat{f} \in L_p(\mathbb{R})$ such that $f_n \to \widetilde{f}$ and $N_{\omega^+}(f_n) \to \widehat{f}$. Clearly, $P^\pm f_n \to P^\pm\widetilde{f}$. Furthermore, $N_{\omega^+}(f_n) \to \widehat{f}$ implies that $f_n - \mathcal{Y}_{\omega^+}(f_n) \to \omega^+\widehat{f}$. Applying the projector P^-, we have $P^-f_n \to P^-\omega^+\widehat{f}$, and so $P^-\widetilde{f} = P^-\omega^+\widehat{f}$, because the limit is unique. Applying now the projector P^+ we get $P^+f_n - \mathcal{Y}_{\omega^+}(f_n) \to P^+\omega^+\widehat{f}$, whence $\mathcal{Y}_{\omega^+}(f_n) \to P^+\widetilde{f} - P^+\omega^+\widehat{f}$. This implies that $\widehat{f_2^+} = P^+\widetilde{f} - P^+\omega^+\widehat{f} \in \mathcal{M}(h^+)$, because all the elements of the sequence $\{\mathcal{Y}_{\omega^+}(f_n)\}$ lie in the subspace $\mathcal{M}(h^+)$. It follows that $\widetilde{f} = P^+\widetilde{f} + P^-\widetilde{f} = P^+\omega^+\widehat{f} + f_2^+ + P^-\omega^+\widehat{f} = \omega^+\widehat{f} + f_2^+ \in X(\omega^+)$ and $N(\widetilde{f}) = \widehat{f}$. Finally,

$$\begin{aligned}
\|f_n - \widetilde{f}\| &= \|f_n - \widetilde{f}\|_p + \|N_{\omega^+}(f_n) - N_{\omega^+}(\widetilde{f})\|_p \\
&= \|f_n - f\|_p + \|N_{\omega^+}(f_n - \widehat{f})\|_p \to 0
\end{aligned}$$

as $n \to \infty$, i.e., $f_n \to \widetilde{f}$ in $X(\omega^+)$. The proof for the case of the space $X(\omega^-)$ is analogous. □

Let us remark that the norm $\|f\|$ given by (6.16) is equivalent to the norm

$$\|f\|_1 = \|f_1\|_p + \|f_2^\pm\|_p. \tag{6.17}$$

Indeed, $\|f\|_p = \|\omega^\pm f_1 + f_2^\pm\|_p \le c(\|f_1\|_p + \|f_2^\pm\|_p)$, whence $\|f\| \le c_1\|f\|_1$. Similarly, $\|f_2^\pm\|_p = \|f - \omega^\pm f_1\|_p \le c(\|f\|_p + \|f_1\|_p)$ and thus $\|f\|_1 \le c_2\|f\|$. This shows that $\|f\| \asymp \|f\|_1$.

We say that the function a is a *multiplier* in the space $X(\omega^\pm)$, and write $a \in \operatorname{mult} X(\omega^\pm)$, if $af \in X(\omega^\pm)$ for all $f \in X(\omega^\pm)$. From the definition of the space $X(\omega^\pm)$ it follows that $a \in \operatorname{mult} X(\omega^\pm)$ if and only if $af^\pm \in X(\omega^\pm)$ for all $f^\pm \in \mathcal{M}(h^\pm)$.

Proposition 6.11 *Suppose that the conditions*

$$h^- = (h^+)^{-1}, \qquad \omega^- = h^-\omega^+. \tag{6.18}$$

are satisfied. Then the following assertions are equivalent:

(a) $X(\omega^+) = X(\omega^-)$;

(b) $h^+ \in \operatorname{mult} X(\omega^+)$ *and* $h^- \in \operatorname{mult} X(\omega^-)$.

Proof. (*a*) $\Longrightarrow$ (*b*). Let $f \in X(\omega^+) = X(\omega^-)$. Then $f = \omega^+ f_1 + f_2^+ = \omega^- g_1 + g_2^-$, where $f_1, g_1 \in L_p(\mathbb{R})$, $f_2^+ \in \mathcal{M}(h^+)$, and $g_2^- \in \mathcal{M}(h^-)$. It follows that $h^+ f = \omega^+ g_1 + h^+ g_2^- \in X(\omega^+)$, because $h^+ g_2^- \in \mathcal{M}(h^+)$. Analogously, $h^- f = \omega^- f_1 + h^- f_2^+ \in X(\omega^-)$, because $h^- f_2^+ \in \mathcal{M}(h^-)$. Therefore, $h^\pm \in \operatorname{mult} X(\omega^\pm)$.

(*b*) $\Longrightarrow$ (*a*). Let $f \in X(\omega^+)$. Then $f = \omega^+ f_1 + f_2^+$, where $f_1 \in L_p(\mathbb{R})$ and $f_2^+ \in \mathcal{M}(h^+)$. Now $h^+ f_2^+ \in X(\omega^+)$, because $h^+ \in \operatorname{mult} X(\omega^+)$. This yields $h^+ f_2^+ = \omega^+ N_{\omega^+}[h^+ f_2^+] + \mathcal{Y}_{\omega^+}[h^+ f_2^+]$ (see Proposition 6.8).

Let us solve the equation $\omega^- g_1 + g_2^- = f$ with the unknowns g_1 and g_2^-. Multiplying both sides of this equation by h^+, we obtain

$$\omega^+ g_1 + h^+ g_2^- = h^+\omega^+ f_1 + h^+ f_2^+,$$

or, equivalently,

$$\omega^+ g_1 + h^+ g_2^- = \omega^+(h^+ f_1 + N_{\omega^+}[h^+ f_2^+]) + \mathcal{Y}_{\omega^+}[h^+ f_2^+].$$

This yields the desired solution

$$g_1 = h^+ f_1 + N_{\omega^+}[h^+ f_2^+] \in L_p(\mathbb{R}), \quad g_2^- = h^- \mathcal{Y}_{\omega^+}[h^+ f_2^+] \in \mathcal{M}(h^-).$$

We conclude that $X(\omega^+) \subset X(\omega^-)$. The opposite inclusion is established in much the same manner. □

Proposition 6.12 *Suppose that condition* (6.18) *is satisfied and* $X(\omega^+) = X(\omega^-)$. *Then the projectors* $P^\pm$, *and together with them the operator* $S_\mathbb{R}$, *are continuous in the space* $X = X(\omega^+) = X(\omega^-)$. *Moreover,*

$$X^\pm = P^\pm X = \omega^\pm I(L_p^\pm(\mathbb{R})) \oplus \mathcal{M}(h^\pm). \tag{6.19}$$

Proof. Consider, say, the subspace X^+. Let $f \in X$. Then $f = \omega^+ f_1 + f_2^+$, where $f_1 \in L_p(\mathbb{R})$, $f_2^+ \in \mathcal{M}(h^+)$, and consequently $P^+ f = \omega^+ P^+ f_1 + f_2^+ + P^+\omega^+ P^- f_1 \in \omega^+ I(L_p^+(\mathbb{R})) + \mathcal{M}(h^+)$, because $P^+\omega^+P^- f_1 \in \mathcal{M}(h^+)$. Therefore, $X^+ \subset \omega^+ I(L_p^+(\mathbb{R})) + \mathcal{M}(h^+)$. The opposite inclusion is obvious. Further,

$$\begin{aligned} \|P^+ f\| &\leq c(\|P^+ f_1\|_p + \|f_2^+\|_p + \|P^+\omega^+P^- f_1\|_p) \\ &\leq c_1(\|f_2^+\|_p + \|f_1\|_p) = c_1\|f\|_1 \leq c_2\|f\|. \end{aligned}$$

Since the space X is complete in the norm (6.17), the fact that the linear manifolds $\omega^+ I(L_p^+(\mathbb{R}))$ and $\mathcal{M}(h^+)$ are closed in X follows from the isomorphisms

$$\omega^+ I(L_p^+(\mathbb{R})) \cong L_p^+(\mathbb{R}), \quad f \mapsto f_1 = N_{\omega^+}(f),$$

$$\mathcal{M}(h^+) \cong l_p(\mathbb{Z}), \quad f \mapsto f|_{\zeta(\omega^+)}.$$

At the same time, from the interpolation property of the set $\zeta(\omega^+)$ in the space $\mathcal{M}(h^+)$ it follows that

$$\omega^+ I(L_p^+(\mathbb{R})) \cap \mathcal{M}(h^+) = \{0\}.$$

Therefore, the sum in (6.19) is direct. The case of the subspace X^- is dealt with similarly. □

Now let us address the description of the spaces $X(\omega^\pm)$ in terms of the local behavior of functions in the vicinity of the set $\zeta(\omega^\pm)$. We will assume that $\zeta(\omega^\pm)$ satisfies the following three requirements:

(1) the set $\{x_j\}_{j\in\mathbb{Z}} = \zeta(\omega^\pm)$ is separated;

(2) $(\omega^\pm)'(x_j) \geq c > 0$, $j \in \mathbb{Z}$;

(3) the function $\omega^\pm(x)$ is uniformly bounded away from zero outside each sufficiently small ε-neighborhood V_ε of the set $\zeta(\omega^\pm)$, i.e., for any $\delta > 0$ there exists an $\varepsilon > 0$ such that $|\omega^\pm(x)| \geq \delta$ whenever $x \in \mathbb{R} \setminus V_\varepsilon$.

We denote by $\widetilde{\mathcal{K}}^\pm(\mathbb{R})$ the subclass of all functions of the class $\mathcal{K}^\pm(\mathbb{R})$ for which conditions (1)–(3) are satisfied.

Let $\zeta = \{x_j\}_{j\in\mathbb{Z}}$ be a separated real set, that is,

$$|x_k - x_j| \geq d > 0, \quad k \neq j, \quad k, j \in \mathbb{Z}, \tag{6.20}$$

and let $DL_p(\mathbb{R})$ denote the linear manifold of the functions $f \in L_p(\mathbb{R})$ with the following properties:

(i) the points x_j, $x_j \in \zeta$, are Lebesgue points for f;

(ii) $\{f(x_j)\} = f|_\zeta \in l_p(\mathbb{Z})$;

(iii) there is an ε, $0 < \varepsilon < \frac{d}{2}$, such that

$$\sum_{j\in\mathbb{Z}} \int_{x_j-\varepsilon}^{x_j+\varepsilon} \left| \frac{f(x) - f(x_j)}{x - x_j} \right|^p dx < \infty.$$

Notice that condition (iii) does not depend on ε. When it is satisfied, it holds for every ε_1 such that $0 < \varepsilon_1 < d/2$.

Denoting by V_ε the real ε-neighborhood of the set ζ, $0 < \varepsilon < d/2$, let us give a simple sufficient condition for membership in the space $DL_p(\mathbb{R})$.

Proposition 6.13 *Let $f \in L_p(\mathbb{R})$ and suppose that at each point $x \in V_\varepsilon$ the function $f(x)$ is continuous and that for almost every point $x \in V_\varepsilon$ it has a derivative $f'(x)$ that belongs to $L_p(V_\varepsilon)$. Then $f \in DL_p(\mathbb{R})$.*

Proof. First let us assume that the function f takes only real values. By the mean value theorem, for each $j \in \mathbb{Z}$ there exists a point $\eta_j \in [x_j - \varepsilon, x_j + \varepsilon]$ such that

$$f(\eta_j) = \frac{1}{2\varepsilon} \int_{x_j-\varepsilon}^{x_j+\varepsilon} f(t)dt.$$

Therefore,

$$\begin{aligned}
\Big(\sum_j |f(x_j)|^p\Big)^{1/p} &\le \Big(\sum_j |f(x_j) - f(\eta_j)|^p\Big)^{1/p} + \Big(\sum_j |f(\eta_j)|^p\Big)^{1/p} \\
&= \Big(\sum_j \Big|\int_{x_j}^{\eta_j} f'(t)dt\Big|^p\Big)^{1/p} + \Big(\sum_j \Big|\frac{1}{2\varepsilon}\int_{x_j-\varepsilon}^{x_j+\varepsilon} f(t)dt\Big|^p\Big)^{1/p} \\
&\le (2\varepsilon)^{p/q}\Big(\sum_j \int_{x_j}^{\eta_j} |f'(t)|^p dt\Big)^{1/p} \\
&\qquad + (2\varepsilon)^{(p-q)/q}\Big(\sum_j \int_{x_j-\varepsilon}^{x_j+\varepsilon} |f(t)|^p dt\Big)^{1/p} \\
&\le c\left(\|f'|_{V_\varepsilon}\|_p + \|f\|_p\right) < \infty.
\end{aligned}$$

Further, let

$$f^*(x) = \sup_{x,y\in[x_j-\varepsilon,x_j+\varepsilon],\, x\ne y} \frac{1}{x-y}\int_y^x |f'(t)|dt$$

be the Hardy-Littlewood maximal function for the function f'. By the Hardy-Littlewood maximal theorem (see, e.g., Garnett [1], Chap. I, Theorem 4.3),

$$\int_{x_j-\varepsilon}^{x_j+\varepsilon} [f^*(t)]^p dt \le A_p^p \int_{x_j-\varepsilon}^{x_j+\varepsilon} |f'(t)|^p dt.$$

Consequently,

$$\begin{aligned}
\sum_j \int_{x_j-\varepsilon}^{x_j+\varepsilon} \left|\frac{f(x)-f(x_j)}{x-x_j}\right|^p dx &\le \sum_j \int_{x_j-\varepsilon}^{x_j+\varepsilon} \left(\frac{1}{x-x_j}\int_{x_j}^{x} |f'(t)|dt\right)^p dx \\
\le \sum_j \int_{x_j-\varepsilon}^{x_j+\varepsilon} [f^*(x)]^p dx &\le c\sum_j \int_{x_j-\varepsilon}^{x_j+\varepsilon} |f'(x)|^p dx \le c\|f'|_{V_\varepsilon}\|_p^p < \infty.
\end{aligned}$$

In the general case, $f(x) = \operatorname{Re} f(x) + i \operatorname{Im} f(x)$, where $\operatorname{Re} f(x), \operatorname{Im} f(x) \in L_p(\mathbb{R})$, and $\operatorname{Re} f'(x)$, $\operatorname{Im} f'(x) \in L_p(V_\varepsilon)$. Since

$$|f(x)| \leq |\operatorname{Re} f(x)| + |\operatorname{Im} f(x)|,$$

$$\left|\frac{f(x) - f(x_j)}{x - x_j}\right| \leq \left|\frac{\operatorname{Re} f(x) - \operatorname{Re} f(x_j)}{x - x_j}\right| + \left|\frac{\operatorname{Im} f(x) - \operatorname{Im} f(x_j)}{x - x_j}\right|,$$

the desired result is obtained via an appropriate modification of the above argument. □

We need one more general assertion concerning smoothness properties of model subspaces on $\mathbb{R}$.

Proposition 6.14 *Let $a \in C_B^\infty(\mathbb{R})$ and*

$$(A_k(f))(x) = \frac{d^k}{dx^k} \int_{\mathbb{R}} \frac{a(t) - a(x)}{t - x} f(t)dt, \tag{6.21}$$

where $f \in L_p(\mathbb{R})$, $1 < p < \infty$. Then

$$\|A_k(f)\|_p \leq c_k \|f\|_p, \quad k = 0, 1, \ldots.$$

Proof. Since

$$A_0 = \pi i [S_{\mathbb{R}} a I - a S_{\mathbb{R}}],$$

the case $k = 0$ requires no proof.

Let $k \geq 1$. Take $f, g \in \mathcal{D}(\mathbb{R})$, where $\mathcal{D}(\mathbb{R})$ is the space of infinitely differentiable functions with compact support, and let us estimate the bilinear form

$$\langle A_k(f), g \rangle = \int_{\mathbb{R}} [A_k(f)](x) g(x) dx.$$

With no loss of generality, we may assume that the function $a(x)$ has compact support. Let us continue a to $\mathbb{C}$ using the formula

$$a(x + iy) = \sum_{m=0}^{M} a^{(m)}(x) \frac{(iy)^m}{m!},$$

where M will be chosen below. Then

$$\left|\frac{\partial a}{\partial \bar{z}}\right| = \left|\frac{1}{2}\left(\frac{\partial a}{\partial x} + i \frac{\partial a}{\partial y}\right)\right| = \left|a^{(M+1)}(x) \frac{(iy)^M}{M!}\right| \leq \|a^{(M+1)}\|_\infty |y|^M.$$

Setting $\zeta = \xi + i\eta$, the Cauchy-Green formula yields

$$a(x) = \frac{1}{2\pi i} \int_{\eta = \pm 1} a(\zeta) \frac{d\zeta}{\zeta - x} - \frac{1}{\pi} \iint_{|\eta| < 1} \frac{\partial a(\zeta)}{\partial \bar{\zeta}} \cdot \frac{d\xi d\eta}{\zeta - x}, \quad x \in \mathbb{R}.$$

But then

$$\begin{aligned}\frac{d^k}{dx^k}\left(\frac{a(t)-a(x)}{t-x}\right) &= \frac{k!}{2\pi i}\int_{\eta=\pm 1}\frac{a(\zeta)d\zeta}{(\zeta-x)^{k+1}(\zeta-t)} \\ &\quad -\frac{k!}{\pi}\iint_{|\eta|<1}\frac{\partial a(\zeta)}{\partial\overline{\zeta}}\cdot\frac{d\zeta d\eta}{(\zeta-x)^{k+1}(\zeta-t)}\end{aligned}$$

and

$$\begin{aligned}\langle A_k(f),g\rangle &= \frac{k!}{2\pi i}\int_{\eta=\pm 1}a(\zeta)d\zeta\int_{\mathbb{R}}\frac{g(x)dx}{(\zeta-x)^{k+1}}\int_{\mathbb{R}}f(t)dt \\ &\quad -\frac{k!}{\pi}\iint_{|\eta|<1}\frac{\partial a(\zeta)}{\partial\overline{\zeta}}\int_{\mathbb{R}}\frac{g(x)dx}{(\zeta-x)^{k+1}}\int_{\mathbb{R}}\frac{f(t)dt}{\zeta-t}.\end{aligned}$$

Let $f\in L_p(\mathbb{R})$, $g\in L_q(\mathbb{R})$, $q=p/(p-1)$. Then for any $\eta\neq 0$,

$$\begin{aligned}&\left|\int_{\mathbb{R}}\frac{\partial a(\xi+i\eta)}{\partial\overline{\zeta}}d\xi\int_{\mathbb{R}}\frac{g(x)dx}{(\xi+i\eta-x)^{k+1}}\int_{\mathbb{R}}\frac{f(t)dt}{\xi+i\eta-t}\right| \\ &\le \|a^{(M+1)}\|_\infty|\eta|^M\left\|\int_{\mathbb{R}}\frac{g(x)dx}{(\xi+i\eta-x)^{k+1}}\right\|_q\left\|\int_{\mathbb{R}}\frac{f(t)dt}{\xi+i\eta-t}\right\|_p \\ &\le c|\eta|^M\|(x-i\eta)^{-k-1}\|_1\|f\|_p\|g\|_q\le c_k\|f\|_p\|g\|_q|\eta|^{M-k},\end{aligned}$$

because (see Kreĭn [1])

$$\left\|\int_{\mathbb{R}}\frac{g(x)dx}{(\xi+i\eta-x)^{k+1}}\right\|_q\le\|(x-i\eta)^{-k-1}\|_1\|g\|_q$$

and (see Dwight [1], pp. 29, 46)

$$\|(x-i\eta)^{-k-1}\|_1=O(|\eta|^{-k}).$$

Now taking $M=k$ we have

$$\begin{aligned}&\left|\iint_{|\eta|<1}\frac{\partial a(\zeta)}{\partial\overline{\zeta}}d\xi d\eta\int_{\mathbb{R}}\frac{g(x)dx}{(\zeta-x)^{k+1}}\int_{\mathbb{R}}\frac{f(t)dt}{\zeta-t}\right| \\ &\le\int_{-1}^{1}d\eta\left|\int_{\mathbb{R}}\frac{\partial a(\xi+i\eta)}{\partial\overline{\zeta}}d\xi\int_{\mathbb{R}}\frac{g(x)dx}{(\xi+i\eta-x)^{k+1}}\int_{\mathbb{R}}\frac{f(t)dt}{\xi+i\eta-t}\right| \\ &\le c_k\|f\|_p\|g\|_q\int_{-1}^{1}|\eta|^{M-k}d\eta=2c_k\|f\|_p\|g\|_q.\end{aligned}$$

Similarly,

$$\begin{aligned}
&\left|\int_{\eta=\pm 1} a(\zeta)d\zeta \int_{\mathbb{R}} \frac{g(x)dx}{(\zeta-x)^{k+1}} \int_{\mathbb{R}} \frac{f(t)dt}{\zeta-t}\right| \\
&= \left|\int_{\eta=\pm 1} \sum_{m=0}^{k} a^{(m)}(\xi)\frac{(i\eta)^m}{m!} \int_{\mathbb{R}} \frac{g(x)dx}{(\xi+i\eta-x)^{k+1}} \int_{\mathbb{R}} \frac{f(t)dt}{\xi+i\eta-t}\right| \\
&\le d_k \sum_{m=0}^{k} \|a^{(m)}\|_\infty \|f\|_p \|g\|_q \le \widetilde{d}_k \|f\|_p \|g\|_q .
\end{aligned}$$

Since $\mathcal{D}(\mathbb{R})$ is dense in $L_p(\mathbb{R})$, $1 < p < \infty$, the theorem on the extension of bounded bilinear functionals defined on a dense linear set yields

$$|\langle A_k(f), g\rangle| \le c_k \|f\|_p \|g\|_q, \quad f \in L_p(\mathbb{R}), \quad g \in L_q(\mathbb{R}),$$

which proves the boundedness of the operator A_k in $L_p(\mathbb{R})$. □

The next result establishes a connection between the classes $\widetilde{\mathcal{K}}^\pm(\mathbb{R})$ and the spaces $X(\omega^\pm)$ and $DL_p(\mathbb{R})$.

Proposition 6.15 *Suppose that $\omega^\pm \in \widetilde{\mathcal{K}}^\pm(\mathbb{R})$ and conditions (6.18) are satisfied. Then*

$$X(\omega^+) = X(\omega^-) = DL_p(\mathbb{R}),$$

where the space $DL_p(\mathbb{R})$ is constructed using the set $\{x_j\} = \zeta(\omega^\pm)$.

Proof. First let us show that $DL_p(\mathbb{R}) \subset X(\omega^+)$. Take $f \in DL_p(\mathbb{R})$. Then from the definition of the class $\mathcal{K}^+(\mathbb{R})$ it follows that $\mathcal{Y}_{\omega^+}(f) \in \mathcal{M}(h^+)$. By conclusion (2) of Proposition 6.7, a function f lies in $\mathcal{M}(h^+)$ if and only if $f = 2\pi i A_0(g^-)$, where $g^- \in L_p^-(\mathbb{R})$ and the operator A_0 has the form (6.21), where $a = h^+$. Further, by Proposition 6.14, $f \in \mathcal{M}(h^+)$ implies that $f' \in L_p(\mathbb{R})$. Hence, applying Proposition 6.13 we conclude that $\mathcal{M}(h^+) \subset DL_p(\mathbb{R})$. Thus, we have shown that $\mathcal{Y}_{\omega^+}(f) \in DL_p(\mathbb{R})$.

Choosing a sufficiently small neighborhood V_ε of the set $\zeta = \{x_j\}_{j\in\mathbb{Z}}$, which is possible because the separation condition (6.20) is satisfied, we obtain

$$\begin{aligned}
\|N_{\omega^+}(f)\|^p &= \|(\omega^+)^{-1}(f - \mathcal{Y}_{\omega^+}(f))\|^p \\
&= \sum_j \int_{x_j-\varepsilon}^{x_j+\varepsilon} \left|\frac{f(x) - [\mathcal{Y}_{\omega^+}(f)](x)}{\omega^+(x)}\right|^p dx \\
&\quad + \int_{\mathbb{R}\setminus V_\varepsilon} \left|\frac{f(x) - [\mathcal{Y}_{\omega^+}(f)](x)}{\omega^+(x)}\right|^p dx.
\end{aligned}$$

By condition (3) in the definition of the class $\widetilde{\mathcal{K}}^+(\mathbb{R})$, the function $|\omega^+|^{-1}$ is uniformly bounded on the set $\mathbb{R} \setminus V_\varepsilon$, while by condition (2) in that definition, for $x \in [x_j - \varepsilon, x_j + \varepsilon]$,

$$\begin{aligned}\left|\frac{x - x_j}{\omega^+(x)}\right| &= \left|\omega^{+(1)}(x_j) + \frac{1}{x - x_j}\int_{x_j}^{x} dt \int_{x_j}^{t} \omega^{+(2)}(\tau)d\tau\right|^{-1} \\ &\leq (|\omega^{+(1)}(x_j)| - |x - x_j|\|\omega^+\|_\infty)^{-1} \leq c(\varepsilon),\end{aligned}$$

where the constant $c(\varepsilon)$ does not depend on j and k. Using this, we obtain

$$\begin{aligned}\|N_{\omega^+}(f)\|^p \leq &\left[\left(\sum_j \int_{x_j-\varepsilon}^{x_j+\varepsilon} \left|\frac{x - x_j}{\omega^+(x)}\right|^p \left|\frac{f(x) - f(x_j)}{x - x_j}\right|^p dx\right)^{1/p}\right. \\ &+ \left.\left(\sum_j \int_{x_j-\varepsilon}^{x_j+\varepsilon} \left|\frac{x - x_j}{\omega^+(x)}\right|^p \left|\frac{[\mathcal{Y}_{\omega^+}(f)](x) - [\mathcal{Y}_{\omega^+}(f)](x_j)}{x - x_j}\right|^p dx\right)^{1/p}\right]^p \\ &+ c_1(\varepsilon)(\|f\|_p + \|\mathcal{Y}_{\omega^+}(f)\|_p)^p \\ \leq\, &c_2(\varepsilon)\left[\left(\sum_j \int_{x_j-\varepsilon}^{x_j+\varepsilon} \left|\frac{f(x) - f(x_j)}{x - x_j}\right|^p dx\right)^{1/p}\right. \\ &+ \left.\left(\sum_j \int_{x_j-\varepsilon}^{x_j+\varepsilon} \left|\frac{[\mathcal{Y}_{\omega^+}(f)](x) - [\mathcal{Y}_{\omega^+}(f)](x_j)}{x - x_j}\right|^p dx\right)^{1/p}\right]^p \\ &+ c_1(\varepsilon)(\|f\|_p + \|\mathcal{Y}_{\omega^+}(f)\|_p)^p < \infty,\end{aligned}$$

because the constants $c_1(\varepsilon)$ and $c_2(\varepsilon)$ are independent of j, and the functions f and $\mathcal{Y}_{\omega^+}(f)$ satisfy condition (iii) of the definition of the space $DL_p(\mathbb{R})$. By Proposition 6.8, $f \in X(\omega^+)$.

It remains to verify that $X(\omega^+) \subset DL_p(\mathbb{R})$. Since $\mathcal{M}(h^+) \subset DL_p(\mathbb{R})$, to do this it suffices to show that $\omega^+ f \in DL_p(\mathbb{R})$ whenever $f \in L_p(\mathbb{R})$. As in Section 1.7, one shows that the set ζ consists of Lebesgue points of the function $\omega^+ f$ and that $\omega^+ f|_\zeta = 0$. At the same time, for any ε, $0 < \varepsilon < d/2$,

$$\left|\frac{\omega^+(x)}{x - x_j}\right| = \frac{1}{|x - x_j|}\left|\int_{x_j}^{x} (\omega^+)'(t)dt\right| \leq \|(\omega^+)^{(1)}\|_\infty$$

uniformly for all $x \in [x_j - \varepsilon, x_j + \varepsilon]$ and $j \in \mathbb{Z}$. Therefore,

$$\begin{aligned}\sum_j \int_{x_j-\varepsilon}^{x_j+\varepsilon} \left|\frac{\omega^+(x)f(x)}{x - x_j}\right|^p dx &\leq \|(\omega^+)'\|_\infty^p \sum_j \int_{x_j-\varepsilon}^{x_j+\varepsilon} |f(x)|^p dx \\ &\leq \|(\omega^+)'\|_\infty^p \|f\|_p^p < \infty.\end{aligned}$$

The case of lower signs can be disposed of in a similar manner. □

As a corollary of Propositions 6.12 and 6.15 we obtain that, under the assumptions of Proposition 6.15,

$$DL_p^{\pm}(\mathbb{R}) \stackrel{\text{def}}{=} DL_p(\mathbb{R}) \cap L_p^{\pm}(\mathbb{R}) = P^{\pm}(X(\omega^{\pm})). \tag{6.22}$$

Now let us consider Toeplitz operators with symbols in $\widetilde{\mathcal{K}}^{\pm}(\mathbb{R})$.

Proposition 6.16 *Suppose that* $\omega^- \in \widetilde{\mathcal{K}}^-(\mathbb{R})$ *and conditions* (6.18) *are satisfied, where* h^+ *is a model function, for which* $\omega^+ = h^+\omega^- \in H^\infty(\Pi_+)$ *and* $h^- = (h^+)^{-1}$. *Then the operator* $T(\omega^-)$ *is unboundedly invertible in the space* $L_p^+(\mathbb{R})$ *and* $\operatorname{im} T(\omega^-) = DL_p^+(\mathbb{R})$. *The unbounded inverse* $[T(\omega^-)]^{-1}$ *acts according to the rule*

$$[T(\omega^-)]^{-1} = N_{\omega^+}(h^+ I), \tag{6.23}$$

where the operator N_{ω^+} *is given by* (6.15).

Proof. First let us show that $\ker T(\omega^-) = \{0\}$. Let $f \in L_p^+(\mathbb{R})$ and $[T(\omega^-)](f) = 0$. Then $P^+\omega^- f = 0$, whence $\omega^- f = P^-\omega^- f$. Multiplying both sides by h^+ we obtain $h^+\omega^- f = h^+P^-\omega^- f$, and consequently

$$P^+h^+\omega^- f = \omega^+ f = P^+h^+P^-\omega^- f \in \mathcal{M}(h^+)$$

(by the definition of $\mathcal{M}(h^+)$). Denoting $f^+ = P^+h^+P^-\omega^- f$, we have $\omega^+ f = f^+$. But then $f^+|_{\zeta(\omega^+)} = 0$, and so $f^+ = 0$ because $\zeta(\omega^+)$ is an interpolation set in the subspace $\mathcal{M}(h^+)$ (see the definition of the class $\mathcal{K}^+(\mathbb{R})$). We conclude that $f = 0$ almost everywhere on $\mathbb{R}$, i.e., $\ker T(\omega^-) = \{0\}$, since f was arbitrary.

Now let us show that $\operatorname{im} T(\omega^-) = DL_p^+(\mathbb{R})$. Let $g \in \operatorname{im} T(\omega^-)$. Then $g = P^+\omega^- f$, where $f \in L_p^+(\mathbb{R})$, and hence $g = \omega^- f - P^-\omega^- f \in X(\omega^-) = DL_p(\mathbb{R})$, because $P^-\omega^- f \in \mathcal{M}(h^-)$ (Proposition 6.15). Therefore, $g \in DL_p^+(\mathbb{R}) = DL_p(\mathbb{R}) \cap L_p^+(\mathbb{R})$, i.e., $\operatorname{im} T(\omega^-) \subset DL_p^+(\mathbb{R})$, since g was arbitrary. Conversely, if $g \in DL_p^+(\mathbb{R})$, then $h^+g \in DL_p^+(\mathbb{R})$ (by Propositions 6.11, 6.12 and 6.15; see also equality (6.22)). It follows that $N_{\omega^+}(h^+g) \in L_p^+(\mathbb{R})$ (Propositions 6.8 and 6.9), and consequently

$$\begin{aligned} T(\omega^-)N_{\omega^+}(h^+g) &= P^+\omega^-(\omega^+)^{-1}(h^+g - \mathcal{Y}_{\omega^+}(h^+g)) \\ &= P^+(h^+)^{-1}(h^+g - \mathcal{Y}_{\omega^+}(h^+g)) = g - P^+h^-\mathcal{Y}_{\omega^+}(h^+g) = g, \end{aligned}$$

because

$$h^-\mathcal{Y}_{\omega^+}(h^+g) \in \mathcal{M}(h^-) \subset L_p^-(\mathbb{R}).$$

Hence, $g \in \operatorname{im} T(\omega^-)$, which proves that $DL_p^+(\mathbb{R}) \subset \operatorname{im} T(\omega^-)$, since g was arbitrary.

Finally, $[T(\omega^-)]^* = T\left(\overline{\omega^-}\right) = P^+\overline{\omega^-}I = \overline{\omega^-}I$, and since $\overline{\omega^-} \in H^\infty(\Pi_+)$, it follows that $\ker[T(\omega^-)]^* = \{0\}$, and so

$$\operatorname{clos} \operatorname{im} T(\omega^-) = \operatorname{clos} DL_p^+(\mathbb{R}) = L_p^+(\mathbb{R}).$$

Therefore, the operator $T(\omega^-)$ is unboundedly invertible. The proof also shows that the operator (6.23) is a right unbounded inverse, and hence a two-sided unbounded inverse for the operator $T(\omega^-)$. □

We are now ready to formulate the analogue of Theorem 6.11 for Toeplitz operators whose symbol degenerates on a separated real set.

Theorem 6.12 *Suppose that the function a belongs to $L_\infty(\mathbb{R})$ and admits a representation*

$$a = \omega^- a_1, \tag{6.24}$$

where $\omega^- \in \widetilde{\mathcal{K}}^-(\mathbb{R})$, $a_1 \in \mathrm{fact}(\infty, p, 1)$, h satisfies condition (2.31), *and* $\operatorname{ind} b = 0$ *($a_1 = bh^{-1}$ or $a_1 = bh$, where $h \in H^\infty(\Pi_+) \cap GL_\infty(\mathbb{R})$, $b = b^+b^- \in \mathrm{fact}(p,1)$).*

Then the operator $T(a)$ belongs to $\mathcal{L}(L_p^+(\mathbb{R}), DL_p^+(\mathbb{R}))$, where the space $DL_p^+(\mathbb{R})$ is constructed using the set $\zeta(\omega^-)$.

If $a_1 = bh^{-1}$, then $T(a)$ is right invertible, $\dim\ker T(a) = \infty$, *and the operator*

$$[T(a)]^{-1} = \frac{h}{b^+}P^+\frac{1}{b^-}N,$$

where N is defined by (6.23), *lies in $\mathcal{L}(DL_p^+(\mathbb{R}), L_p^+(\mathbb{R}))$ and is a right inverse of $T(a)$. Moreover* $\ker T(a) = \ker T(a_1)$.

If $a_1 = bh$, then $T(a)$ is left invertible, $\dim \operatorname{coker} T(a) = \infty$, *and the operator*

$$[T(a)]^{-1} = P^+\frac{1}{b^+h}P^+\frac{1}{b^-}N,$$

where N is again defined by (6.23), *lies in $\mathcal{L}(DL_p^+(\mathbb{R}), L_p^+(\mathbb{R}))$ and is a left inverse of $T(a)$.*

Finally, for f to lie in $\operatorname{im} T(a)$ *it is necessary and sufficient that*

$$\int_{\mathbb{R}} \psi_j(x)[N(f)](x)dx = 0, \quad j = 1, 2, \ldots,$$

where

$$\psi_j = \frac{1}{b^-}P^-\frac{(x+i)^{-j}}{hb^+} \in L_q^-(\mathbb{R}), \quad q = p/(p-1).$$

Proof. This theorem is a simple consequence of Theorem 6.9, Proposition 2.4, and Proposition 6.16 (see also the proof of Theorem 6.11). □

6.5 Basic examples

In this section we consider some sufficiently general examples of symbols in the class $\widetilde{\mathcal{K}}^\pm(\mathbb{R})$. They are special entire functions of exponential type – sine-type functions with real zeros – and special meromorphic functions connected with Blaschke products whose set of zeros lies on lines parallel to the real line.

Example 6.1 *Sine-type functions.*
Let us return to Section 3.7. Let $s(z)$ be a sine-type function of type $\sigma/2$, with a separated set of real zeros $\zeta(s) = \{x_j\}_{j\in\mathbb{Z}}$. Then

$$\omega^{\pm}(x) = \exp(\pm i\sigma x/2)s(x). \tag{6.25}$$

By definition, $\omega^{\pm} \in B_{\sigma}^{\pm} \subset H_{\infty}(\Pi_{\pm})$, and Bernstein's inequality yields

$$\|\omega^{\pm(k)}\|_{\infty} \leq \sigma^k \|\omega^{\pm}\|_{\infty}, \quad k = 1, 2, \ldots.$$

Therefore, $\omega^{\pm} \in C_B^{\infty}(\mathbb{R})$. The model function here is

$$h^{\pm}(x) = \exp(\pm i\sigma x) \in \mathcal{U}^{\pm}(\mathbb{R}) \cap C_B^{\infty}(\mathbb{R}), \quad (h^+)^{-1} = h^-,$$

$$\omega^{\pm} = h^{\pm}\omega^{\mp}.$$

Consequently, conditions (6.18) are satisfied. By Theorem 3.4, the set $\zeta(\omega^{\pm}) = \zeta(s)$ is an interpolation set in the model subspace $\mathcal{M}(h^{\pm})$ of $L_p^{\pm}(\mathbb{R})$. Hence, $\omega^{\pm} \in \mathcal{K}^{\pm}(\mathbb{R})$. Moreover, thanks to inequalities (3.27) and (3.26) the function $\omega^{\pm}$ satisfies conditions (2) and (3) of the definition of the classes $\widetilde{\mathcal{K}}^{\pm}(\mathbb{R})$, and so $\omega^{\pm} \in \widetilde{\mathcal{K}}^{\pm}(\mathbb{R})$. We see that if the function ω^- is given by (6.25), then the Toeplitz operator $T(a)$ with a symbol a of the form (6.24) satisfies all the assumptions of Theorem 6.12.

Let us give concrete examples of symbols of the form (6.25) that are encountered in the literature.

(1) Let $\sigma, \sigma_0 \in \mathbb{R}$, $\sigma > 0$, $\sigma_0 > 0$. Then

$$\omega^+(x) = \exp(i\sigma_0) - \exp(i\sigma x) \in \widetilde{\mathcal{K}}^+(\mathbb{R}). \tag{6.26}$$

Since

$$\begin{aligned} \omega^+(x) &= (\cos\sigma_0 - \cos\sigma x) + i(\sin\sigma_0 - \sin\sigma x) \\ &= (-2i)\sin\frac{\sigma x - \sigma_0}{2}\exp(i(\sigma x + \sigma_0)/2) \\ &= c\cdot\sin\frac{\sigma x - \sigma_0}{2}\exp(i\sigma x/2), \end{aligned}$$

where $c = (-2i)\exp(i\sigma_0/2)$, the symbol ω^+ given by (6.26) is generated by the "pure" sine function

$$s(z) = \sin\left[\frac{\sigma}{2}\left(z - \frac{\sigma_0}{\sigma}\right)\right].$$

Symbols of the form (6.26) were considered above in Section 1.7 and are of importance in the theory of integro-difference equations of convolution type.

(2) The function $\sin((\pi - \sigma z)/2)$ can be written as an infinite product (3.28):

$$\sin\frac{\pi - \sigma z}{2} = \prod_{j\in\mathbb{Z}}\left(1 - \frac{z}{(2j+1)\pi\sigma^{-1}}\right).$$

If $\{x_j\} \in l_r(\mathbb{Z})$, where $x_j \in \mathbb{R}$ and $1 < r < \infty$, then

$$\omega^+(x) = \exp(i\sigma x/2) \prod_{j\in\mathbb{Z}} \left(1 - \frac{x}{(2j+1)\pi\sigma^{-1} + x_j}\right) \tag{6.27}$$

is a sine-type function of type σ, and $\zeta(\omega^+) = \{(2j+1)\pi\sigma^{-1}+x_j\}_{j\in\mathbb{Z}}$ is a separated real set for an appropriate choice of the points $\{x_j\}$.

(3) Let $\nu \in (1/p, 1/p+1)$, $1 < p < \infty$, $\sigma > 0$, and let

$$E_1(i\sigma x, \nu) = \sum_{k=0}^{\infty} \frac{(i\sigma x)^k}{\Gamma(\nu+k)}$$

be a function of Mittag-Leffler type. The sine-type function

$$\omega^+(x) = \frac{1}{2i} \exp(i\sigma x/2)[E_1(i\sigma x/2, \nu) - E_1(-i\sigma x/2, \nu)] \tag{6.28}$$

of type σ is encountered in the theory of differential equations of fractional order (see Dzhrbash-yan [3], were one can find also a description of its set of zeros).

Example 6.2 *Meromorphic functions connected with a Blaschke product.*
This example is based on the following sharper form of Proposition 3.20 for Blaschke products with an arbitrary separated set of simple zeros that lie on a line.

Proposition 6.17 *Let*

$$B(z) = \prod_{k\in\mathbb{Z}} \nu_k \frac{z - z_k}{z - \overline{z}_k}, \tag{6.29}$$

where $z_k = x_k + ir$, $r > 0$, $|x_k - x_j| \geq d > 0$, $k \neq j$, $\nu_k = \overline{z}_k / z_k$. Further, given $\delta > 0$, denote $c_\delta(\overline{z}_k) = \{z : |z - \overline{z}_k| \leq \delta\}$. Then for any integer $m \geq 0$ and any real number $\delta > 0$ the function $(d^m/dz^m)B(z)$ is uniformly bounded on the set $\mathbb{C}_\delta = \mathbb{C} \setminus \bigcup_{k\in\mathbb{Z}} c_\delta(\overline{z}_k)$.

Proof. First of all, notice that the set of zeros $\{z_k\}$ of the function $B(z)$ lies on the line $\mathbb{R} + ir$ and is separated. Therefore, the Blaschke condition (1.19), which guarantees the convergence of the infinite product (6.29), is satisfied.

By Proposition 3.20, the product (6.29) is an analytic and uniformly bounded function in $\mathbb{C}_\delta$. The uniform convergence of this infinite product in a neighborhood of a point $z \in \mathbb{C}_\delta \setminus \{z_k\}$ is equivalent to the uniform convergence of the series

$$\log B(z) = \sum_{k\in\mathbb{Z}} \log \left(\frac{\overline{z}_k}{z_k} \cdot \frac{z - z_k}{z - \overline{z}_k} \right)$$

in that neighborhood for an appropriate choice of branch of the function $\log z$. Differentiating this equality, we obtain

$$(\log B(z))' = \sum_{k\in\mathbb{Z}} \frac{2ir}{(z - z_k)(z - \overline{z}_k)}.$$

It follows that in a neighborhood of the point in question the function $B'(z)$ has an expansion in a uniformly convergent series of functions:

$$B'(z) = B(z)(\log B(z))' = \sum_{k\in\mathbb{Z}} \frac{2irB_k(z)}{(z-\overline{z}_k)^2}, \tag{6.30}$$

where we put $B_k(z) = B(z)\frac{z-\overline{z}_k}{z-z_k}$. We claim that even more is true, namely, that the above series converges at any point $z \in \mathbb{C}_\delta$ and that its sum is uniformly bounded in $\mathbb{C}_\delta$. This in turn will imply the uniform boundedness of the function $B'(z)$ in $\mathbb{C}_\delta$.

Indeed, $|B_k(z)| \leq 1$ for all $z \in \overline{\Pi}_+$ and all $k \in \mathbb{Z}$, whereas $|B_k(z)| \leq |B(z)|$ for all $z \in \mathbb{C}_\delta \setminus \Pi_+$ and all $k \in \mathbb{Z}$. Hence, by Proposition 3.20, the functions $B_k(z)$ are uniformly bounded for $z \in \mathbb{C}_\delta$ and $k \in \mathbb{Z}$. It follows that

$$\left|\sum_{k\in\mathbb{Z}} \frac{2irB_k(z)}{(z-\overline{z}_k)^2}\right| \leq c\sum_{k\in\mathbb{Z}} \frac{1}{|z-\overline{z}_k|^2} = c\sum_{k\in\mathbb{Z}} \frac{1}{(x-x_k)^2+(y+r)^2},$$

where $(x-x_k)^2+(y+r)^2 > \delta^2$.

Given any $x \in \mathbb{R}$, one can find a $j \in \mathbb{Z}$ such that

$$\frac{1}{2}(x_j - x_{j-1}) < x \leq \frac{1}{2}(x_j + x_{j+1}).$$

Moreover, thanks to the separation property,

$$|x - x_{j\pm n}| \geq \frac{nd}{2}, \qquad n \in \mathbb{N}.$$

Therefore,

$$\begin{aligned}\sum_{k\in\mathbb{Z}} \frac{1}{(x-x_k)^2+(y+r)^2} &\leq \frac{1}{\delta^2} + \sum_{|k-j|\geq 1} \frac{1}{(x-x_{j+(k-j)})^2} \\ &\leq \frac{1}{\delta^2} + 2\sum_{n=1}^{\infty}\left(\frac{nd}{2}\right)^{-2} = \frac{1}{\delta^2} + \frac{8}{d^2}\sum_{n=1}^{\infty}\frac{1}{n^2} < \infty.\end{aligned}$$

If now $m \geq 2$, we use the identity

$$\frac{d^m}{dz^m}B(z) = \sum_{k=1}^{m} c_k B^{(k-1)}(z)(\log B)^{(m-k+1)}(z),$$

where

$$(\log B)^{(m)}(z) = (-1)^m(m-1)!\sum_{j\in\mathbb{Z}}\left[\frac{1}{(z-\overline{z}_j)^m} - \frac{1}{(z-z_j)^m}\right]$$

and $c_k \in \mathbb{N}$, which is obtained by differentiating identity (6.30) $m-1$ times. The proof of the proposition in the general case is carried out by induction. □

Starting with a function $B(z)$ of the form (6.28), consider the function

$$\omega^{\pm}(x) = \exp(\pm i\theta) \prod_{k\in\mathbb{Z}} \nu_k^{\pm 1} \frac{x - x_k}{x - x_k \pm 2ir}, \quad \theta \in \mathbb{R}. \tag{6.31}$$

Observe that

$$\omega^{+}(x) = \exp(i\theta)B(x + ir), \quad \omega^{-}(x) = \overline{\exp(i\theta)B(x + ir)}.$$

By Proposition 6.17, $\omega^{\pm} \in C_B^{\infty}(\mathbb{R})$. Therefore, $\omega^{\pm} \in \mathcal{K}^{\pm}(\mathbb{R})$. The corresponding model functions $h^{\pm}$ are

$$h^{\pm}(x) = \left(\prod_{k\in\mathbb{Z}} \mu_k \frac{x - x_k - 2ir}{x - x_r + 2ir}\right)^{\pm 1}, \quad \mu_k = \frac{x_k - 2ir}{x_k + 2ir}, \tag{6.32}$$

with $(h^{+})^{-1} = h^{-}$, $h^{-}\omega^{+} = \omega^{-}$, $\exp(-2i\theta) = \prod_{k\in\mathbb{Z}} \overline{\mu_k} \nu_k^2$. We see that conditions (6.18) are satisfied. The fact that $\zeta(\omega^{\pm}) = \{x_k\}_{k\in\mathbb{Z}}$ is an interpolation set in $\mathcal{M}(h^{\pm})$ follows from Theorem 3.5.

Let us verify that the functions $\omega^{\pm}$ satisfy conditions (2) and (3) in the definition of the classes $\widetilde{\mathcal{K}}^{\pm}(\mathbb{R})$. Since

$$\exp(-i\theta)(\omega^{+})'(x) = \exp(-i\theta)\omega^{+}(x)(\log \omega^{+}(x))' = B(x + ir)(\log B(x + ir))',$$

equality (6.30) yields

$$\exp(-i\theta)(\omega^{+})'(x_j) = \sum_{k\in\mathbb{Z}} \frac{2irB_k(x_j + ir)}{(x_j - x_k + 2ir)^2} = \frac{B_j(x_j + ir)}{2ir},$$

where, as one recalls, $B_j(z) = B(z)\frac{z - \overline{z}_j}{z - z_j}$. Further, using inequality (3.43) for the functions $B_j(z)$, we have

$$|B_j(z)| \ge \exp\left[\left(1 + 2\log\frac{1}{a}\right) \sum_{k\ne j} \frac{-2yr}{|z - \overline{z}_k|^2}\right], \tag{6.33}$$

where

$$a = \inf_{k\ne j} \left|\frac{z - z_k}{z - \overline{z}_k}\right| > 0.$$

Let $\delta_j = \inf_{k\ne j} |x_j - x_k|$. Obviously, $\delta_j \ge d$ for all $j \in \mathbb{Z}$. Then for any $k \in \mathbb{Z}$, $k \ne j$,

$$1 > \frac{|x_j - x_k|}{|x_j - x_k + 2ir|} \ge \frac{\delta_j}{\sqrt{\delta_j^2 + 4r^2}} \ge \frac{d}{\sqrt{d^2 + 4r^2}} > 0.$$

It follows that for any $j \in \mathbb{Z}$,

$$1 \ge a_j = \inf_{k\ne j} \left|\frac{x_j - x_k}{x_j - x_k + 2ir}\right| \ge \frac{d}{\sqrt{d^2 + 4r^2}} > 0 \qquad \text{and} \qquad 1 + 2\log\frac{1}{a_j} \ge 1.$$

Setting $z = x_j + ir$ in (6.33), we conclude that for any $j \in \mathbb{Z}$,

$$|B_j(x_j + ir)| \geq \exp\left(\sum_{k \neq j} \frac{-2r^2}{(x_j - x_k)^2 + 4r^2}\right) > \exp\left(\sum_{n=1}^{\infty} \frac{-2r^2}{(nd)^2}\right) = c > 0,$$

where the constant c does not depend on j. But then

$$|(\omega^+)'(x_j)| > \frac{c}{2r} > 0$$

for all $j \in \mathbb{Z}$. Thus, condition (2) of the definition of the class $\widetilde{\mathcal{K}}^+(\mathbb{R})$ holds for $\omega^+(x)$. Turning our attention to condition (3) of that definition, we remark that $x \in \mathbb{R} \setminus V_\varepsilon$ implies that $x + ir \in \mathbb{C} \setminus \bigcup_{j \in \mathbb{Z}} c_\varepsilon(z_j)$, and therefore

$$|\omega^+(x)| = |B(x + ir)| \geq M_3\varepsilon \qquad \text{for } x \in \mathbb{R} \setminus V_\varepsilon,$$

thanks to inequality (3.46).

In a similar manner one can verify that ω^- belongs to $\widetilde{\mathcal{K}}^-(\mathbb{R})$.

Thus, if the function ω^- has the form (6.31), then the operator $T(a)$ with a symbol a of the form (6.24) satisfies all the assumptions of Theorem 6.12.

Let us make two remarks concerning Examples 6.1 and 6.2.

Remark 6.6 In general, not every separated set is the set of zeros of some sine-type function (see Levin and Ostrovskiĭ [1]). For that reason, in Example 6.2 we considered a more general distribution of the zeros of the symbol – an arbitrary separated real set.

The study of symbols of the form (6.25) is of independent interest for applications (integro-difference equations, diffraction theory, and so on).

Remark 6.7 Functions of the form (6.31) play an important role in the theory of Toeplitz operators with infinitely differentiable symbols. The finite products of the form (6.31) provide the largest class of sets of zeros for which an operator $T(a)$ whose symbol degenerates on such a set remains a Φ-operator in the appropriate space of infinitely differentiable functions.

References and comments

6.1–6.2 For Theorems 6.1–6.4 the reader is referred to Gohberg and Krupnik [3]. The normalization principle for linear operators was formulated in the case of convolution-type operators by Dybin in 1967 [2], [3]. However, already in 1956, Khvedelidze [1] has in fact carried out the normalization of a singular integral operator in the space L_p. The history of the development of this research direction in the following twenty years is reflected in the monographs of Prössdorf [1] and Dybin [13].

The first investigations of singular integral operators whose coefficients are allowed to have infinitely many zeros were carried out in the papers of Gaponenko and Dybin [1], Gaponenko [1], and Dybin and Dzhirgalova [1], [2].

N.V. Govorov's approach to this problem is discussed by Zhuravleva [1], [2], Rogozin [1], and Rogozin and Tolochko [1].

For acquaintance with the modern development of normalization theory and its applications the reader is referred to the works of Volevich and Gindikin [1], Vladimirov [1]–[3], Kravchenko [1], Kravchenko and Litvinchuk [1], Drekova and Kravchenko [1], and Kravchenko, Lebre, Litvinchuk, and Teixeira [1].

Theorem 6.10 is based on Khaĭkin's paper [1], which deals with the finite-index case.

6.3 The case of polynomial degeneracy of the symbol in the Riemann boundary value problem, which is closely related to the operator $T(a)$, was studied already in the 1940s by F.D. Gahov; for the history of the problem, see Gahov [2]. This case, in the setting of one-dimensional integral equations and equations of convolution type, is the object of study of a large part of Prössdorf's monograph [1].

6.4–6.5 The first investigations of Toeplitz operators whose symbols have locally-polynomial degeneracies of general form were carried out in the works of Dybin [5], [7], [8], [11]. The treatment in these sections follows his paper [11], in which he also studied the operators Π and $\Pi^\top$ in $L_p(\Gamma, \varrho)$, $1 < p < \infty$, under the assumption that the coefficients may have locally-polynomial degeneracies.

The proof of Proposition 6.14 was communicated to us by E.M. Dynkin.

Example 6.1 is from Dybin's work [7], where it is also shown how to transfer the corresponding results from the real line $\mathbb{R}$ to the case of a smooth closed contour.

Example 6.2 is taken from Dybin's paper [11].

In 1966, Dybin and Karapetyants [1] constructed the countably-normed space $\mathcal{L}_2^\infty(\mathbb{R})$ ($f \in \mathcal{L}_2^\infty(\mathbb{R}) \iff f^{(k)} \in L_2(\mathbb{R}), k = 0, 1, \ldots$), in which the singular integral operators Π and $\Pi^\top$ with infinitely differentiable coefficients with polynomial degeneracies on the real line $\mathbb{R}$ have one-sided continuous inverses. At the same time there arose the question of which sets of zeros are the coefficients of the operators Π and $\Pi^\top$ allowed to have so that properties like continuous invertibility, Fredholmness, semi-Fredholmness, or normal solvability are preserved. Pioneering work in this direction was done by Silbermann [1], who showed that the coefficients of a singular integral operator that is a Φ-operator in the algebra $C^\infty(\Gamma_0)$ are allowed to have only polynomial degeneracy.

The paper of Gorodetskiĭ and Dybin [1] sharpened Silbermann's results in the case of the algebra $\mathbb{C} + W^\infty(\mathbb{R})$ by showing that no zero is allowed at infinity.

Further progress in this area is connected with symbols of the form (6.31). A set of points on the real line is said to be *weakly nonseparated* if it can be written as a union of finitely many separated sets. The generators of weakly nonseparated real sets are finite products of functions of the form (6.31).

Dybin [15] has shown that an operator $T(a)$ with symbol a in the class $C_B^\infty(\mathbb{R})$ that has a weakly nonseparated set of real zeros, the set of multiplicities of which is bounded, has a continuous inverse in the space $\mathcal{L}_2^{\infty+}(\mathbb{R}) = P^+(\mathcal{L}_2^\infty(\mathbb{R}))$, at least from one side. From the paper of Dybin and Smirnova [1] it follows that the indicated restrictions, imposed on the set of zeros of the symbol a, are necessary for $T(a)$ to be a Φ_+-operator in $\mathcal{L}_2^{\infty+}(\mathbb{R})$.

Dybin [14] has shown that the operator $N_{\omega^\pm}$ given by (6.15) enables one to solve the regularization problem for integrals that diverge on countable sets, which plays an important role in the theory of distributions. This regularization method is an alternative to the method given in Gel′fand and Shilov [1], Chap. I, §3, which uses infinite partitions of unity, and in contrast to the latter it preserves the analyticity properties of functions.

References

Abrahamse, M.B.

1. *The spectrum of a Toeplitz operator with a multiplicatively periodic symbol,* J. Funct. Anal. **31** (1979), no. 2, 224–233. [MR **82b:47029**]

Akhiezer, N.I.

1. *Lectures on Approximation Theory*, Second, revised and enlarged edition, "Nauka," Moscow, 1965. (Russian) [MR **32** #6108]; English transl. of first edition *Theory of Approximation*, Frederick Ungar Publishing Co., New York, 1956 [MR **20 #1872**]

Aiena, P. and González, M.

1. *On the perturbation classes of semi-Fredholm and Fredholm operators.* Proceedings of the Workshop "Functional Analysis: Methods and Applications" (Cosenza 1995); Rend. Circ. Math. Palermo (2) Suppl. No. 40 (1996), 37–46. [MR **97f:47010**]

Aĭrapetyan, G.M.

1. *A basis of rational functions in subspaces of the Hardy classes* H_p $(1 < p < \infty)$, Izv. Akad. Nauk Armjan. SSR Ser. Mat. **8** (1973), no. 6, 429–450. (Russian) [MR **49 #7451**]
2. *Multiple interpolation and the property of being a basis for certain biorthogonal systems of rational functions in the Hardy* H_p *classes*, Izv. Akad. Nauk Armjan. SSR Ser. Mat. **12** (1977), no. 4, 262–277. (Russian) [MR **58** #11419]

Alehno, A.G.

1. *On a Riemann boundary value problem with a finite number of whirl points,* Dokl. Akad. Nauk BSSR **23** (1971), no. 12, 87–112. (Russian)
2. *On a Riemann boundary value problem with a coefficient that has an arbitrary finite oscillation of the argument,* Vestn. Beloruss. Gos. Univ. Ser. 1 Fiz. Mat. Inform. no. 3 (1997) 46–49. (Russian) [MR **99m:30082**]
3. *The inhomogeneous Riemann boundary value problem with an infinite index in the case of multilateral whirling,* Dokl. Akad. Nauk Belarusi **41** (1997), no. 5, 35–43. (Russian) [MR **99m:30081**]

Alekna, P.Yu.

1. *The homogeneous Riemann boundary value problem with infinite index of logarithmic order for the half-plane*, Litovsk. Mat. Sb. **13** (1973), no. 3, 5–13. (Russian) [MR **48 #6437**]
2. *Necessary and sufficient conditions for the solvability of an inhomogeneous Riemann boundary value problem with a negative infinite index of logarithmic order* $\min(\alpha, \beta) > 1$, Litovsk. Mat. Sb. **18** (1978), no. 3, 5–14. (Russian) [MR **80b:30032**]

Aleman, A., Richter S. and Sundberg, C.

1. *Beurling's theorem for the Bergman space*, Acta Math. **177** (1996), no. 2, 275–310. [MR **98a:46034**]

Arov, D.Z.

1. *On strictly positive compressions of block Toeplitz operators.* Indiana Univ. Math. J. **46** (1997), no. 2, 357–373. [MR **98j:47058**]

Axler, S., Chang, S-Y.A. and Sarason, D.

1. *Products of Toeplitz operators*, Integral Equations Operator Theory **1** (1977), no. 3, 285–309. [MR **80d:47039**]

Babeshko, V.A.

1. *Generalized Factorization Method in Three-dimensional Dynamic Mixed Problems of Elasticity Theory*, "Nauka," Moscow, 1984. (Russian) [MR **87b:73012**]

Bantsuri, R.D. and Dzhanashiya, G.A.

1. *Equations of convolution type for the semi-axis*, Dokl. Akad. Nauk SSSR **155** (1964) 251–253. (Russian) [MR **28 #3301**]

Bart, H., Gohberg, I. and Kaashoek, M.A.

1. *Minimal Factorization of Matrix and Operator Functions*, Operator Theory: Advances and Applications, 1. Birkhäuser Verlag, Basel and Boston, 1979. [MR **81a:47001**]

Bastos, M.A., Karlovich Yu.I., dos Santos, A.F. and Tishin, P.M.

1. *The corona theorem and the existence of canonical factorization of triangular AP matrix functions*, J. Math. Anal. Appl. **223** (1998), no. 2, 494–522. [MR **99f:47022a**]
2. *The corona theorem and the canonical factorization of triangular matrix functions – effective criteria and explicit formulas*, J. Math. Anal. Appl. **223** (1998), no. 2, 523–550. [MR **99f:47022b**]

Bastos, B.A., Karlovich, Yu.I., Spitkovsky, I.M. and Tishin, P.M.

1. *On a new algorithm for almost-periodic factorization*, in: Recent Progress in Operator Theory (Regensburg, 1995), 53–74, Oper. Theory Adv. Appl., 103, Birkhäuser, Basel, 1998. [MR **99g:47034**]

Bastos, M.A., dos Santos, A.F. and Duduchava, R.

1. *Finite interval convolution operators on the Bessel potential spaces* H_p^s, Math. Nachr. **173** (1995), 49–63. [MR **96e:47032**]

Ben-Artzi, A., Gohberg, I. and Kaashoek, M.

1. *A time varying generalization of the canonical factorization theorem for Toeplitz operators*, Indag. Math. (N.S.) **4** (1993), no. 4, 385–405. [MR **94j:47040**]

Berkovich, F.D.

1. *Convolution-type integral equations with infinitely many solutions* in: Mathematical Analysis and its Applications, 27–33, Rostov-on-Don, 1969. (Russian)

Berkovich, F.D. and Konyshkova, E.M.

1. *On a case of the Riemann boundary value problems with infinite index*, Soobshch. Nauch. Mat. Obs., Rostov-on-Don, 158–164, 1968. (Russian)

Bernard, A., Garnett, J.B. and Marshall, D.E.

1. *Algebras generated by inner functions.* J. Functional Anal., **25** (1977), no. 3, 275–285. [MR **56** #1072]

Birger, I.I. and Dybin, V.B.

1. *Homogeneous Riemann boundary value problem whose coefficients have power-logarithmic singularities*, in: Integral and Differential Equations and Approximate Solutions, 11–23, Kalmytsk. Gos. Univ., Elista, 1985. (Russian) [MR **88e:30105**]

Bonsall, F.F. and Gillespie, T.A.

1. *Hankel operators with PC symbols and the space* $H^\infty + PC$, Proc. Roy. Soc., Edinburg **A86** (1981), no. 1–2, 17–24. [MR **83b:47038**]

Böttcher, A.

1. *On Toeplitz operators generated by symbols with three essential cluster points*, Preprint P–MATH–04/86, Akad. Wiss. DDR, Inst. Math., Berlin, 1986.
2. *Scalar Toeplitz operators, distance estimates, and localization over subalgebras of* $C + H^\infty$, in: Operator Equations and Numerical Analysis (Seminar Analysis 1985/86), 1–17, Akad. Wiss. DDR, Inst. Math., Berlin, 1986. [MR **88e:47046**]

Böttcher, A. and Grudsky, S.M.

1. *Toeplitz operators with discontinuous symbols: phenomena beyond piecewise continuity*, in: Singular Integral Operators and Related Topics (Tel Aviv, 1995), 55–118, Oper. Theory Adv. Appl., 90, Birkhäuser, Basel, 1996. [MR **97k:47025**]
2. *On the composition of Muckenhoupt weights and inner functions*, J. London Math. Soc., (2) **58** (1998), no. 1, 172–184. [MR **99j:47033**]

Böttcher, A., Grudsky, S.M. and Spitkovsky, I.M.

1. *The spectrum is discontinuous on the manifold of Toeplitz operators*, Arch. Math. (Basel) **75** (2000), no. 1, 46–52. [MR 2001c:47034]
2. *On the Fredholm indices of associated systems of Wiener–Hopf equations*, J. Integral Equations Appl. **12** (2000), no. 1, 1–29. [MR **2001c:47052**]
3. *Matrix functions with arbitrarily prescribed left and right partial indices*, Integral Equations Operator Theory **36** (2000), no. 1, 71–91. [MR **2000j:47033**]
4. *Toeplitz operators with frequency modulated semi-almost-periodic symbols*, J. Fourier Anal. Appl. **7** (2001), no. 5, 523–535.
5. *On the essential spectrum of Toeplitz operators with semi-almost-periodic symbols*, in Oper. Theory Adv. Appl. (to appear).

Böttcher, A. and Karlovich, Yu.I.

1. *Carleson Curves, Muckenhoupt Weights, and Toeplitz Operators*, Progress in Math., Vol. 154, Birkhäuser Verlag, Basel, 1997. [MR **98m:47031**]

Böttcher, A., Karlovich, Yu.I. and Rabinovich, V.S.

1. *Emergence, persistence, and disappearance of logarithmic spirals in the spectra of singular integral operators.* Integral Equations and Operator Theory **25** (1996), 405–444.
2. *Mellin pseudodifferential operators with slowly varying symbols and singular integrals on Carleson curves with Muckehoupt weights.* Manuscripta math. **95** (1998), 363–376.
3. *The method of limit operators for one-dimensional singular integrals with oscillating data.* J. Operator Theory **43** (2000), 171–198.

Böttcher, A., Karlovich, Yu.I. and Spitkovsky, I.M.

1. *Toeplitz operators with semi-almost-periodic symbols on spaces with Muckenhoupt weight*, Integral Equations Operator Theory **18** (1994), no. 3, 261–276. [MR **95e:47034**]
2. *Toeplitz operators with semi-almost-periodic matrix symbols on Hardy spaces*, Acta Appl. Math. **65** (2001), no. 1–3, 115–136.
3. *Convolution Operators and Factorization of Almost Periodic Matrix Functions*, Birkhäuser Verlag, Basel, 2002.

Böttcher, A. and Seybold, M.

1. *Discrete Wiener-Hopf operators on spaces with Muckenhoupt weight*, Studia Math. **143** (2000), no. 2, 121–144. [MR **2001m:47044**]

Böttcher, A. and Silbermann, B.

1. *Toeplitz operators in l^p with $C + H^\infty$ symbols*, Zap. Nauchn. Sem. Leningrad. Otdel. Mat. Inst. Steklov. (LOMI) **157** (1987), Issled. po Linein. Operator. i Teorii Funktsii. XVI, 124–128. (Russian); English transl. in J. Soviet Math. **44** (1989), no. 6, 834–836. [MR **88g:47055**]
2. *Analysis of Toeplitz Operators*, Springer-Verlag, Berlin, 1990. [MR **92e:47001**]
3. *Introduction to Large Truncated Toeplitz Matrices*, Universitext, Springer-Verlag, New York, 1999. [MR **2001b:47043**]

Böttcher, A., Silbermann, B. and Spitkovsky I.M.

1. *Toeplitz operators with piecewise quasisectorial symbols*, Bull. London Math. Soc. **22** (1990), no. 3, 281–286. [MR **90m:47042**]

Böttcher, A. and Spitkovsky, I.M.

1. *Toeplitz operators with PCQ symbols on weighted Hardy spaces*, J. Funct. Anal. **97** (1991), no. 1, 194–214. [MR **92k:47046**]
2. *Wiener-Hopf integral operators with PC symbols on spaces with Muckenhoupt weight*, Rev. Mat. Iberoamericana **9** (1993), no. 2, 257–279. [MR **94e:45004**]

Boutet de Monvel, L. and Guillemin, V.

1. *The Spectral Theory of Toeplitz Operators*, Annals of Mathematics Studies, 99, Princeton University Press, Princeton, NJ; University of Tokyo Press, Tokyo, 1981. [MR **85j:58141**]

Calderón, A.P.

1. *Commutators of singular integral operators*, Proc. Nat. Acad. Sci. U.S.A. **53** (1965), 1092–1099. [MR **31 #1575**]
2. *Cauchy integrals on Lipschitz curves and related operators*, Proc. Nat. Acad. Sci. U.S.A. **74** (1977), no. 4, 1324–1327. [MR **57 #6445**]
3. *Commutators, singular integrals on Lipschitz curves and applications,* in: Proceedings of the International Congress of Mathematicians (Helsinki, 1978), 85–96, Acad. Sci. Fennica, Helsinki, 1980. [MR **82f:42016**]

Calderón, A.P., Calderón, C.P., Fabes, E., Jodeit, M. and Rivière, N.M.

1. *Applications of the Cauchy integral on Lipschitz curves*, Bull. Amer. Math. Soc. **84** (1978), no. 2, 287–290. [MR **57 #649**]

Carleson, L.

1. *An interpolation problem for bounded analytic functions*, Amer. J. Math. **80** (1958), 921–930. [MR **22 #8129**]

Chibrikova, L.I.

1. *A special case of the Riemann problem on an unbounded contour.* I., Trudy Sem. Kraev. Zadacham no. 15 (1978), 139–167. (Russian) [MR **81j:30064**]
2. *A special case of the Riemann problem on an unbounded contour.* II., Trudy Sem. Kraev. Zadacham no. 16 (1979), 185–201. (Russian) [MR **82d:30041**]

Clancey, K. and Gohberg, I.

1. *Factorization of Matrix Functions and Singular Integral Operators*, Operator Theory: Advances and Applications, 3. Birkhäuser Verlag, Basel and Boston, 1981. [MR **84a:47016**]

Clancey, K.F. and Gosselin, J.A.

1. *On the local theory of Toeplitz operators*, Illinois J. Math. **22** (1978), no. 3, 449–458. [MR **58 #2419**]

Clancey, K.F. and Morrel, B.B.

1. *The essential spectrum of some Toeplitz operators*, Proc. Amer. Math. Soc. **44** (1974), 129–134. [MR **49 #5912**]

Coburn, L.A.

1. *The C^*-algebra generated by an isometry*, Bull. Amer. Math. Soc. **73** (1967), 722–726. [MR **35 #4760**]
2. *The C^*-algebra generated by an isometry*, II, Trans. Amer. Math. Soc. **137** (1969), 211–217. [MR **38 #5015**]

Coburn, L.A. and Douglas, R.G.

1. *Translation operators on the half-line*, Proc. Nat. Acad. Sci. U.S.A. **62** (1969), 1010–1013. [MR **43 #985**]

Coburn, L.A., Douglas, R.G., Schaeffer, D. and Singer, I.

1. *On the C^*-algebras of operators on a half-space.* II. *Index theory*, Inst. Hautes Études Sci. Publ. Math. no. 40 (1971), 69–79. [MR **50 #10884**]

Coifman, R., David, G. and Meyer, Y.

1. *La solution des conjectures de Calderón*, Adv. in Math. **48** (1983), no. 2, 144–148. [MR **84i:42025**]

Danilov, E.A.

1. *Dependence of the number of solutions of a homogeneous Riemann problem on the contour and the modulus of the coefficient*, Dokl. Akad. Nauk SSSR **264** (1982), no. 6, 1305–1308; correction: Dokl. Akad. Nauk SSSR **267** (1982), no. 5, 1032. (Russian) [MR **84g:30044a,b**]

Danilyuk, I.I.

1. *Nonregular Boundary Value Problems in the Plane*, "Nauka," Moscow, 1975. (Russian) [MR **58 #6269**]

David, G.

1. *Opérateurs intégraux singuliers sur certaines courbes du plan complexe*, Ann. Sci. École Norm. Sup. (4) **17** (1984), no. 1, 157–189. [MR **85k:42026**]

Devinatz, A.

1. *Toeplitz operators on H^2 spaces*, Trans. Amer. Math. Soc. **112** (1964), 304–317. [MR **29 #477**]

Douglas, R.G.

1. *Toeplitz and Wiener-Hopf operators in $H^\infty + C$*, Bull. Amer. Math. Soc. **774** (1968), 895–899. [MR **37 #4648**]
2. *Banach Algebra Techniques in Operator Theory*, Second edition. Graduate Texts in Mathematics, 179, Springer-Verlag, New York, 1998. [MR **99c:47001**]
3. *Banach Algebra Techniques in the Theory of Toeplitz Operators*, Conference Board of the Mathematical Sciences Regional Conference Series in Mathematics, No. 15. American Mathematical Society, Providence, R.I., 1973. [MR **50 #14336**]
4. *Local Toeplitz operators*, Proc. London Math. Soc. (3) **36** (1978), no. 2, 243–272 [MR **58 #2421**]

Douglas, R.G. and Rudin, W.

1. *Approximation by inner functions*, Pacific J. Math. **31** (1969), 313–320. [MR **40 #7814**]

Douglas, R.G. and Sarason, D.

1. *Fredholm–Toeplitz operators*, Proc. Amer. Math. Soc. **26** (1970), 117–120. [MR **41 #4275**]
2. *A class of Toeplitz operators*, Indiana Univ. Math. J. **20** (1971), no. 10, 891–895. [MR **53 #11414**]

Douglas, R.G. and Widom, H.

1. *Toeplitz operators with locally sectorial symbols*, Indiana Univ. Math. J. **20** (1970/1971), 385–388. [MR **41 #9024**]

Drekova, G.V. and Kravchenko, V.G.

1. *On the theory of the normalization of the Riemann problem*, Izv. Vyssh. Uchebn. Zaved. Mat., (1991), no. 9, 20–28. (Russian); English transl. in Soviet Math. (Iz. VUZ) **35** (1991), no. 9, 19–27. [MR **93h:47082**]

Duduchava, R.

1. *Integral Equations in Convolution with Discontinuous Presymbols, Singular Integral Equations with Fixed Singularities, and their Applications to some Problems of Mechanics*, Teubner-Texte zur Mathematik, B. G. Teubner Verlagsgesellschaft, Leipzig, 1979. [MR **82c:45001b**]

Duduchava, R.V. and Saginashvili, A.I.

1. *Convolution integral operators on the half-line with semi-almost-periodic presymbols*, Soobshch. Akad. Nauk Gruzin. SSR **98** (1980), no. 1, 21–24. (Russian) [MR **82k:47041**]
2. *Integral equations of convolution on the half-line with semi-almost-periodic presymbols*, Differentsial′nye Uravneniya **17** (1981), no. 2, 301–312. (Russian) [MR **82g: 45010**]

Dunford, N. and Schwartz, J.T.

1. *Linear operators.* I*: General Theory*, With the assistance of W. G. Bade and R. G. Bartle. Pure and Applied Mathematics, Vol. 7, Interscience Publishers, Inc., New York; Interscience Publishers, Ltd., London, 1958. [MR **22 #8302**]

Duren, P.L.

1. *Theory of H^p Spaces*, Pure and Applied Mathematics, Vol. 38, Academic Press, New York and London, 1970. [MR **42 #3552**]

Dwight, H.B.

1. *Tables of Integrals and Other Mathematical Data*, 4th ed. The Macmillan Company, New York, 1961. [MR **23 #B2613**]

Dybin, V.B.

1. *An exceptional case of integral equations of convolution type in the class of generalized functions*, Dokl. Akad. Nauk SSSR **161** (1965), 753–756. (Russian) [MR **31 #1528**]
2. *An exceptional case of dual integral equation of convolution type*, Dokl. Akad. Nauk SSSR **176** (1967), 251–254. (Russian); English transl. in Soviet Math. Dokl., **8** (1967), no. 5, 1073–1077. [MR **37 #1917**]
3. *The Wiener-Hopf integral operator in classes of functions with polynomial behavior at infinity*, Izv. Akad. Nauk Armjan. SSR Ser. Mat. **2** (1967), no. 4, 250–270. (Russian) [MR **37 #4513**]
4. *A singular integral operator on the real axis with almost-periodic coefficients*, in: Theory of Functions. Differential Equations and their Applications, no. 1, 98–108, Kalmyck. Gos. Univ., Elista, 1976. (Russian) [MR **58 #30498**]
5. *The Riemann boundary value problem on a closed contour with an infinite number of zeros in its coefficient*, in: Problems and Methods of Mathematical

Physics (Karl-Marx-Stadt, 1983), 27–31, Teubner-Texte Math., 63, Teubner, Leipzig, 1984. [MR **86e:30040**]

6. *Correct formulations of the Riemann boundary value problem with standard almost-periodic discontinuities in its coefficients*, Math. Nachr. **122** (1985), 301–338. (Russian) [MR **88f:47020**]
7. *Singular integral equations with coefficients that vanish on countable sets*, Math. Nachr. **124** (1985), 65–84. (Russian) [MR **87e:45001**]
8. *Correct formulation of the Riemann problem in the case of an infinite number of zeros in its coefficient*, Dokl. Akad. Nauk SSSR **280** (1985), no. 4, 785–788. (Russian); English transl. in Soviet Math. Dokl., **31** (1985), no. 1, 147–150. [MR **87a:30065**]
9. *Singular integral equations with a countable set of almost-periodic discontinuities in their coefficients*, Trudy Tbiliss. Mat. Inst. Razmadze Akad. Nauk Gruzin. SSR **82** (1986), 90–105. (Russian) [MR **88f:45005**]
10. *On bases in spaces of analytic functions with a countable set of essential singular points*, Izv. Akad. Nauk Armyan. SSR Ser. Mat., **21** (1986), no. 6, 566–582. (Russian) [MR **88j:30098**]
11. *One-dimensional singular integral equations with coefficients that vanish on countable sets.* Izv. Akad. Nauk SSSR Ser. Mat. **51** (1987), no. 5, 936–961 (Russian); English transl. in Math. USSR-Izv. **31** (1988), no. 2, 245–271. [MR **89f:45005**]
12. *The Wiener-Hopf equation with a vanishing symbol, and the limiting absorption principle*, Dokl. Akad. Nauk SSSR **300** (1988), no. 6, 1305–1308 (Russian); English transl. in Soviet Math. Dokl. **37** (1988), no. 3, 813–816. [MR **89h:45008**]
13. *Well-Posed Problems for Singular Integral Equations*, Rostov. Gos. Univ., Rostov-on-Don, 1988. (Russian) [MR **91a:45006**]
14. *Multiple interpolation in weighted Hardy classes, and regularization of divergent integrals*, Izv. Akad. Nauk Armyan. SSR Ser. Mat. **25** (1990), no. 3, 236–260 (Russian); English transl. in Soviet J. Contemporary Math. Anal. **25** (1990), no. 3, 22–44. [MR **92k:30056**]
15. *The Wiener-Hopf equation and the Blaschke product*, Mat. Sb. **181** (1990), no. 6, 779–803 (Russian); English transl. in Math. USSR-Sb. **70** (1991), no. 1, 205–230. [MR **91g:47018**]
16. *The Wiener-Hopf equation with a symbol that vanishes on a countable set*, Dokl. Rassh. Zased. Seminara Inst. Priklad. Mat. im I. N. Vekua **5** (1990), no.1, 54–58. (Russian)

Dybin, V.B. and Dodokhova, G.V.,

1. *Correct formulation of the Riemann boundary value problem on the line with almost-periodic discontinuous coefficient*, in: Mathematical Analysis and its

Applications, 12–22, 116, Rostov. Gos. Univ., Rostov-on-Don, 1983. (Russian)

2. *Correct formulation of the Riemann boundary value problem on a closed contour in the case of almost-periodic discontinuities in its coefficient*, Izv. Akad. Nauk Armyan. SSR Ser. Mat. **18** (1983), no. 5, 380–393. (Russian) [MR **85g:30072**]

Dybin, V.B. and Karapetyants, N.K.

1. *Integral equations of the convolution type in a class of generalized functions*, Sibirsk. Mat. Zh., **7** (1966), no. 9, 531–545. (Russian) [MR **33 #4634**]

2. *Application of the normalization method to a class of infinite systems of linear algebraic equations*, Izv. Vyssh. Uchebn. Zav. Mat., no. 10 (1967), 39–49. (Russian) [MR **39 #4643**]

Dybin, V.B. and Smirnova, I.Yu.

1. *One-dimensional integro-differential convolution-type equation in the class of function with power-exponential behavior at infinity. Power decay*, Rostov-on-Don, Deposited VINITI 30.10.96, N 3154–B96, 49 pages, 1996. (Russian)

2. *A convolution-type integro-differential operator on the line in the class* $\{\alpha, \beta\}_2$, in: Integro-Differential Operators and their Applications, no. 2, 63–68, Rostov-on-Don, Izdat. DGTU, 1997. (Russian)

Dyn′kin, E.M.

1. *Methods of the theory of singular integrals (the Hilbert transform and Calderón-Zygmund theory)*, in: Current Problems in Mathematics. Fundamental Directions, Vol. 15, 197–292, Itogi Nauki i Tekhniki, Akad. Nauk SSSR, Vsesoyuz. Inst. Nauchn. i Tekhn. Inform., Moscow, 1987 (Russian); English trannsl. in Commutative Harmonic Analysis, I, 167–259, Encyclopaedia Math. Sci., 15, Springer, Berlin, 1991. (see [MR **91b:42021**])

2. *Methods of the theory of singular integrals.* II. *The Littlewood-Paley theory and its applications*, in: Current Problems in Mathematics. Fundamental Directions, Vol. 42, 105–198, Itogi Nauki i Tekhniki, Akad. Nauk SSSR, Vsesoyuz. Inst. Nauchn. i Tekhn. Inform., Moscow, 1989 (Russian); English transl. in Commutative Harmonic Analysis, IV, 97–194, Encyclopaedia Math. Sci., 42, Springer, Berlin, 1992. [see MR **91j:42015**]

Dyn′kin, E.M. and Osilenker, B.P.

1. *Weighted estimates for singular integrals and their applications*, in: Mathematical Analysis, Vol. 21, 42–129, Itogi Nauki i Tekhniki, Akad. Nauk SSSR, Vsesoyuz. Inst. Nauchn. i Tekhn. Inform., Moscow, 1983. (Russian) [MR **85j:42034**]

Dzhirgalova, S.B. and Dybin, V.B.

1. *A Riemann boundary value problem on the circle with almost-periodic discontinuities in its coefficient in the nonelliptic case*, in: Differential and Integral Equations and their Applications, 44–53, Kalmytsk. Gos. Univ., Elista, 1983. (Russian) [MR **85m:30028**]
2. *A Wiener-Hopf integral-difference equation in the exceptional case*, Izv. Vyssh. Uchebn. Zaved. Mat. (1985) no. 5, 64–67. (Russian) [MR **87d:47036**]

Dzhrbashyan, M.M.

1. *Expansions in systems of rational functions with fixed poles*, Izv. Nauk Armjan. SSR Ser. Mat., no. 2 (1967), 3–51. (Russian) [MR **35 #4438**]
2. *The basis property of some biorthogonal systems, and the solution of a multiple interpolation problem in H^p classes in a half plane*, Izv. Akad. Nauk SSSR Ser. Mat. **42** (1978), no. 6, 1322–1384. (Russian) [MR **81d:30008**]
3. *Interpolation and spectral expansions associated with differential operators of fractional order*, Izv. Akad. Nauk Armyan. SSR Ser. Mat. **19** (1984), no. 2, 81–181. (Russian) [MR **85k:30060**]

Dzhrbashjan, M.M. and Rafaelyan, S.G.

1. *Entire functions of exponential type of weighted L_p classes*, Akad. Nauk Armyan. SSR Dokl. **73** (1981), no. 1, 29–36. (Russian) [MR **83h:30024**]

Ehrhardt, T., Roch, S. and Silbermann, B.

1. *Symbol calculus for singular integrals with operator-valued PCQ coefficients*, in: Singular Integral Operators and Related Topics (Tel Aviv, 1995), 182–203, Oper. Theory Adv. Appl., 90, Birkhäuser Verlag, Basel, 1996. [MR **98d:47107**]

Feldman, I., Krupnik, N. and Spitkovsky, I.

1. *Norms of the singular integral operator with Cauchy kernel along certain contours*, Integral Equations Operator Theory **24** (1996), no. 1, 68–80. [MR **96j:47046**]

Gahov, F.D.

1. *On Riemann's boundary problem for a system of n pairs of functions*, Dokl. Akad. Nauk SSSR (N.S.) **67** (1949), 601–606. (Russian) [MR **11,169d**]
2. *Boundary Value Problems*, Third edition, revised and augmented, "Nauka," Moscow, 1977 (Russian); English transl. Dover Publications, Inc., New York, 1990. [MR **58 #6270**]

Gahov, F.D. and Cherskiĭ, Yu.I.

1. *Equations of Convolution Type*, "Nauka," Moscow, 1978. (Russian) [MR **80j:45001**]

Gaponenko, V.N.

1. *A dual integro-difference operator with vanishing symbol in the space* $L_p(-\infty,\infty)$, Izv. Vyshch. Uchebn. Zaved. Matematika (1976), no. 5(168), 108–111. (Russian) [MR **58 #23768**]

Gaponenko, V.N. and Dybin, V.B.

1. *The Riemann boundary value problem with a quasiperiodic degeneracy of the coefficients*, Dokl. Akad. Nauk SSSR **212** (1973), 1046–1049. (Russian) [MR **49 #3158**]
2. *A Wiener-Hopf integro-difference equation with a vanishing symbol*, Mat. Issled. **7** (1972), no. 4(26), 50–59 (Russian) [MR **48 #12133**]

Garnett, J.

1. *Bounded Analytic Functions*, Pure and Applied Mathematics, 96. Academic Press, Inc. [Harcourt Brace Jovanovich, Publishers], New York and London, 1981. [MR **83g:30037**]

Garnett. J. and Nicolau, A.

1. *Interpolating Blaschke products generate* H^∞, Pacif. J. Math., **173** (1966), no. 2, 501–510. [MR **97f:30050**]

Gelfand, I.M., Raikov, D.A. and Shilov, G.E.

1. *Commutative Normed Rings*, Translated from the Russian, with a supplementary chapter, Chelsea Publishing Co., New York, 1964. [MR **34 #4940**]

Gelfand, I.M. and Shilov, G.E.

1. *Generalized Functions.* Vol. 1. *Properties and Operations*, Translated from the Russian, Academic Press [Harcourt Brace Jovanovich, Publishers], New York and London, 1964. [MR **55 #8786a**]

Gelfand, M. and Spitkovsky, I.M.

1. *Almost periodic factorization: applicability of the division algorithm*, in: Voronezh Winter Mathematical Schools, 97–109, Amer. Math. Soc. Transl. Ser. 2, 184, Amer. Math. Soc., Providence, RI, 1998. [MR **2000k:47019**]

Gohberg, I.

1. *On an application of the theory of normed rings to singular integral equations*, Uspekhi Matem. Nauk (N.S.) **7** (1952), no. 2 (48), 149–156. (Russian). [MR **14,54a**]
2. *The factorization problem in normed rings, functions of isometric and symmetric operators, and singular integral equations*, Uspekhi Matem. Nauk **19** (1964), no. 1 (115), 71–124. (Russian) [MR **29 #487**]

Gohberg, I. and Feldman, I.A.

1. *Wiener-Hopf integro-difference equations*, Dokl. Akad. Nauk SSSR **183** (1968), 25–28 (Russian); English transl. in Soviet Math. Dokl. **9** (1968), 1312–1316. [MR **44 #3096**]
2. *Integro-difference Wiener-Hopf equations*, Acta Sci. Math., **30** (1969), 119–137. (Russian) [MR **40 #7880**]
3. *Convolution Equations and Projection Methods for their Solution*, "Nauka," Moscow, 1971 (Russian); English transl. Translations of Mathematical Monographs, Vol. 41, American Mathematical Society, Providence, R.I., 1974. [MR **50 #8148, #8149**]

Gohberg, I., Kaashoek, M.A. and van Shagen, F.

1. *Similarity of operator blocks and canonical forms. II. Infinite dimensional case and Wiener-Hopf factorization*, in: Topics in Modern Operator Theory (Timişoara/Herculane, 1980), 121–170, Operator Theory: Adv. Appl., 2, Birkhäuser Verlag, Basel and Boston, 1981. [MR **84a:47032**]

Gohberg, I. and Kreĭn, M.G.

1. *The basic propositions on defect numbers, root numbers and indices of linear operators*, Uspekhi Mat. Nauk (N.S.) **12** (1957), no. 2(74), 43–118 (Russian); English transl. in Amer. Math. Soc. Transl. (2) **13** (1960), 185–264. [MR **20 #3459, 22 #3984**]
2. *Systems of integral equations on the half-line with kernels depending on the difference of the arguments*, Uspekhi Mat. Nauk (N.S.) **13** (1958), no. 2 (80), 3–72 (Russian); English transl. in Amer. Math. Soc. Transl. (2) **14** (1960), 217–287. [MR **21 #1506, 22 #3954**]

Gohberg, I. and Krupnik, N.Ya.

1. *The algebra generated by one-dimensional singular integral operators with piecewise continuous coefficients*, Funktsional. Anal. i Priložen. **4** (1970), no. 3, 26–36. (Russian) [MR **42 #5057**]
2. *Singular integral operators with piecewise continuous coefficients and their symbols*, Izv. Akad. Nauk SSSR Ser. Mat. **35** (1971), 940–964 (Russian); English transl. in Math. USSR Izv. **5** (1971), 955–979 (1972). [MR **45 #981**]
3. *Introduction to the Theory of One-dimensional Singular Integral Operators*, "Shtiintsa," Kishinev, 1973 (Russian); English transl. *One-dimensional Linear Singular Integral Equations*. Vol. I. *Introduction*, Vol. II. *General Theory and Applications*, Translated from the 1979 German translation. Operator Theory: Advances and Applications, 53 and 54. Birkhäuser Verlag, Basel, 1992. [MR **53 #8971, 93c:47061, 93k:47059**]

Gohberg, I. and Sementsul, A.A.

1. *Toeplitz matrices composed of the Fourier coefficients of functions with discontinuities of almost-periodic type*, Mat. Issled. **5** (1970), 4(18), 63–83. (Russian) [MR **44 #7379**]

Gorodetskiĭ, M.B. and Dybin, V.B.

1. *Operators of the convolution type in some classes of topological spaces*, I., in: Mathematical Analysis and its Applications, 17–23, Rostov-on-Don, 1978. (Russian)

Govorov, N.V.

1. *An example of a solvable homogeneous Riemann boundary value problem with a coefficient that has negative index*, Uchen. Zapiski Kabardino-Balkarskogo Univ. **19** (1963), 141–144. (Russian)
2. *On the Riemann boundary value problem with infinite index*, Dokl. Akad. Nauk **154** (1964), no. 6, 1247–1249. (Russian) [MR **28 #3165**]
3. *A non-homogeneous Riemann boundary value problem with infinite index*, Dokl. Akad. Nauk SSSR **159** (1964), no. 5, 961–964. (Russian) [MR **30 #260**]
4. *The Riemann boundary value problem with infinite index of power order smaller than* 1/2, Teor. Funktsiĭ Funktsional. Anal. i Prilozhen., no. 6 (1968), 151–176. (Russian) [MR **41 #7123**]
5. *On the bounded solutions of a Riemann boundary value problem with infinite index of power type*, Dokl. Akad. Nauk SSSR **182** (1968), 750–753. (Russian) [MR **39 #1666**]
6. *Bounded solutions of a homogeneous Riemann boundary value problem with infinite index of power type*, Teor. Funktsiĭ Funktsional. Anal. i Prilozhen., no. 11 (1970), 3–34. (Russian) [MR **45 #3727**]
7. *The solution in a class of functions of completely regular growth of a homogeneous Riemann boundary value problem with infinite index*, Teor. Funktsiĭ Funktsional. Anal. i Prilozhen., no. 15 (1972), 213–244. (Russian) [MR **46 #7523**]
8. *The Riemann Boundary Value Problem with an Infinite Index*, "Nauka," Moscow, 1986 (Russian); English transl. *Riemann's Boundary Problem with Infinite Index*, Operator Theory: Advances and Applications, 67, Birkhäuser Verlag, Basel, 1994. [MR **88k:30001, 95d:30001**]

Grigoryan, S.A.

1. *Generalized analytic functions*, Uspekhi Mat. Nauk **49** (1994), no. 2(296), 3–42 (Russian); English transl. in Russian Math. Surveys **49** (1994), no. 2, 1–40. [MR **95h:43006**]

Grozev, C.R. and Rahman, Q.I.

1. *Entire functions of exponential type belonging to $L_p(\mathbf{R})$*. J. London Math. Soc., (2) **50** (1994), no. 1, 121–138 [MR **95d:30045**]

Grudskaya, O.N. and Grudsky, S.M.

1. *Wave propagation in the air half-space above an absorbing earth surface crossed by a straight motorway*, in: Integral and Differential Operators and their Applications, Rostov-on-Don, no. 3, 47–53, 1998.

Grudsky, S.M.

1. *On the compactness of a certain integral operator*, Rostov-on-Don, Deposited VINITI 18.10.80, N 4856–80 (1980).
2. *The Riemann boundary value problem with discontinuities of almost-periodic type in the class $L_\infty(\Gamma)$*, in: Differential and Integral Equations and their Applications, 30–41, Kalmytsk. Gos. Univ., Elista, 1982. (Russian) [MR **85e:30070**]
3. *The Riemann boundary value problem with semi-almost-periodic discontinuities in the space $L_p(\Gamma, \rho)$*, in: Integral and Differential Equations and Approximate Solutions 54–68, Kalmytsk. Gos. Univ., Elista, 1985. (Russian) [MR **88i:30069**]
4. *Singular integral equations and the Riemann boundary value problem with an infinite index in the space $L_p(\Gamma, \omega)$*, Izv. Akad. Nauk SSSR Ser. Mat. **49** (1985), no. 1, 55–80. (Russian) [MR **86h:30075**]
5. *Singular integral operators with infinite index, and Blaschke products*, Math. Nachr. **129** (1986), 313–331. (Russian) [MR **87m:47067**]
6. *Factorization of u-periodic matrix functions and problems with infinite index*, Dokl. Akad. Nauk SSSR **295** (1987), no. 6, 1298–1302 (Russian); English transl. in Soviet Math. Dokl. **36** (1988), no. 1, 180–184. [MR **88m:30103**]
7. *Matrix singular integral operators with infinite index*, in: Integral Operators and Equations, 49–64, Kuban. Gos. Univ., Krasnodar, 1987. (Russian) [MR **90e:47043**]
8. *Convolution equations on a finite interval with a small parameter multiplying the growing part of the symbol*, Izv. Vyssh. Uchebn. Zaved. Mat. (1990), no. 7, 7–17 (Russian); English transl. in Soviet Math. (Iz. VUZ) **34** (1990), no. 7, 7–18. [MR **92e:47039**]
9. *Matrix singular integral operators with infinite index.* II, Izv. Vyssh. Uchebn. Zaved. Mat. (1991), no. 6, 69–72 (Russian); English transl. in Soviet Math. (Iz. VUZ) **35** (1991), no. 6, 70–72. [MR **92j:47096**]
10. *u-factorization and Toeplitz operators with infinite index*, in: Problems and Methods in Mathematical Physics (Chemnitz, 1993), 59–70, Teubner, Stuttgart, 1994. [MR **95e:47037**]
11. *Toeplitz operators with symbols that have discontinuities of infinite index type*, Dokl. Akad. Nauk **342** (1995), no. 3, 307–309 (Russian); English transl. in Dokl. Math. **51** (1995), no. 3, 356–358. [MR **96j:47023**]

12. *The convolution equation on the finite interval. The Riemann boundary value matrix problem method*, in: Proceedings of the International Symposium on Electromagnetic Theory, St.-Petersburg, May 23–26, pp. 351–353, 1995.
13. *Toeplitz operators and the modelling of oscillating discontinuities with the help of Blaschke product*, in: Problems and Methods in Mathematical Physics (Chemnitz, 1999), Oper. Theory Adv. Appl. **121** (2001), 162–193.

Grudsky, S.M. and Dybin, V.B.

1. *A Riemann boundary value problem with discontinuities of almost-periodic type in its coefficient*, Dokl. Akad. Nauk SSSR **237** (1977), no. 1, 21–24 (Russian); English transl. in Soviet Math. Dokl. **18** (1977), no. 6, 1383–1387. [MR **57 #6451**]
2. *A Riemann boundary value problem in the space $L_p(\Gamma, \rho)$ with almost-periodic discontinuities in its coefficient*, in: Linear Operators, Mat. Issled., no. 54 (1980), 36–49. (Russian) [MR **81h:30044**]

Grudsky, S.M. and Khevelev, A.B.

1. *Invertibility in $L_2(R)$ of singular integral operators with periodic coefficients and shift*, Dokl. Akad. Nauk SSSR **269** (1983), no. 6, 1303–1306 (Russian); English transl in Sov. Math., Dokl. **27** (1983), 486–489. [MR **85g:45010**]

Grudsky, S.M. and Silbermann, B.

1. *Approximate identities, almost-periodic functions and Toeplitz operators*, Acta Appl. Math. **65** (2001), no. 1–3, 237–271.

Hartman, Ph. and Wintner, A.

1. *The spectra of Toeplitz's matrices*, Amer. J. Math. **76** (1954), 867–882. [MR **17,499a**]

Hayashi, Eric

1. *The kernel of a Toeplitz operator*, Integral Equations Operator Theory **9** (1986), no. 4, 588–591. [MR **87m:47068**]

Haykin, M.I.

1. *The regularization of operators with a nonclosed range*, Izv. Vyssh. Uchebn. Zaved. Mat. (1970), no. 8 (99), 118–123. (Russian) [MR **57 #1175**]

Heinig, G. and Silbermann, B.

1. *Factorization of matrix functions in algebras of bounded functions*, in: Spectral Theory of Linear Operators and Related Topics (Timişoara/Herculane, 1983), 157–177, Oper. Theory Adv. Appl., 14, Birkhäuser Verlag, Basel, 1984. [MR **86m:47017**]

Helson, H. and Sarason, D.

1. *Past and future*, Math. Scand **21** (1967), 5–16, see also Sarason, D., *An addendum to: "Past and future"*, Math. Scand. **30** (1972), 62–64. [MR **38 #5282**, **52 #6849**].

Hlgatyan, R.G.

1. *Some problems in the theory of systems of Wiener-Hopf difference equations,* Kazan Univ., Deposited VINITI 14.01.80 N 227, 26 p., 1980. (Russian)

Hoffman, K.

1. *Banach Spaces of Analytic Functions*, Prentice-Hall Series in Modern Analysis, Prentice-Hall, Inc., Englewood Cliffs, N. J., 1962. [MR **24 #A2844**]

Hruschev, S.V., Nikol'skiĭ, N.K. and Pavlov, B.S

1. *Unconditional bases of exponentials and reproducing kernels*, in: Complex Analysis and Spectral Theory (Leningrad, 1979/1980), 214–335, Lecture Notes in Math., 864, Springer, Berlin and New York, 1981. [MR **84k:46019**]

Hunt, R., Muckenhoupt, B. and Wheeden, R.

1. *Weighted norm inequalities for the conjugate function and Hilbert transform*, Trans. Amer. Math. Soc. **176** (1973), 227–251 (1973). [MR **47 #701**]

Jahnke, E., Emde, F. and Lösch, F.

1. *Tables of Higher Functions*, 6th ed. Revised by Friedrich Lösch, McGraw-Hill Book Co., Inc., New York-Toronto-London; B. G. Teubner Verlagsgesellschaft, Stuttgart 1960. [MR **22 #5140**]

Kantorovich, L.V. and Akilov, G.P.

1. *Functional Analysis*, Second edition, revised. Izdat. "Nauka," Moscow, 1977 (Russian); English transl., Second edition. Pergamon Press, Oxford-Elmsford, N.Y., 1982. [MR **58 #23465**, **83h:46002**]

Karlovich, Yu.I.

1. *On the Haseman problem*, Demonstratio Math. **26** (1993), no. 3–4, 582–595. [MR **95a:47048**]

Karlovich, Yu.I. and Litvinchuk, G.S.

1. *On some classes of semi-Noetherian operators*, Izv. Vyssh. Uchebn. Zaved. Mat. (1990), no. 2, 3–16 (Russian); English transl. in Soviet Math. (Iz. VUZ) **34** (1990), no. 2, 1–15 [MR **91k:47122**]

Karlovich, Yu.I. and Spitkovsky, I.M.

1. *On the Noethericity of some singular integral operators with matrix coefficients of class SAP and systems of convolution equations on a finite interval associated with them*, Dokl. Akad. Nauk SSSR **269** (1983), no. 3, 531–535 (Russian); English transl. in Soviet. Math. Dokl. **27**, 358–363 (1983). [MR **85d:47053**]
2. *Factorization of almost-periodic matrix functions and the Noether theory of certain classes of equations of convolution type*, Izv. Akad. Nauk SSSR Ser. Mat. **53** (1989), no. 2, 276–308 (Russian); English transl. in Math. USSR-Izv. **34** (1990), no. 2, 281–316. [MR **90f:47034**]
3. *Factorization of almost-periodic matrix functions and Fredholm theory of Toeplitz operators with semi almost-periodic matrix symbols,* in: Linear and Complex Analysis Problem Book: 199 research problems, 279–282, Lecture Notes in Mathematics, 1043, Springer-Verlag, Berlin, 1984.
4. *Vector Riemann and Haseman boundary value problems with oscillating coefficients*, Dokl. Rash. Zased. Inst. Priklad. Mat. im I. N. Vekua **5** (1990), 86–89. (Russian)
5. *(Semi)-Fredholmness of convolution operators on the spaces of Bessel potentials*, in: Toeplitz Operators and Related Topics (Santa Cruz, CA, 1992), 122–152, Oper. Theory Adv. Appl., 71, Birkhäuser Verlag, Basel, 1994. [MR **95h:47034**]
6. *Factorization of almost-periodic matrix functions*, J. Math. Anal. Appl. **193** (1995), no. 1, 209–232. [MR **96m:47047**]
7. *Almost periodic factorization: an analogue of Chebotarev's algorithm*, in: Harmonic Analysis and Operator Theory (Caracas, 1994), 327–352, Contemp. Math., 189, Amer. Math. Soc., Providence, RI, 1995. [MR **96h:47024**]
8. *Semi-Fredholm properties of certain singular integral operators*, in: Singular Integral Operators and Related Topics (Tel Aviv, 1995), 264–287, Oper. Theory Adv. Appl., 90, Birkhäuser Verlag, Basel, 1996. [MR **97k:47046**]

Kats, B.A.

1. *A Riemann boundary value problem with an oscillating coefficient*, Trudy Sem. Kraev. Zadacham No. 14 (1977), 110–120. (Russian) [MR **80a:30047**]
2. *A Riemann boundary value problem with a coefficient that allows discontinuities of oscillating type*, Dokl. Akad. Nauk SSSR **244** (1979) (Russian); English transl. in Sov. Math., Dokl. **20** (1979), 77–81. [MR **80b:30036**]
3. *Boundary value problems with infinite indices and not one-to-one shifts*, Mat. Zametki **30** (1981), no. 6, 847–856 (Russian); English transl. in Math. Notes **30** (1982), 925–929. [MR **83d:30041**]
4. *The Riemann problem on a closed Jordan curve*, Izv. Vyssh. Uchebn. Zaved. Mat. (1983), no. 4, 68–80 (Russian); English transl. in Sov. Math. (Iz. VUZ) **27** (1983), no. 4, 83–98. [MR **84h:30066**] (Russian)

Kats, B.A. and Zapuskalova, T.A.

1. *A Riemann boundary value problem on a spiral-shaped contour*, Trudy Sem. Kraev. Zadacham No. 15 (1978), 53–61. (Russian) [MR **81k:30054**]

Khvedelidze, B.V.

1. *Linear discontinuous boundary problems in the theory of functions, singular integral equations and some of their applications*, Akad. Nauk Gruzin. SSR. Trudy Tbiliss. Mat. Inst. Razmadze **23** (1956), 3–158. (Russian) [MR **21 #5873**]
2. *The method of Cauchy type integrals in discontinuous boundary value problems of the theory of holomorphic functions of a complex variable*, in: Current Problems in Mathematics, Vol. 7, 5–162, Akad. Nauk SSSR VINITI, Moscow, 1975 (Russian); English transl. in J. Soviet Math. **7** (1977), 309–414. [MR **56 #12276**]
3. *Boundary value problems of the theory of functions of a complex variable and singular integral equations*, Akad. Nauk Gruzin. SSR. Trudy Tbiliss. Mat. Inst. Razmadze LXXVII 39–55 (1985).

Kolmogorov, A.N. and Fomin, S.V.

1. *Elements of the Theory of Functions and Functional Analysis*, Fourth edition, revised. Izdat. "Nauka," Moscow, 1976 (Russian); English transl. of the first edition, *Elements of the Theory of Functions and Functional Analysis.* Vol. 1. *Metric and Normed Spaces* and Vol. 2 *Measure. The Lebesgue Integral. Hilbert space*, Graylock Press, Rochester, N. Y., 1957 and 1961. [MR **55 #8728, 19,44d, 22 #9566a**]

Konyshkova, E.M.

1. *On a characteristic singular integral equation with infinite index*, Izv. Vyssh. Uchebn. Zaved. Matematika (1973), no. 8(135), 43–53 (Russian) [MR **48 #6843**]

Koosis, P.

1. *Introduction to H^p Spaces*, Second edition. With two appendices by V. P. Havin. Cambridge Tracts in Mathematics, 115, Cambridge University Press, Cambridge, 1998. [MR **2000b:30052**]

Kravchenko, V.G.

1. *Normalization of singular integral operators*, Dokl. Akad. Nauk SSSR **285** (1985), no. 6, 1314–1317 (Russian); English transl in Sov. Math. Dokl. **32** (1985), 880–883. [MR **87d:47064**]

Kravchenko, V.G., Lebre, A.B., Litvinchuk, G.S. and Teixeira, F.S.

1. *A normalization problem for a class of singular integral operators with Carleman shift and unbounded coefficients*, Integral Equations Operator Theory **21** (1995), no. 3, 342–354. [MR **95m:47095**]

Kravchenko, V.G. and Litvinchuk, G.S.

1. *On the problem of normalization of singular integral operators with shift*, Uspekhi Mat. Nauk **39** (1984), no. 4, 126–127. (Russian)

Kreĭn, M.G.

1. *Integral equations on a half-line with a kernel depending upon the difference of the arguments.* Uspehi Mat. Nauk **13** (1958), no. 5, 3–120 (Russian); English transl. in Amer. Math. Soc. Transl. **22**(2) (1962), 163–288. [MR **21 #1507**]
2. *On some new Banach algebras and Wiener-Levy type theorems for Fourier series and integrals*, Matem. Issled. **1** (1966), 82–109 (Russian); Engl. transl. in Amer. Math. Soc. Transl. **93** (1970), 177–199. [MR **34 #3365**]

Krupnik, N.Ya.

1. *Singular integral operators with matrix coefficients*, in: Spectral Properties of Operators, Mat. Issled. Vyp. 45 (1977), 93–100. (Russian) [MR **58 #12525**]
2. *Some corollaries of a theorem of Hunt, Muckenhoupt and Wheeden*, in: Operators in Banach Spaces, Mat. Issled. No. 47, (1978), 64–70. (Russian) [MR **80a:42011**]
3. *Banach algebras with Symbol and Singular Integral Operators*, "Shtiintsa," Kishinev, 1984 (Russian); English transl. Operator Theory: Advances and Applications, 26 Birkhäuser Verlag, Basel, 1987. [MR **86j:47071, 90g:47092**]

Krupnik, N.Ya. and Nyaga, V.I.

1. *Singular integral operators in the case of a nonsmooth contour*, Mat. Issled. **10** (1975), no. 1(35), 144–164. (Russian) [MR **52 #4035**]

Kulagina, M.F.

1. *The Riemann boundary value problem for almost-periodic functions*, Izv. Vyssh. Uchebn. Zaved. Mat. (1983), no. 4, 95–96 (Russian); English transl. in Sov. Math. **27** (1983), no. 4, 116–118. [MR **85c:30044**]
2. *Almost periodic solutions of characteristic singular equations on the line*, in: Current Problems in the Theory of Boundary Value Problems and their Applications, 82–88, Chuvash. Gos. Univ., Cheboksary, 1988. (Russian)

Lee, M. and Sarason, D.

1. *The spectra of some Toeplitz operators*, J. Math. Anal. Appl. **33** (1971), 529–543. [MR **43 #960**]

Levin, B.Ya.

1. *Interpolation by entire functions of exponential type*, in: Mat. Fiz. i Funktsional. Anal. Vyp. 1 (1969), 136–146. (Russian) [MR **57 #6439**]

Levin, B.Ya. and Ostrovskiĭ, I.V.

1. *Small perturbations of the set of roots of sine-type functions*, Izv. Akad. Nauk SSSR Ser. Mat. **43** (1979), no. 1, 87–110 (Russian); English transl. in Math. USSR-Izv. **14** (1980), no. 1, 79–101. [MR **82b:30022**]

Levitan, B.M.

1. *Almost Periodic Functions*, Gosudarstv. Izdat. Tekhn.-Teor. Lit., Moscow, 1953 (Russian). [MR **15,700a**]

Li, Kin Y.

1. *Interpolating Blaschke products and the left spectrum of multiplication operators on the Bergman space*, Hokkaido Math. J. **21** (1992), no. 2, 295–304. [MR **93h:30053**]

Litvinchuk, G.S.

1. *Boundary Value Problems and Singular Integral Equations with Shift*, Izdat. "Nauka," Moscow, 1977. (Russian) [MR **58 #28532**]

Litvinchuk, G.S. and Spitkovsky, I.M.

1. *Factorization of Measurable Matrix Functions*, Mathematical Research, 37, Akademie-Verlag, Berlin, 1987; and Operator Theory: Advances and Applications, 25, Birkhäuser Verlag, Basel, 1987. [MR **89e:47022, 90g:47030**]

Lopes, P.A. and dos Santos, A.F.

1. *A new approach to the convolution operator on a finite interval*, Integral Equations Operator Theory **26** (1996), no. 4, 460–475. [MR **97m:47028**]

Lyubarskii, Yu. and Seip, K.

1. *Complete interpolating sequences for Paley-Wiener spaces and Muckenhoupt's (A_p) condition*, Rev. Mat. Iberoamericana **13** (1997), no. 2, 361–376. [MR **99e:42004**]

Martirosjan, V.M.

1. *The closure and basis property of some biorthogonal systems and the solution of a multiple interpolation problem in angular domains*, Izv. Akad. Nauk Armyan. SSR Ser. Mat. **13** (1978), no. 5–6, 490–531. (Russian) [MR **81e:30007**]
2. *Closure and the basis property of certain biorthogonal systems and the solution of a multiple interpolation problem in $H_p[\alpha;\, \omega]$*, Dokl. Akad. Nauk SSSR **245** (1979), no. 1, 24–27 (Russian); English transl. in Soviet Math. Dokl. **20** (1979), no. 2, 260–263. [MR **81g:30049**]

Mikhlin, S.G.

1. *Singular integral equations*, Uspehi Mat. Nauk **3** (1948), no. 3, 29–112 (Russian); English transl. in Amer. Math. Soc. Translation (1950), no. 24, 116 pp. [MR **10,305a, 12,107d**]

Mikhlin, S.G. and Prössdorf, S.

1. *Singuläre Integraloperatoren*, Mathematische Lehrbücher und Monographien, II. Abteilung: Mathematische Monographien, 52 Akademie-Verlag, Berlin, 1980 (German); English transl. *Singular Integral Operators*, Springer-Verlag, Berlin, 1986. [MR **82c:45002**, **88e:47097**]

Monakhov, V.N. and Semenko, E.V.

1. *Singular integral equations with an infinite index in L_p spaces*, Dinamika Sploshn. Sredy No. 65 (1984), 149–151. (Russian)
2. *Well-posedness classes of boundary value problems of conjugation of analytic functions with infinite index*, Dokl. Akad. Nauk SSSR **286** (1986), no. 1, 27–30 (Russian); English transl. in Soviet Math. Dokl. **33** (1986), no. 1, 17–20. [MR **87j:30099**]
3. *Boundary Value Problems on Compact Riemann Surfaces*, Izdatel′stvo Rossiĭskoĭ Akademii Nauk, Sibirskoe Otdelenie, Novosibirsk, 1996, 151 pp. (Russian)

Muskhelishvili, N.I.

1. *Singular Integral Equations. Boundary Value Problems of Function Theory and their Application to Mathematical Physics*, Third, corrected and augmented edition. With an appendix by B. Bojarski. Izdat. "Nauka," Moscow, 1968 (Russian); English transl. edited by J. R. M. Radok, reprint of the 1958 edition, Noordhoff International Publishing, Leyden, 1977. [MR **50 #7969**, **55 #10978**]

Nakazi, T.

1. *Kernels of Toeplitz operators*, J. Math. Soc. Jap. **38** (1986), no. 4, 607–616. [MR **87k:47061**]
2. *Intersection of two invariant subspaces*, Canad. Math. Bull. **30** (1987), no. 2, 129–133. [MR **88e:47062**]
3. *The spectra of Toeplitz operators with unimodular symbols*, Proc. Edinburgh Math. Soc. (2) **41** (1998), no. 1, 133–139. [MR **99m:47032**]

Nicolau, A.

1. *Interpolating Blaschke products solving Pick-Nevanlinna problems*, J. Anal. Math. **62** (1994), 199–224. [MR **94m:30071**]

Nikol′skiĭ, N.K.

1. *Lectures on the Shift Operator*, "Nauka," Moscow, 1980 (Russian); English transl. *Treatise on the Shift Operator. Spectral Function Theory*, with an appendix by S. V. Hruščev and V. V. Peller, Grundlehren der Mathematischen Wissenschaften, 273, Springer-Verlag, Berlin, 1986. [MR **82i:47013**, **87i:47042**]

2. Operators, Functions, and Systems: An Easy Reading. Volume 1: Hardy, Hankel, and Toeplitz. Mathematical Surveys and Monographs, Volume 92, American Mathematical Society, Providence, RI, 2002, 461 pp.
3. Operators, Functions, and Systems: An Easy Reading. Volume 2: Model Operators and Systems. Mathematical Surveys and Monographs, Volume 93, American Mathematical Society, Providence, RI, 2002, 439 pp.

Nitievskiĭ, V.S.

1. *Some classes of functions that admit factorization with infinite index*, Kuban. Univ., Krasnodar, Deposited VINITI 18.04.86, N 2851–B, 16 pp, 1986.
2. *Factorization with infinite index of a certain class of functions*, Kuban. Univ., Krasnodar, Deposited VINITI 09.10.87, N 7192–B87, 9 pp, 1987.

Noble, B.

1. *Methods Based on the Wiener-Hopf Technique for the Solution of Partial Differential Equations*, International Series of Monographs on Pure and Applied Mathematics, Vol. 7, Pergamon Press, New York-London-Paris-Los Angeles, 1958. [MR **21 #1505**]

Nordgren, E.

1. *Composition operators*, Canadian J. Math. **20** (1968), 442–449. [MR **36#6961**]

Novokshenov, V.Yu.

1. *Convolution equations on a finite segment and factorization of elliptic matrices*, Mat. Zametki **27** (1980), no. 6, 935–946 (Russian); English transl. in Math. Notes **27** (1980), no. 5–6, 449–455. [MR **82d:45010**]

Ostrovskiĭ, I.V.

1. *Conditions for the solvability of a homogeneous Riemann boundary value problem with infinite index*, Dokl. Akad. Nauk Ukrain. SSR Ser. A (1990), no. 5, 24–27. (Russian) [MR **91g:30046**]
2. *Conditions for the solvability of the homogeneous Riemann boundary value problem with an infinite index*, Teor. Funktsiĭ Funktsional. Anal. i Prilozhen. no. 55 (1991), 3–23 (Russian); English transl. in J. Soviet Math. **59** (1992). [MR **93g:30058, 94d:30075**]
3. *The homogeneous Riemann boundary value problem with infinite index on a curvilinear contour*, I, Teor. Funktsiĭ Funktsional. Anal. i Prilozhen. no. 56 (1991), 95–105 (Russian); English transl. in J. Math. Sci. **76** (1995), no. 4, 2517–2524. [MR **94d:30076**]

Paltsev, B.V.

1. *On a method for constructing the canonical matrix of solutions for a Hilbert problem that arises in the solution of convolution equations on a finite interval*, Izv. Akad. Nauk SSSR Ser. Mat. **45** (1981), no. 6, 1332–1390 (Russian);

English transl. in Math. USSR, Izv. **19** (1982), 559–610. [MR **83c:30042**]; see also Izv. Akad. Nauk SSSR Ser. Mat. **46** (1982), no. 3, 668 and Math. USSR, Izv. **20** (1983), 627.

Pavlov, B.S.

1. *The basis property of a system of exponentials and the Muckenhoupt condition*, Dokl. Akad. Nauk SSSR **247** (1979), no. 1, 37–40 (Russian); English transl. in Soviet Math. Dokl. **20** (1979), no. 4, 655–659. [MR **84j:42042**]

Plaksa, S.A.

1. *The Riemann boundary value problem with an oscillating coefficient, and singular integral equations on a rectifiable curve*, Ukrain. Mat. Zh. **41** (1989), no. 1, 116–121 (Russian); English transl. in Ukrainian Math. J. **41** (1989), no. 1.
2. *The Riemann boundary value problem with infinite index of logarithmic order on a spiral-like contour*, I. Ukrain. Mat. Zh. **42** (1990), no. 11, 1509–1517 (Russian); English transl. in Ukrainian Math. J. **42** (1990), no. 11, 1351–1358. [MR **92g:30040**]

Pousson, H.R.

1. *Systems of Toeplitz operators on H^2*, Proc. Amer. Math. Soc. **19** (1968), 603–608. [MR **37 #783**]

Power, S.C.

1. *Fredholm–Toeplitz operators and slow oscillation*, Canad. J. Math. **32** (1980), no. 5, 1058–1071. [MR **82b:47032**]

Privalov, I.I.

1. *Boundary Properties of Analytic Functions*, 2nd ed. Gosudarstv. Izdat. Tehn.-Teor. Lit., Moscow-Leningrad, 1950. [MR **13,926h**]

Prössdorf, S.

1. *Einige Klassen singulärer Gleichungen*, Mathematische Reihe, Band 46. Birkhäuser Verlag, Basel and Stuttgart, 1974; English transl. *Some Classes of Singular Equations*, North-Holland Mathematical Library, 17, North-Holland Publishing Co., Amsterdam and New York, 1978. [MR **58 #17715, 80d:47044**]
2. *Linear integral equations*, in: Current Problems in Mathematics. Fundamental Directions, Vol. 27, 5–130, Itogi Nauki i Tekhniki, Akad. Nauk SSSR, Vsesoyuz. Inst. Nauchn. i Tekhn. Inform., Moscow, 1988 (Russian); English transl. in *Analysis*, IV, 1–125, Encyclopaedia Math. Sci., 27, Springer, Berlin, 1991. [MR **93e:45001**]

Prudnikov, A.P., Brychkov, Yu.A. and Marichev, O.I.

1. *Integrals and Series. Elementary Functions*, "Nauka," Moscow, 1981 (Russian); English transl. *Integrals and Series*. Vol. 1. *Elementary functions*, Gordon and Breach Science Publishers, New York, 1986. [MR **83b:0000,88f:00013**]

Quint, D., Rodman, L. and Spitkovsky, I.M.

1. *New cases of almost-periodic factorization of triangular matrix functions*, Michigan Math. J. **45** (1998), no. 1, 73–102. [MR **99c:47020**]

Rabindranathan, M.

1. *On the inversion of Toeplitz operators*, J. Math. Mech. **19** (1969/1070), no. 3, 195–206. [MR **40 #4785**]

Rabinovich, V.S.

1. *Singular integral operators on a composite contour with an oscillating tangent, and Mellin pseudodifferential operators*, Dokl. Akad. Nauk SSSR **321** (1991), no. 4, 692–696 (Russian); English transl. in Soviet Math. Dokl. **44** (1992), no. 3, 791–796. [MR **93c:47065**]
2. Algebras of singular integral operators with whirl points. Functionalniy Analis i ego priloj, Vol. 30, No. 3, 1996, p. 85–87. (English translation in Functional analysis and its applications, Vol. 30, No. 3, 1996, p. 213–215.)
3. Algebras of singular integral operators on complicated contours with nodes being of logarithmic whirl points. Izvestia AN Rossii, ser. mathem., v. 60, No. 6, 1996, 169–200. (In Russian) English Translation: Izvestia: Mathematics 60:6, 1996, 1261–1292.

Rafaelyan, L.I.

1. *Interpolation and the basis property in weighted classes of entire functions of exponential type*, Izv. Akad. Nauk Armyan. SSR Ser. Mat. **18** (1983), no. 3, 167–186 (Russian); English transl. in Soviet J. Contemporary Math. Anal. **18** (1983), no. 3, 1–21. [MR **85g:30063**]

Rakowski, M. and Spitkovsky, I.

1. *On normal solvability of the Riemann problem with singular coefficient*, Proc. Amer. Math. Soc. **125** (1997), no. 3, 815–826. [MR **97e:47026**]

Riesz, F. and Sz.-Nagy, B.

1. *Leçons d'Analyse Fonctionnelle*, 2ème éd. Akadémiai Kiadó, Budapest, 1953; English transl. *Functional Analysis*, Frederick Ungar Publishing Co., New York, 1955. [MR **17,175i**]

Roch, S. and Silbermann, B.

1. *Algebras of Convolution Operators and Their Image on the Calkin Algebra*, Report Math 90-05, Karl-Weierstrass-Inst. f. Math., Berlin, 1990 [MR **92d:47067**].

Rochberg, R.

1. *Toeplitz operators on weighted H^p spaces*, Indiana Univ. Math. J. **26** (1977), no. 2, 291–298. [MR **56 #16431**]

Rogozin, S.V.

1. *An inhomogeneous Riemann boundary value problem with infinite index in an exceptional case for a half plane*, Vestnik Beloruss. Gos. Univ. Ser. I (1983), no. 2, 60–62. (Russian) [MR **85c:30045**]

Rogozin, S.V. and Tolochko, M.E.

1. *A homogeneous Riemann boundary value problem with an infinite index for a half-plane in the exceptional case*, Vesci̅ Akad. Navuk BSSR Ser. Fi̅z.-Mat. Navuk (1978), no. 3, 5–10. (Russian) [MR **58 #22589**]

Rosenblum, M. and Rovnyak, J.

1. *Hardy Classes and Operator Theory*, Oxford Mathematical Monographs, Oxford Science Publications, The Clarendon Press, Oxford University Press, New York, 1985. [MR **87e:47001**]

Saginashvili, A.I.

1. *Singular integral operators with coefficients having semi-almost-periodic type discontinuities*, Soobshch. Akad. Nauk Gruzin. SSR **94** (1979), no. 2, 289–291. (Russian) [MR **81d:45012a**]
2. *Singular integral operators with coefficients having semi-almost-periodic discontinuities*, Soobshch. Akad. Nauk Gruzin. SSR **95** (1979), no. 3, 541–543. (Russian) [MR **81d:45012b**]
3. *Singular integral equations with coefficients that have discontinuities of semi-almost-periodic type*, in: Theory of Analytic Functions and Harmonic Analysis, Akad. Nauk Gruzin. SSR Trudy Tbiliss. Mat. Inst. Razmadze **66** (1980), 84–95 (Russian); English transl. in Amer. Math. Soc. Transl. II **127** (1985), 49–59. [MR **82j:45006**]

Sakhnovich, L.A.

1. *Integral Equations with Difference Kernels on Finite Intervals*, Operator Theory: Advances and Applications, 84, Birkhäuser Verlag, Basel, 1996. [MR **97e:45001**]

Sandrygailo, I.E.

1. *The Riemann boundary value problem with infinite index for the half-plane*, Dokl. Akad. Nauk BSSR **19** (1975), no. 10, 872–875. (Russian) [MR **52 #5986**]
2. *The Riemann boundary value problem with infinite index for the half-plane in a class of functions of completely regular growth*, Vesci̅ Akad. Navuk BSSR Ser. Fi̅z.-Mat. Navuk (1976), no. 1, 21–24. (Russian) [MR **56 #635**]

dos Santos, A.F.

1. *General Wiener-Hopf operators and representation of their generalized inverses*, in: The Gohberg Anniversary Collection, Vol. II (Calgary, AB, 1988), 473–483, Oper. Theory Adv. Appl., 41, Birkhäuser Verlag, Basel, 1989. [MR **91f:47044**]

Sarason, D.

1. *On products of Toeplitz operators*, Acta Sci. Math. (Szeged) **35** (1973), 7–12. [MR **48 #9443**]
2. *Functions of vanishing mean oscillation*, Trans. Amer. Math. Soc. **207** (1975), 391–405. [MR **51 #13690**]
3. *Toeplitz operators with semi-almost-periodic symbols*, Duke Math. J. **44** (1977), no. 2, 357–364. [MR **56 #12965**]
4. *Toeplitz operators with piecewise quasicontinuous symbols*, Indiana Univ. Math. J. **26** (1977), no. 5, 817–838. [MR **57 #3906**]
5. *Function Theory on the Unit Circle*, Notes for lectures given at a Conference at Virginia Polytechnic Institute and State University, Blacksburg, Va., June 19–23, 1978, Virginia Polytechnic Institute and State University, Department of Mathematics, Blacksburg, VA., 1978. [MR **80d:30035**]
6. *Shift-invariant spaces from the Brangesian point of view*, in: The Bieberbach Conjecture (West Lafayette, Ind., 1985), 153–166, Math. Surveys Monogr., 21, Amer. Math. Soc., Providence, RI, 1986. [MR **88d:47014a**]

Semenko, E.V.

1. *Boundary value problems with infinite index for elliptic systems of equations*, Candidate Dissertation, Novosibirsk, 1984.

Semencul, A.A.

1. *Singular integral equations with coefficients that have discontinuities of almost-periodic type*, Mat. Issled. **6** (1971), no. 3(21), 92–114. (Russian) [MR **45 #988**]

Shabat, B.V.

1. *Introduction to Complex Analysis*, Third edition, "Nauka," Moscow, 1985. (Russian) [MR **86m:30001**]

Shamoyan, F.A.

1. *Toeplitz operators and division by an inner function in some spaces of analytic functions*, Akad. Nauk Armyan. SSR Dokl. **76** (1983), no. 3, 109–113 (Russian); English transl. in Amer. Math. Soc. Transl. Ser. II. Ser **133** (1986), 5–9. [MR **84j:30093**]

Shapiro, J.H.

1. *Composition Operators and Classical Function Theory*, Universitext, Springer-Verlag, New York, 1993. [MR **94k:47049**]

Shapiro, H.S. and Shields, A.L.

1. *On some interpolation problems for analytic functions*, Amer. J. Math. **83** (1961), no. 3, 513–532. [MR **24 #A3280**]

Shargorodskiĭ, E.M.

1. *Singular integral operators with coefficients in P_nC*, Trudy Tbiliss. Mat. Inst. Razmadze Akad. Nauk Gruzin. SSR **93** (1990), 52–66. (Russian) [MR **92e:47043**]

Shirokov, N.A.

1. *Analytic Functions Smooth up to the Boundary*, Lecture Notes in Mathematics, 1312. Springer-Verlag, Berlin, 1988. [MR **90h:30087**]

Silbermann, B.

1. *Singular integral operators in spaces of infinitely differentiable and generalized functions*, Mat. Issled. **6** (1971), no. 3 (21), 168–179. (Russian) [MR **44 #5833**]

Simonenko, I.B.

1. *Riemann's boundary value problem with a continuous coefficient*, Dokl. Akad. Nauk SSSR **124** (1959) 278–281. (Russian) [MR **22 #3805**]
2. *Riemann's boundary problem with a measurable coefficient*, Dokl. Akad. Nauk SSSR **135** (1960), no. 3, 538–541 (Russian); English transl. in Soviet Math. Dokl. **1** (1960) 1295–1298. [MR **25 #4112**]
3. *The Riemann boundary value problem for n pairs of functions with continuous coefficients*, Izv. Vyssh. Uchebn. Zaved. Matematika (1961), no. 1 (20), 140–145 (Russian) [MR **24 #A838**]
4. *The Riemann boundary value problem for n pairs of functions with measurable coefficients and its application to the study of singular integrals in weighted L_p spaces*, Dokl. Akad. Nauk SSSR **141** (1961) 36–39 (Russian); English transl. in Sov. Math. Dokl. **2** (1961), 1391–1394. [MR **29 #4899**]
5. *The Riemann boundary value problem for n pairs of functions with measurable coefficients and its application to the investigation of singular integrals in the space L^p with weight.* Izv. Akad. Nauk SSSR, Mat., **28** (1964), 277–306. (Russian) [MR **29 #253**]
6. *A new general method of investigating linear operator equations of singular integral equation type*, I. Izv. Akad. Nauk SSSR Ser. Mat. **29** (1965), 567–586, and II. Izv. Akad. Nauk SSSR Ser. Mat. **29** (1965), 757–782. (Russian) [MR **31 #3876**, **32 #6174**]
7. *Some general questions of the theory of the Riemann boundary value problem*, Izv. Akad. Nauk SSSR Ser. Mat. **32** (1968), 1138–1146 (Russian); English transl. in: Math. USSR Izv. **2** (1968), 1091–1099. [MR **38 #3447**]
8. *Factorizations and local factorization of measurable functions*, Dokl. Akad. Nauk SSSR **250** (1980), no. 5, 1063–1066 (Russian); English transl. in Soviet Math. Dokl. **21** (1980), no. 1, 271–275. [MR **81d:30075**]
9. *On the relation between local factorizability and the local Noethericity property*, Soobshch. Akad. Nauk Gruzin. SSR **98** (1980), no. 2, 281–283. (Russian) [MR **82f:45004**]

10. *An example of function that satisfies the Muckenhoupt condition, but is not a weight function for the Cauchy singular operator, in the case of a contour with a turning point* Rostov. Gos. Univ., Deposited VINITI 18.05.83, N 2659 (1982).
11. *Some properties of a singular integral with continuous density in the case of a contour of type R*, Izv. Severo-Kavkaz. Nauchn. Tsentra Vyssh. Shkoly Estestv. Nauk. (1982), no. 2, 17–20. (Russian) [MR **84d:30072**]
12. *Stability of weight properties of functions with respect to the Cauchy singular integral*, Mat. Zametki **33** (1983), no. 3, 409–416 (Russian); English transl. in Math. Notes **33** (1983), no. 3–4, 208–212. [MR **84i:30053**]
13. *Global and local factorizability of a measurable matrix function, and the Noether property of the singular operator generated by it*, Izv. Vyssh. Uchebn. Zaved. Mat. (1983), no. 4, 81–87 (Russian); English transl. Soviet Math. (Iz. VUZ) **27** (1983), no. 4, 99–106. [MR **85c:45010**]
14. *The theory of operators of local type*. I, Deposited VINITI 23.01.96, N 275–B96, 74 pp.

Simonenko, I.B. and Minh, Trinh Ngok

1. *A local method in the theory of one-dimensional singular integral equations with piecewise continuous coefficients. Noethericity*, Rostov. Gos. Univ., Rostov-on-Don, 1986. (Russian) [MR **89h:45006**]

Skomaha, L.N.

1. *A singular integral equation with infinite index for the case of an analytic kernel*, Litovsk. Mat. Sb. **16** (1976), no. 3, 161–177. (Russian) [MR **54 #8188**]

Speck, F.-O.

1. *General Wiener-Hopf Factorization Methods*, Research Notes in Mathematics, 119, Pitman (Advanced Publishing Program), Boston, MA, 1985. [MR **87a:47045**]

Spitkovsky, I.M.

1. *Multipliers that do not affect factorizability*, Dokl. Akad. Nauk SSSR **231** (1976), no. 6, 1300–1303 (Russian); English transl. in Soviet Math. Dokl. **17** (1976), no. 6, 1733–1738. [MR **55 #6232**]
2. *Multipliers that do not affect factorizability*, Mat. Zametki **27** (1980), no. 2, 291–299 (Russian); English transl. in Math. Notes. **27** (1980), no. 1–2, 145–149. [MR **82g:42008**]
3. *Factorization of matrix functions from the classes $\widetilde{A_n}(p)$ and TL*, Ukrain. Mat. Zh. **35** (1983), no. 4, 455–460 (Russian); English transl. in Ukrainian Math. J. **35** (1983), no. 4, 383–388. [MR **84k:30045**]
4. *Factorization of certain classes of semi-almost-periodic matrix functions and its applications to systems of equations of convolution type*, Izv. Vyssh.

Uchebn. Zaved. Mat. (1983), no. 4, 88–94 (Russian); English transl. in Soviet Math. (Iz. VUZ) **27** (1983), no. 4, 107–115. [MR **85h:45009**]

5. *Generalized factorization of matrix functions and the Riemann boundary value problem with infinite partial indices*, Dokl. Akad. Nauk SSSR **286** (1986), no. 3, 559–562 (Russian); English transl. in Soviet Math. Dokl. **33** (1986), no. 1, 145–149. [MR **88d:30052**]
6. *The vector Riemann boundary value problem with infinite deficiency numbers and the factorization of matrix functions connected with it*, Mat. Sb. (N.S.) **135(177)** (1988), no. 4, 533–550 (Russian); English transl. in Math. USSR-Sb. **63** (1989), no. 2, 521–538. [MR **89j:30058**]
7. *Factorization of almost-periodic matrix functions*, Mat. Zametki **45** (1989), no. 6, 74–82 (Russian); English transl. in Math. Notes **45** (1989), no. 5–6, 482–488. [MR **90k:47033**]
8. *Singular integral operators with PC symbols on the spaces with general weights*, J. Funct. Analysis **105** (1992), no. 1, 129–143 [MR **93d:47057**].

Spitkovsky, I.M. and Tishin, P.M.

1. *Factorization of new classes of almost-periodic matrix functions*, in: Reports of the Extended Sessions of a Seminar of the I. N. Vekua Institute for Applied Mathematics **3** (1989), no. 1, 170–173. (Russian)

Spitkovsky, I.M. and Yatzko, S.I.

1. *A Riemann boundary value problem whose matrix coefficient admits infinite partial indices*, Izv. Vyssh. Uchebn. Zaved. Mat. (1985), no. 6, 45–53 (Russian); English transl. in Soviet Math. (Iz. VUZ) **29** (1985), no. 6, 55–65. [MR **87b:30068**]

Tolochko, M.E.

1. *The Riemann boundary value problem with infinite index for the half plane*, Vescī Akad. Navuk BSSR Ser. Fīz.-Mat. Navuk (1969), no. 4, 52–59. (Russian) [MR **41 #3785**]
2. *On the solvability of the Riemann boundary value problem with infinite index for the half-plane*, Izv. Akad. Nauk BSSR, Ser. Fiz.-Mat. Nauk (1971), no. 3, 31–38. (Russian)
3. *The homogeneous Riemann problem with infinite index for the half-plane*, Vescī Akad. Navuk BSSR Ser. Fīz.-Mat. Navuk (1972), no. 5, 34–41. (Russian) [MR **48 #8813**]

Treil, S.R.

1. *Invertibility of a Toeplitz operator does not imply its invertibility by the projection method*, Dokl. Akad. Nauk SSSR **292** (1987), no. 3, 563–567 (Russian); English transl. in Soviet Math. Dokl. **35** (1987), no. 1, 103–107. [MR **88d:47041**]

Tumarkin, G.C.

1. *The expansion of analytic functions in a series of rational fractions with a given set of poles*, Izv. Akad. Nauk Armjan. SSR Ser. Fiz.-Mat. Nauk **14** (1961), no. 1, 9–31 (Russian) [MR **26 #333**]

Vekua, N.P.

1. *Systems of Singular Integral Equations and Certain Boundary Value Problems*, Second revised and supplemented edition, Izdat. "Nauka," Moscow 1970 (Russian); English transl. *Systems of Singular Integral Equations*, P. Noordhoff, Ltd., Groningen 1967. [MR **42 #4979, 35 #2102**]

Vladimirov, V.S.

1. *The Wiener-Hopf equation on the half-axis in Nevanlinna and Smirnov algebras*, Dokl. Akad. Nauk SSSR **293** (1987), no. 2, 278–283 (Russian); English transl. in Soviet Math. Dokl. **35** (1987), no. 2, 287–292. [MR **88g:45008**]
2. *The Wiener-Hopf equation in Nevanlinna and Smirnov algebras*, Izv. Akad. Nauk SSSR Ser. Mat. **51** (1987), no. 4, 767–784 (Russian); English translation in Math. USSR-Izv. **31** (1988), no. 1, 77–94. [MR **89a:45012**]
3. *The Wiener-Hopf equation on the half-line in the Nevanlinna and the Smirnov algebras*, Zap. Nauchn. Sem. Leningrad. Otdel. Mat. Inst. Steklov. (LOMI) **170** (1989), Issled. Linein. Oper. Teorii Funktsii. **17** (1989), 67–81 (Russian); English transl. in J. Soviet Math. **63** (1993), no. 2, 149–158. [MR **91d:47023**]

Volberg, A.

1. *Matrix A_p weights via S-functions.* J. Amer. Math. Soc. **10** (1997), no. 2, 445–466. [MR **98a:42013**]

Volevich, L.P. and Gindikin, S.G.

1. *The Wiener-Hopf equation in generalized functions*, Trudy Moskov. Mat. Obshch. **35** (1976), 165–214 (Russian); English transl. in Trans. Mosc. Math. Soc. **35** (1979), 167–216. [MR **58 #30189**]
2. *Generalized Functions and Convolution Equations*, Fizmatlit "Nauka," Moscow, 1994 (Russian); see also *Distributions and Convolution Equations*, Translated from the Russian, Gordon and Breach Science Publishers, Philadelphia, PA, 1992. [MR **93c:46061, 97g:46041**]

Widom, H.

1. *Inversion of Toeplitz matrices,* II, Illinois J. Math. **4** (1960), 88–99. [MR **24 #A432**]
2. *Inversion of Toeplitz matrices,* III, Notices Amer. Math. Soc. **7**, p. 63 (1960).
3. *A theorem on translation kernels in n dimensions*, Trans. Amer. Math. Soc. **94** (1960), 170–180. [MR **22 #1793**]

4. *Singular integral equations in L_p spaces,* Trans. Amer. Math. Soc. **97** (1960), 131–160. [MR **22 #9830**]
5. *On the spectrum of a Toeplitz operator*, Pacific J. Math. **14** (1964), 365–375. [MR **29 #476**]
6. *Toeplitz operators on H^p*, Pacific J. Math. **19** (1966), 573–582. [MR **34 #1859**]

Xia, Jing Bo

1. *Piecewise continuous almost-periodic functions and mean motion*, Trans. Amer. Math. Soc. **288** (1985), no. 2, 801–811. [MR **86e:47040**]
2. *Wiener-Hopf operators with piecewise continuous almost-periodic symbol*, J. Operator Theory **14** (1985), no. 1, 147–171. [MR **87a:47046**]
3. *The K-theory and the invertibility of almost-periodic Toeplitz operators*, Operator Theory **11** (1988), no. 2, 267–286. [MR **89h:47039**]

Yamamoto, T.

1. *Invertibility of some singular integral operators and a lifting theorem*, Hokkaido Math. J. 22 (1993), no. 2, 181–198. [MR **94e:47038**]

Yurov (Jurov), P.G.

1. *The homogeneous Riemann boundary value problem with an infinite index of logarithmic type*, Izv. Vyssh. Uchebn. Zaved. Mat. (1966), no. 2(51), 158–163. (Russian) [MR **33 #7568**]
2. *The inhomogeneous Riemann boundary value problem with infinite index of logarithmic order $\alpha \geq 1$*, in: Proceedings of a Conference on Boundary Value Problems (Kazan, 1969), 279–284. Izdat. Kazan. Univ., Kazan, 1970. (Russian) [MR **50 #10284**]

Zhuravleva M.I.

1. *A homogeneous Riemann boundary value problem with infinite index whose coefficients have a countable set of discontinuities*, Trudy Tbilis. Mat. Inst. Razmadze **43** (1973), 53–71. (Russian)
2. *An inhomogeneous Riemann boundary value problem with infinite index whose coefficient has a countable set of zeros and poles*, Dokl. Akad. Nauk SSSR **214** (1974), no. 4, 755–758 (Russian); English transl. Sov. Math., Dokl. **15** (1974), 228–232. [MR **51 #3469**]

Index

MIX
Papier aus verantwortungsvollen Quellen
Paper from responsible sources
FSC® C105338

If you have any concerns about our products,
you can contact us on
ProductSafety@springernature.com

In case Publisher is established outside the EU,
the EU authorized representative is:
Springer Nature Customer Service Center GmbH
Europaplatz 3, 69115 Heidelberg, Germany

Printed by Libri Plureos GmbH
in Hamburg, Germany